Basic Biochemical Laboratory Procedures and Computing

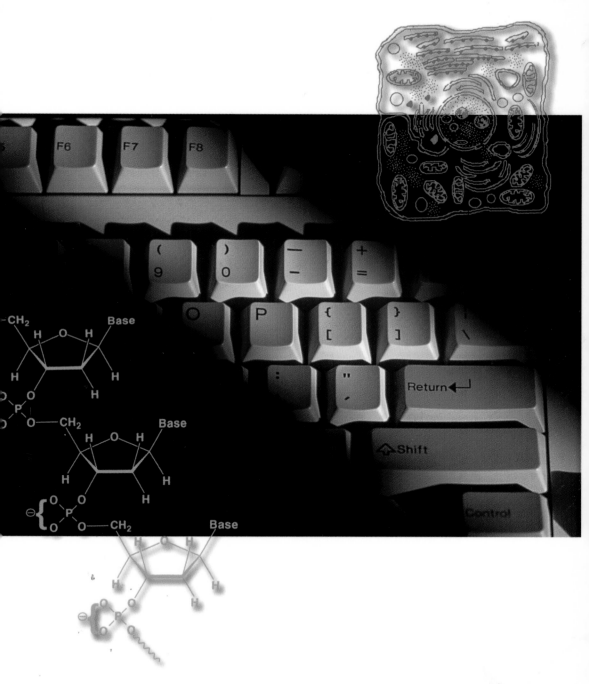

R. Cecil Jack

BASIC BIOCHEMICAL
LABORATORY PROCEDURES
AND COMPUTING

TOPICS IN BIOCHEMISTRY

Series Editor, David L. Nelson

C. Jack, *Basic Biochemical Laboratory Procedures and Computing*

BASIC BIOCHEMICAL LABORATORY PROCEDURES AND COMPUTING

With Principles, Review Questions, Worked Examples, and Spreadsheet Solutions

R. Cecil Jack

New York   Oxford
OXFORD UNIVERSITY PRESS
1995

Oxford University Press

Oxford   New York
Athens Auckland Bangkok Bombay
Calcutta Cape Town Dar es Salaam Delhi
Florence Hong Kong Istanbul Karachi
Kuala Lumpur Madras Madrid Melbourne
Mexico City Nairobi Paris Singapore
Taipei Tokyo Toronto

and associated companies in
Berlin   Ibadan

Library of Congress Cataloging-in-Publication Data
Jack, R. Cecil.
Basic biochemical laboratory procedures and computing /
R. Cecil Jack.
p.   cm. Includes index.
ISBN 0-19-507897-7
1. Biochemistry—Laboratory manuals.
2. Biochemistry—Data processing.
3. Electronic spreadsheets. I. Title.
QD415.5.J33   1994   547.7'078—dc20   94-36165

9 8 7 6 5 4 3 2 1
Printed in the United States of America
on acid-free paper

To Valerie and Marcy

Preface

The discovery of the structure of DNA in 1953, and the development of the computer microchip in 1958, unleashed a torrent of activity which has transformed the modern biomedical sciences into molecular, multidisciplinary, quantitative and informational disciplines.

Because of the quantitative and informational characteristics of today's biomedical sciences, students and scientists will want to use computers for calculations, for sequence analyses, for model-building, and for testing hypotheses. But, even in this computer age, biomedical workers should know the steps used to solve biomolecular calculations and understand the theoretical bases of experimental procedures. Developing such proficiency can, however, be time-consuming, if one attempts to do so by studying detailed textbooks in the traditional manner. For basic biochemical computing, an alternative goal is to develop broad (in contrast to expert) understanding of important multidisciplinary facts and principles. Then, equipped with such understanding, users can realize much of the potential of computers for increased efficiency in basic biochemical computing.

The computer, therefore, should not be a substitute for students or scientists who know how to analyze and interpret their results. Rather, the computer should enhance the problem-solving skills of knowledgeable investigators who understand how to analyze and interpret their data. Hence, although the book of *Basic Biochemical Laboratory Procedures and Computing* is, in large part, a book about performing biochemical calculations with computers, it tries to do more than provide a catalog of such calculations: in addition to showing how to perform certain calculations, the book emphasizes mastery of basic theoretical and laboratory biomedical principles through the choice of topics and the inclusion of many review questions and problems. This approach is consistent with the view that, to achieve good results both in the laboratory, and in computing, modern biomedical scientists must be able to use modern methods and apply broad knowledge of the disciplines or subdisciplines upon which the methods are based.

The book has four sections, each with an interdisciplinary theme depicting the types of procedures and computations used by practicing biomedical scientists. The four themes are: (a) Bioanalytical Chemistry; (b) Macromolecular structure and assay; (c) Physicochemical approaches; and (d) Data analysis. These sections are subdivided into eleven chapters, nine on procedures from specific areas of research and two on data analysis. Some topics discussed may be found in textbooks of biology, general chemistry, mathematics, biochemistry, immunology, physical chemistry, statistics, or computer science. Other topics may be found in the primary biomedical research journals, annual review publications and monographs. Whenever possible, recent applications of the procedures and computations in clinical, academic, industrial, and basic research laboratories are cited. Thus, *Basic Biochemical Laboratory Procedures and Computing* provides theoretical background and spreadsheet solutions in selected areas of the biomedical sciences; it also reflects the thinking, methods, instruments and experimental strategies of today's practicing biomedical scientists. This approach is based on the premise that, to achieve good results in the laboratory and in computing, modern biomedical scientists must be able both to use modern methods and apply broad knowledge of the disciplines or subdisciplines upon which the methods are based. Indeed, it is now well known that the solution of modern research problems often requires the participation of teams of scientists with skills from several traditional disciplines.

Basic Biochemical Laboratory Procedures and Computing may be used in various ways. It may be used to teach biochemical computing. It also may be used for courses on laboratory methods for undergraduate biochemistry or biophysics majors, or it may be used for lecture courses on laboratory techniques for first year graduate students in biochemistry, biophysics or the biological sciences. The book also may serve as a guide for graduate students and laboratory assistants in a broad range of biomedical research activities; it may be used as a tool in corporate or government training programs for biomedical employees; or it may be put to use as a self-study package for graduate students and biomedical workers.

It is perhaps inevitable in a small book that some readers would choose to delete some topics and substitute different ones. Other readers would change the emphasis placed on one or another subject. For example, some readers may prefer that greater prominence be given to laboratory procedures and less to computing; others would prefer less emphasis on methodology and more on computing; and others still might like to stress theoretical principles to a greater extent than we have done. Clearly, personal preference and experience influence the choices one makes in attempting to achieve a balanced presentation which reflects the nature of modern biomedical research. But we hope that, our selections in *Basic Biochemical Laboratory Procedures and Computing* will, to some degree, meet the needs and interests of some readers.

It is a pleasure to acknowledge the help and encouragement of Drs. Ed Miranda, Larry Parker, Marcel Sierra, Charles Traina and Alex Tzagoloff, the comments, suggestions and comprehensive review of the manuscript by Dr. David J. Nelson and the editorial support of Robert L. Rogers and his staff at Oxford University Press, New York.

TABLE OF CONTENTS

2   Acids and Bases

3   Countercurrent Distribution and Chromatography

PART II MACROMOLECULAR STRUCTURE AND ASSAY

4 The Separation and Characterization of Macromolecules

5 Kinetics and Binding Equilibria

6 Molecular Immunology and Immunochemical Assays

PART III   PHYSICOCHEMICAL APPROACHES

7 Biomolecular Spectroscopy

8 Radioactivity

9 Oxidation Reduction

PART IV   DATA ANALYSIS

10 One- and Two-sample Analyses

11   Analysis of Variance

Appendices

PART I.

BIOANALYTICAL CHEMISTRY

CHAPTERS 1 TO 3

Chapter 1

Solutions: Preparation and Concentration

1-1.   It is important to understand quantities, units, and dimensions

The Système International d'Unités, the International System of Units (abbreviated SI), was introduced in 1960 by the General Conference of Weights and Measures to standardize units of measurements based on the metric system. Thus, SI was designed to permit scientists in different disciplines, and different countries, to communicate with each other without having to convert units of measurement. But, even today, SI has not been adopted universally, and since knowledge of quantities, units, and dimensions is essential for understanding how to perform biomolecular calculations, a brief review of terms is in order. Yet even as we review SI, we should recognize that it has been modified since its introduction and, undoubtedly, it will continue to be changed as, for example, tools of measurement are improved. Thus, like science itself, units of measurement may be expected to evolve constantly.

A *quantity* is a measurable property of a defined system; examples are mass, amount of substance, temperature, and energy. A *unit* is a quantity of defined size which is used as a reference for other quantities of the same type; examples are kilogram and mole, the units of mass and amount of substance, respectively. A *dimension* is a physical quantity described by a product of symbols with appropriate exponents; the product of symbols expresses the relation of the given physical quantity and its units to SI base quantities (next paragraph). Examples are m^3 and $mol\, dm^{-3} \equiv mol/dm^3$, the dimensions of volume and molarity, respectively, where m (meter) is the symbol for length and mol is the symbol for amount of substance.

Seven base quantities are recognized in SI. They are length, mass, time, electric current, thermodynamic temperature, luminous intensity, and amount of substance. Each base quantity is assigned a base unit. Base quantities (as well as base units) are multiplied or divided by each other or themselves to form *derived quantities or derived units*. Derived quantities and units are divided into two classes: coherent and noncoherent. *Coherent derived quantities* are obtained by multiplication or division of base quantities; *noncoherent derived quantities* are obtained by multiplication or division of multiples or fractions of base quantities. Thus, the cubic meter, m^3 is a coherent derived unit of volume but the liter $10^{-3} m^3$ or $m^3/1000$ is a noncoherent derived unit of volume because it contains a fraction. Table 1-1 reviews base quantities and base units, Table 1-2 reviews derived quantities, units and, dimensions of biomedical interest and, Table 1-3 lists the prefixes which are used with SI units of measurement. When the SI base units and derived units are too large or too small to be convenient, they may be multiplied by powers of 10 as indicated in Table 1-3.

Some biomedical scientists have, traditionally, expressed their data in mass units such as g/100 ml. Other biomedical scientists have reported their data in molar units. SI has standardized both methods of expression. In SI, permitted methods of expressing concentrations on the basis of mass *(mass concentration)* are $kg\ m^{-3}$ and $kg\ l^{-1}$ or $kg\ l^{-1}$. SI allows construction of smaller units of mass concentration by use of factors of base units or factors of derived units, for example g/l, mg/l, μg/l and ng/l. Acceptable methods of expressing *amount of substance concentration* are $mol\ m^{-3}$, $mol\ l^{-1}$, $mmol\ l^{-1}$, $μmol\ l^{-1}$, $nmol\ l^{-1}$ etc. Concentrations expressed as $kg\ m^{-3}$ and $mol\ m^{-3}$ are coherent whereas concentrations expressed as $kg\ l^{-1}$ and $mol\ l^{-1}$ are noncoherent. In this book, concentration usually will be expressed on the basis preferred by biochemists i.e. mass of substance or amount of substance per liter. This will be referred to as *volume-based concentration.*

Table 1-1.   Base Quantities and Base Units

BASE QUANTITY		BASE UNIT	
Name	Dimension	Name	Symbol
Length	L	meter	m
Mass	M	kilogram	kg
Amount of substance	N	mole	mol
Time	T	second	s
Electric current	I	ampere	A
Thermodynamic temperature	Ø	kelvin	K
Luminous intensity	J	candela	cd

Table 1-2. Derived Quantities and Derived Units of Biomedical Interest

DERIVED QUANTITY		DERIVED UNIT		
Name	Dimension	Name of SI Unit	Expression in SI Units	Symbol and Definition of Unit
Volume	L^3	Liter	$10^{-3}\ m^3$	$l\ =\ m^3/1000$
Energy	L^2MT^{-2}	Joule	$m^2\ kg\ s^{-2}$	$J\ =\ $ newton meters
Force	LMT^{-2}	Newton	$m\ kg\ s^{-2}$	$N\ =\ $ kg meter per sec
Pressure	$L^{-1}MT^{-2}$	Pascal	$m^{-1}\ kg\ s^{-2}$	$Pa\ =\ $ newton per m^2
Power	L^2MT^{-3}	Watt	$m^2\ kg\ s^{-3}$	$W\ =\ $ joules per second
Electric charge	T	Coulomb	$A\ s$	$C\ =\ $ ampere seconds
Electric potential	L^2MT^{-3}	Volt	$m^2\ kg\ s^{-3}\ A^{-1}$	$V\ =\ $ watts per ampere
Electric resistance	L^2MT^{-3}	Ohm	$m^2\ kg\ s^{-3}\ A^{-2}$	$\Omega\ =\ $ volts per ampere
Electric capacitance	$L^{-2}M^{-1}T^4$	Farad	$m^{-2}\ kg^{-1}s^4\ A^2$	$F\ =\ $ coulombs per volt
Activity of radionucleotide	T^{-1}	Becquerel	s^{-1}	$Bq\ =\ $ 60 dpm
Enzyme activity		Katal		$kat\ =\ $ mol per sec

In SI, $1\ l = 1\ (dm)^3 = 10^{-3}\ m^3$ or $m^3/1000$. Also in SI, $1\ ml = 1\ (cm)^3 = 10^{-6}\ m^3$ or $m^3/10^6 = 10^{-3}\ dm^3$. Use of the symbol M for *molar* is not recommended in SI because M is assigned to the prefix *mega* (Table 1-3). Thus, what previously would have been reported as a 1 M solution, would, in SI, be $1\ mol\ l^{-1} = 1$ mol/dm^3 or $1\ mol\ dm^{-3}$. However, in spite of this recommendation, M is still used commonly as a free-standing symbol or with prefixes such as mM or µM, because (a) it is convenient and (b) it does not cause confusion. Concentration, expressed on the basis of amount of substance per kg, i.e. *weight-based concentration* is not in common use in the biomedical sciences and will generally not be used in this book. The reason why life scientists prefer volume-based to weight-based units of concentration is that it is easier to measure the volume than the weight of a solution. Disadvantages of using volume-based concentration are that it is temperature dependent (volume increases with temperature) and that it does not provide information on the volume of solvent used.

Table 1-3. Prefixes Used with Quantities and Units

Prefix	Multiple or Fraction	Symbol
exa	10^{18}	E
peta	10^{15}	P
tera	10^{12}	T
giga	10^9	G
mega	10^6	M
kilo	10^3	k
hecto	10^2	h
deci	10^{-1}	d
centi	10^{-2}	c
milli	10^{-3}	m
micro	10^{-6}	μ
nano	10^{-9}	n
pico	10^{-12}	p
femto	10^{-15}	f
atto	10^{-18}	a

1-2.   Simple rules apply to the correct use of units and dimensions

Adherence to the following rules will help to ensure that units and dimensions are used correctly in biomolecular calculations.
*   Report all values with units; but report the ratios of values which have the same units as pure numbers since the units cancel out.
*   Add or subtract only those values which have the same units.
*   Do not make direct comparison of values with different dimensions.
*   Assign a given variable the same units throughout a calculation.
*   Multiply or divide units when values are multiplied or divided; then include the product or quotient of the units with the calculated result.
*   Multiply or divide units at each step of a calculation; do not wait until the end of the calculation to deduce what the units should be.
*   Assign units of x to lengths along the x-axis, units of y along the y-axis.
*   The slope of a curve should have the dimensions of y divided by those of x.
*   Multiply tabulated or plotted values by a suitable power of 10 to avoid having inconveniently large or small numbers.
*   The dimensions on the left side of an equation must equal those on the right.

Dimensional analysis is the process of checking units at each step in a calculation to learn whether the final values have the correct units. Dimensional analysis helps investigators to detect errors made during calculations.

1-3. Calculations should be reported only with significant digits

Calculations carried out with computers can yield results with 10 or more digits. Although few persons would report biomedical results with 10 digits, observant readers of the biomedical literature will notice that it is not uncommon to find that too many digits are communicated in research reports. The inclusion of more digits than warranted implies that the results are more accurate than they actually are. To avoid misleading one's colleagues, report measurements and calculated results only with significant digits.

The significant digits in a measurement are those which were measured with some certainty plus one additional digit which contains some uncertainty. Adherence to the following rules will help to ensure that results are computed and reported with the appropriate number of digits:
•   During addition and subtraction, keep only as many decimal places in the final value as there are in the measurement which has the fewest decimal places.
•   In multiplication and division, report individual measurements and calculated results with a number of digits no greater than those in the measurement with the greatest uncertainty.
•   In logarithms, the characteristic is not a significant number but all digits of the mantissa generally are significant.
•   To round off a value, increase the last significant digit by 1 if the number which follows it is 5 or greater. Do not change the last significant digit if the number which follows it is less than 5.

1-4. Amount of substance concentration is used often in bioresearch

In SI, use of mol l^{-1} and related units is acceptable for expressing concentration on the basis of amount of substance per unit volume. *A mole is defined as the amount of substance which contains as many elementary entities as there are atoms in 0.012 kg of carbon-12.* It has been shown experimentally that 0.012 kg of carbon-12 contains 6.02217×10^{23} atoms. This number is the Avogadro number and it is used for defining the mole of any elementary unit such as atom, molecule, ion, proton or electron. A mole also may be defined as being equivalent to the molecular weight (more correctly molar mass in g) of a substance. Therefore, the weight of a substance (its mass) and moles may be interconverted as shown in equations (1) to (4):

$$\text{number of moles} = \frac{\text{wt in g}}{\text{molecular wt}}, \tag{1}$$

$$\text{molecular weight} = \frac{\text{wt in g}}{\text{number of mols}}, \tag{2}$$

$$\text{number of moles} = \frac{\text{wt in g}}{(\text{wt in g/\# of mols})}, \tag{3}$$

$$\text{number of m mols} = \frac{\text{wt in mg}}{(\text{wt in mg/\# of mols})}. \tag{4}$$

Molar concentration, mol l^{-1}, the number of moles per liter, (abbreviated M for convenience in some equations below), can be expressed as a function of related quantities according to the following relationships:

$$\text{Molarity, M} = \frac{\text{mols of solute}}{\text{l of solution}}, \tag{5}$$

$$\text{number of moles} = \text{vol, l} * \text{mol l}^{-1}, \tag{6}$$

$$\text{number of m mols} = \text{vol, ml} * \text{mol l}^{-1}, \tag{7}$$

$$\text{volume in liters} = \frac{\text{number of mols}}{\text{mol l}^{-1}}, \tag{8}$$

$$\text{weight in grams} = \text{number of mols} * \text{Mol wt}, \tag{9}$$

$$\text{molecular weight} = \frac{\text{wt in g}}{\text{number of mols}}. \tag{10}$$

Equivalents and normality are not recommended in SI; but, because they are still in common use, we will include them. A *normal solution* contains one equivalent per l. For acids and bases, one equivalent is the weight which contains 1 mole of replaceable H$^+$ or OH$^-$. Thus:

$$\text{normality, N} = \frac{\text{number of equivalents}}{\text{volume in l}}, \tag{11}$$

$$\text{normality, N} * \text{vol, l} = \text{number of equivalents}, \tag{12}$$

$$\text{normality, N} * \text{vol, ml} = \text{number of milliequivalents}, \tag{13}$$

$$\text{normality, N} = \frac{\text{number of milliequivalents}}{\text{volume in ml}}. \tag{14}$$

1-5.   Many related variables are used to express concentration

Given any three of the four variables (a) volume in liters (b) molarity (c) weight in grams or (d) molecular weight, the remaining variable may be calculated from equations (15) to (19):

molarity, M	= number mols/l	= $(wt, g/Mol\ wt)\ l^{-1}$,	(15)
volume, l	= number mols/M	= $(wt, g/Mol\ wt)\ M^{-1}$,	(16)
weight, g	= number mols * Mol wt	= $l*M*Mol\ wt$,	(17)
number of mols	= volume, l * M		(18)
mol wt	= wt, g/number mols	= $wt, g/(l*M)$.	(19)

Trace quantities of solutes are often expressed on the basis of *parts per million,* abbreviated ppm; but use of this unit is not recommended in SI. Again, because this manner of expressing results is firmly entrenched in some areas of the life sciences, it is given here. *Ppm* refers to the *number of parts of a given component in 1 million parts of a mixture* and it may be expressed either on the basis of weight or volume. On a volume basis

$$ppm\ (wt/vol) \quad = \quad g\ ml^{-1} * 10^6, \quad (20)$$
$$1\ ppm\ (wt/vol) \quad = \quad 1\ mg\ l^{-1} \quad = \quad 1\ \mu g\ ml^{-1}. \quad (21)$$

Ppm and mol l^{-1} are related *approximately* as follows:

$$ppm \quad = \quad Mol\ wt * Molarity *1000, \quad (22)$$
$$Molarity \quad = \quad ppm\ /\ (Mol\ wt * 1000). \quad (23)$$

To calculate the weight of a compound required for preparing a desired concentration of a constituent metal ion as ppm, the following relationship may be used:

$$\frac{Weight\ of}{compound\ (mg)} \quad = \quad \frac{Mol\ wt\ of\ compound * ppm\ of\ metal\ ion * vol\ (ml)}{(Atomic\ wt\ of\ metal * \#\ of\ metal\ ions\ in\ 1\ molecule * 1000)} \quad (24)$$

Similarly, in those disciplines in which the use of equivalents and normality is common, amount of solute expressed as meq liter^{-1} may be calculated from:

$$meq\ liter^{-1} \quad = \quad \frac{ppm}{(mg/meq)}, \quad (25)$$

$$= \quad \frac{mg/liter}{(mg/meq)}. \quad (26)$$

1-6. Making dilutions is guided by a single principle

Dilutions are guided by one principle: the *total amount* of solute in the concentrated solution is equal to the total amount of solute in the dilute solution; however, the *concentration* of solute is lower in the dilute solution than in the concentrated solution. This principle is expressed in the equation:

$$vol_{concd} * C_{concd} \quad = \quad vol_{dil} * C_{dil}, \quad (27)$$

where	vol	=	volume,
	C	=	concentration,
	concd	=	concentrated solution,
	dil	=	dilute solultion.

Thus, the number of moles (or equivalents) in the dilute solution must equal the number of moles or equivalents taken from the concentrated solution. Appendix 1 provides data on the molarities and normalities of concentrated acids and ammonia. These data are useful for calculating the amounts of concentrated acids or ammonia needed for preparing dilute solutions of approximate molarity or normality.

1-7.   Percent solutions are common in the life sciences

In the life sciences, concentration is often expressed as *percent weight/volume*, abbreviated % wt/vol and defined as *weight of solute in grams per 100 ml of solution.* Similar units are percent weight/weight (% wt/wt) defined as the weight of solute in grams per 100 grams of solution and percent volume/volume (% vol/vol) defined as the volume of the active liquid (the solute) dissolved in a second liquid to give 100 ml of solution. Another unit, milligram percent, defined as weight of solute in mg per 100 ml of solution or body fluid, was used frequently at one time in clinical laboratories but is now looked at unfavorably by many scientists. The reason for the frequent use of percent concentrations is that biomedical scientists often work with substances which are of unknown molecular weight and of doubtful purity. The use of molar units is impossible under these circumstances. Despite their wide acceptance, units of percent concentration are not recommended in SI. Calculations of percent (wt/vol), percent (vol/vvol) and percent (wt/wt) may be made readily from the following relationships:

Percent (weight/volume)

Wt of solute (g)	=	vol, ml * % as decimal (g ml $^{-1}$),	(28)
Percent (wt/vol)	=	(wt solute / vol, ml) * 100,	(29)
Volume of soln, ml	=	wt solute / % (wt/vol) as decimal ;	(30)

Percent (volume/volume)

Vol of solute (ml)	=	vol soln (ml) * % (vol/vol) as decimal ,	(31)
Percent (vol/vol)	=	(ml solute / vol soln) * 100 ,	(32)
Volume of soln, ml	=	ml solute / % (vol/vol) as decimal;	(33)

Percent (weight/weight)

Wt of solute (g)	=	wt soln, g * % (wt/wt) as decimal,	(34)
Percent (wt/wt)	=	(wt solute / wt of soln) * 100,	(35)
Weight of soln, g	=	wt solute / % (wt/wt) as decimal.	(36)

For density gradient solutions (e.g. in centrifugation), a useful relation is:

$$\% \text{ wt/vol} \quad = \quad \rho \ (\% \text{ vol/vol}), \tag{37}$$

where ρ is the density of the solution.

1-8.   Weight-based concentration is sometimes used in bioresearch

In physicochemical studies, concentrations often are expressed in terms of weight instead of volume because, unlike volume-based concentration in which volume of the solvent is temperature dependent, weight-based concentrations are not influenced by temperature. We devote the remainder of this short section to a discussion of weight-based concentration since this method of expressing concentration sometimes appears in the biomedical literature. An occasion during which it would be appropriate to use weight-based concentration is one in which high concentrations of solutes cause unavoidable changes in the volumes and temperatures of solutions.

Molality, designated by m prior to SI, refers to the number of moles of solute per 1000 g of solvent, i.e., mol kg^{-1} of solvent. Molal units, in addition to their use with high concentrations of solutes, are used with procedures carried out at different temperatures. Examples of such procedures are measurements of colligative properties of solutions such as depression of the freezing point and elevation of the boiling point.

In the chemical industry, weight-based concentration units are used by manufacturers of nitric, sulfuric, hydrochloric and other strong acids. *Percent weight/ weight* (% wt/wt), defined as the weight of solute in grams per 100 grams of solution, and the *density, defined as weight per unit volume*, are given with the technical information provided by the manufacturers. These two pieces of information permit calculation of normalities and molarities of the strong acids.

1-9.   Worked examples

Example 1-1. How many micromoles are there in 3.4 mg of NADH, molecular weight 664.44?

From equation (1):

$$\text{Number of moles} \quad = \quad \frac{\text{weight in g}}{\text{molecular weight}} \, ,$$

$$= \quad \frac{0.0034 \text{ g}}{664.44 \text{ g mole}^{-1}} \, ,$$

$$= \quad 5.117 * 10^{-6} \text{ mol} = 5.12 \text{ } \mu \text{ mol}.$$

Example 1-2. How many grams of D-ribose 5-phosphate, molecular weight 230.11, would you have to weigh out to obtain 0.005 moles?

From equation (9):

weight in grams   =   number of mols * Mol wt, g mol^{-1},
     =   0.005 mol * 230.11 g mol^{-1},
     =   1.1506 g.

Example 1-3. How many equivalents are there in 15 ml of concentrated HCl which is approximately 12 molar in concentration?

From equation (12):

number of equivalents $\quad = \quad$ normality, N * vol, liters.

One mole of HCl contains 1 replaceable H^+; $\therefore$ 12 M HCl is also 12 N;

| umberof equivalents | $=$ | 12 equiv l^{-1} * 0.015 l , |
| | $=$ | 0.18 equivalents. |

Example 1-4. A nutrient medium for growing the fungus *Neurospora crassa* contains 4 grams of L-asparagine per liter. What is the molarity of asparagine in the medium? The molecular weight of asparagine is 132.13.

From equations (1) and (5):

numberof moles $\qquad = \qquad \dfrac{\text{weight in grams}}{\text{Molecular weight}}$,

molarity, M $\qquad = \qquad \dfrac{\text{moles of solute}}{\text{liter of solution}}$,

$\therefore$ moles asparagine $\qquad = \qquad \dfrac{4 \text{ grams}}{132.13 \text{ grams mol}^{-1}}$,

| | $=$ | 0.0303 mol , |
| 4 g asparagine l^{-1} | $=$ | $3.03 * 10^{-2}$ M. |

Example 1-5. What weight of each component is required for preparing 500 ml of a solution containing: 0.05 M KH_2PO_4; 0.3 M D-mannitol; 25 mM $MgCl_2$ and 1.5 mM EDTA? The relevant molecular weights are: KH_2PO_4, 136.09; D-mannitol, 182.18; $MgCl_2.6\ H_2O$, 203.31 and EDTA 292.24.

From equation (9):

Wt, g liter $^{-1}$	$=$	number moles l^{-1} * mol wt,
KH_2PO_4: wt, g l^{-1}	$=$	0.05 mol l^{-1} * 136.09 g mol^{-1} ,
	$=$	6.8045 g l^{-1} $\quad = \quad$ 3.4023 g/500 ml ,
Mannitol: wt, g l^{-1}	$=$	0.3 mol l^{-1} * 182.18 g mol^{-1} ,
	$=$	54.654 g l^{-1} $\quad = \quad$ 27.327 g/500 ml ,
$MgCl_2$: wt, g l^{-1}	$=$	0.025 mol l^{-1} * 203.31 g mol^{-1} ,
	$=$	5.0823 g l^{-1} $\quad = \quad$ 2.5412 g/500 ml ,
EDTA: wt, g l^{-1}	$=$	0.0015 mol l^{-1} * 292.24 g mol^{-1} ,
	$=$	0.4384 g l^{-1} $\quad = \quad$ 0.2192 g/500 ml .

Example 1-6. How many grams of KOH are required to prepare 300 ml of 0.25 M solution? Express your answer as $g \, l^{-1}$; % wt/vol and normality. The molecular weight of KOH is 56.11.

From equations (5) and (1):

molarity, M	=	$\dfrac{\text{mols of solute}}{\text{l of solution}}$,
Concn. in $g \, l^{-1}$	=	moles l^{-1} * molecular wt ,
0.25 M	=	$0.25 \text{ mol } l^{-1}$,
$G \, l^{-1}$	=	$0.25 \text{ mol } l^{-1}$ * 56.11 g mol^{-1} ,
	=	$14.03 \text{ g } l^{-1}$ = 4.21 g/300 ml ,
%wt/vol: $14.03 \text{ g } l^{-1}$	≡	1.4 % (wt/vol).

Normality: One mol of KOH has one replaceable OH^-

$$\therefore 0.25 \text{ M KOH} \qquad \equiv \qquad 0.25 \text{ N.}$$

Example 1-7. You have 0.5 ml of a chloroform solution containing 25 mg ml^{-1} each of cholesterol oleate, trilinolenin, oleic acid and ergosterol. To what volume should this chloroform solution be diluted so that the dilute solution contains 5 μg of each component per μl of solution?

From equation (27):

Quantity in dilute solution = Quantity in concentrated solution ,
    Or $v_2 * c_2$ = $v_1 * c_1$,
where subscript 2 refers to dilute solution, 1 to concentrated solution ,

$$v_2 \text{ ml} * 5 \text{ mg ml}^{-1} \qquad = \qquad 0.5 \text{ ml} * 25 \text{ mg ml}^{-1} ,$$

$$v_2 \qquad = \qquad \frac{0.5 \text{ ml} * 25 \text{ mg ml}^{-1}}{5 \text{ mg ml}^{-1}} ,$$

$$v_2 \qquad = \qquad 12.5/5 \text{ ml} = 2.5 \text{ ml} .$$

Example 1-8. What is the concentration, expressed as % (wt/vol), of a solution which contains 250 mg of bovine serum albumin in 250 ml of solution?

From equation (29):

% (wt/vol)	=	grams solute/100 ml solution ,
250 mg/250 ml	=	0.25 g/250 ml ,
	=	0.1 g/100 ml ,
$\therefore$ 250 mg/250 ml	=	0.1 % (wt/vol).

Example 1-9. Calculate the molality of concentrated sulfuric acid, molecular weight 98.1 and concentration 96% (wt/wt).

$$\text{number of mols} = \frac{\text{wt in g}}{\text{molecular wt}},$$

$$\text{molality} = \frac{\text{number of moles}}{1000 \text{ g of solvent}},$$

$$\text{wt of solvent} = \text{wt. of solution - wt. of solute,}$$

$$\text{mol } H_2SO_4/100 \text{ g} = \frac{96 \text{ g}}{98.1 \text{ g mol}^{-1}} = 0.979 \text{ mol,}$$

$$\text{wt of solvent } (H_2O) = 100 \text{ g - 96 g} = 4 \text{ g,}$$

$$\text{molality} = \frac{0.979 \text{ mol}}{4 \text{ g}} * 1000 = 244.75 \text{ mol kg}^{-1}.$$

1-10. Spreadsheets are underused in biomedical work and training

A *spreadsheet* is a computer program which permits a computer user to enter numerical data into a *worksheet* consisting of a rectangular grid of columns and rows. Spreadsheets allow a user to perform arithmetic, algebraic, trigonometric, scientific, statistical, or other calculations on entered, numerical data. After the user tells the program how to act upon one set of data, any change in the entered data causes automatic recalculation of the previous result.

Columns on worksheets are identified by letters, whereas rows are identified by numbers. The vertical and horizontal lines which delineate the columns and rows intersect to form small rectangles known as cells. Thus, *a cell is the smallest working unit in a spreadsheet; columns and rows consist of cells;* and *one locates a cell by its column and row identification number.* For example, the cell in *column 2, row 5,* is *B5* and, as indicated above, the *entire set of columns and rows* into which data are entered, stored, and transformed is the *worksheet*.

In addition to numerical data, text also may be entered into cells of a spreadsheet to label rows, columns, sets of data or results. The specific arrangement or specific pattern of labels, entered data and formulas by which a given problem is solved with spreadsheets is known as a *template*. Once a template has been set up, similar problems may be solved over and over again simply by entering new data into the appropriate cells.

Spreadsheets, therefore, may be regarded as sophisticated programmable calculators. But, unlike calculators, spreadsheets allow investigators to view data and results simultaneously; to change entered data readily; to rearrange data and results as circumstances dictate; to copy or use parts of the spreadsheet without having to reenter data; and to use programmed steps without having to write programs. Another advantage of spreadsheets is that a cell may be made to show either a calculated result or the formula which gave the result. In all of the problems solved with a spreadsheet in this book we will present worksheets showing both the calculated values and the formulas which led to the calculated results. In addition, all problems solved with a spreadsheet will be preceded by the same problems worked manually except for the statistical examples. The reader therefore may follow the steps in the worksheet by comparison with those used in the manual calculation - with one exception. Instead of plotting linear functions of y corresponding to values of x (e.g. instead of estimating molecular weights from plots of log molecular weight versus mobility in gel electrophoresis), we will use the method of least squares. In such use of the *method of least-squares*, the parameters b and a of the slope-intercept form of the equations of a straight line, i.e. $y = bx + a$, are chosen so that the sum of the squares of the vertical distances of the points from the straight line is a minimum. We note, however, that pitfalls and debate await such use of least square methods, partly because it may entail linear transformations of nonlinear equations (cf. sections 5 –8, 10 –14 and 10 --17).

Spreadsheets are now used routinely in the business world. But their use is, apparently, not yet commonplace in the biomedical sciences. Nevertheless a scientist may, profitably, use spreadsheets for routine biomolecular computations, and for mathematical modeling, curve fitting, and statistical analyses. In *Basic Biochemical Laboratory Procedures and Computing,* every problem we work out manually we also solve with a spreadsheet and we identify worksheets by their titles and by the numbers of the worked examples. For instance, immediately following this section, worksheets of examples from Chapter 1 (examples 1–1 to 1–9) may be found under the running heads "Spreadsheet Solutions" or "Spreadsheet Formulas." That is, following the running head, we give in **bold type** within the worksheet, the number of the chapter followed by the number of the example(s) within that chapter (e.g. **Example 1–4** is the spreadsheet solution of Example 1–4 in section 1–9 above). Perhaps, the reader who has used spreadsheets infrequently will be convinced of their usefulness by setting up worksheets similar to those shown below. In setting up our worksheets, we had three related goals:

- that they be usable with virtually any spreadsheet program on the market;
- that they be usable for learning how to perform biomolecular computations;
- that they be easier to use than calculators or pencil and paper.

These goals imposed some restrictions. For example, we had to avoid using powerful routines if they varied greatly among the major spreadsheets. Thus, we did not use the regression routines *Linest* in the Paste Function of the Format menu of Microsoft *Excel*®. Neither did we use *Regression* from the Data menu of *Lotus*®*1-2-3*®. Instead, we have written simple least-squares routines which could be used on any spreadsheet and on any computer see section 10-15 for sums required and for equations (2), (3) and (7). For similar reasons, we have not used *macros* and other powerful features which differ among spreadsheets and require the user to have some expertise. As a more automated alternative we have, however, written thirty-seven computer programs which perform more than one hundred different calculations.

The foregoing restrictions notwithstanding, we submit that it is usually advantageous to use the mathematical instructions of spreadsheets since they are powerful and more accurate than building a formula manually. To conclude this section, we define *two classes of instructions*: operators and functions. *Operators* are of four types: arithmetic, text, comparison, and reference. We will be concerned mainly with the *arithmetic operators*:

- addition (+),
- subtraction (-),
- multiplication (*),
- division (/),
- exponentiation (^) and
- percent (%).

Functions are instructions which transform one value to another.

- In *Excel®* examples are: LOG 10, LN(x), EXP(x), SQRT(x), SUM, VAR, STDEV, MINVERSE and MMULT.

- In *Lotus®1-2-3®* equivalent functions are: @LOG, @LN, @EXP, @SQRT, @SUM, @VARS, @STDS, Matrix invert, and multiply, pronounced *"at log"* etc.

In this chapter, we have solved all spreadsheet examples both with *Excel®* and *Lotus®1-2-3®* to demonstrate how both programs may be used to solve the same problems. With examples 1-10 to 1-12, the reader will note that, while it is possible to show an entire worksheet of input and output on one page, the same could not be done when long formulas are shown. To see long formulas on a computer screen, one has to widen the columns and scroll horizontally; but with the printed page the practical choice is to use more than one page to display the long formulas.

Thus, conventions for naming functions differ among spreadsheets; but apart from these differences in names, our worksheets prepared with *Excel* for the Macintosh, should be readily adaptable to *Lotus 1-2-3, Quattro Pro,* and other spreadsheet programs on DOS, Windows, and other operating systems. Users should consult the appropriate manuals for additional information, including how to prepare graphs from least squares computations.

1-11. Spreadsheet solutions, Microsoft Excel

	A	B	C	D	E	F	G
1	EXAMPLE 1-1				EXAMPLE 1-5		
2		INPUT	OUTPUT			INPUT	OUTPUT
3	Weight, g	0.0034			EDTA:		
4	Mol. weight	664.44			Mol/liter	0.0015	
5	# of moles		5.117E-06		Mol. weight	292.24	
6					Wt., (g/liter)		0.43836
7	EXAMPLE 1-2						
8		INPUT	OUTPUT		EXAMPLE 1-6		
9	# of mols	0.005				INPUT	OUTPUT
10	Mol. weight	230.11			Molarity, KOH	0.25	
11	Weight		1.15055		Mol. weight	56.11	
12					Wt., (g/liter)		14.0275
13	EXAMPLE 1-3				g/300 ml		4.20825
14		INPUT	OUTPUT		g/100 ml		1.40275
15	Normality	12					
16	Volume	0.015			EXAMPLE 1-7		
17	# of equivs.		0.18			INPUT	OUTPUT
18					C2, mg/ml	5	
19	EXAMPLE 1-4				V1, ml	0.5	
20		INPUT	OUTPUT		C1, mg/ml	25	
21	Weight, g	4			V2, ml		2.5
22	Mol. weight	132.13					
23	# of mols/l		0.0302732		EXAMPLE 1-8		
24	Molarity		3.03E-02			INPUT	OUTPUT
25					g, solute	0.25	
26	EXAMPLE 1-5				ml solution	250	
27		INPUT	OUTPUT		% (w/v)		0.1
28	K phosphate:						
29	Mol/liter	0.05			EXAMPLE 1-9		
30	Mol. weight	136.09				INPUT	OUTPUT
31	Wt., (g/liter)		6.8045		Wt. solution, g	100	
32	Mannitol:				Wt. solute, g	96	
33	Mol/liter	0.3			Wt. solvent, g		4
34	Mol. weight	182.18			Mol. wt. acid	98.1	
35	Wt., (g/liter)		54.654		Mol. acid/100g		0.97859327
36	Mg chloride:				Molality		244.648318
37	Mol/liter	0.025					
38	Mol. weight	203.31					
39	Wt., (g/liter)		5.08275				
40							

Spreadsheet formulas, Microsoft Excel

	A	B	C	D	E	F	G
1	**EXAMPLE 1-1**				**EXAMPLE 1-5**		
2		INPUT	OUTPUT			INPUT	OUTPUT
3	Weight, g	0.0034			EDTA:		
4	Mol. weight	664.44			Mol/liter	0.0015	
5	# of moles		=B3/B4		Mol. weight	292.24	
6					Wt., (g/liter)		=F4*F5
7	**EXAMPLE 1-2**						
8		INPUT	OUTPUT		**EXAMPLE 1-6**		
9	# of mols	0.005				INPUT	OUTPUT
10	Mol. weight	230.11			Molarity, KOH	0.25	
11	Weight		=B9*B10		Mol. weight	56.11	
12					Wt., (g/liter)		=F10*F11
13	**EXAMPLE 1-3**				g/300 ml		=G12*0.3
14		INPUT	OUTPUT		g/100 ml		=G13/3
15	Normality	12					
16	Volume	0.015			**EXAMPLE 1-7**		
17	# of equivs.		=B15*B16			INPUT	OUTPUT
18					C2, mg/ml	5	
19	**EXAMPLE 1-4**				V1, ml	0.5	
20		INPUT	OUTPUT		C1, mg/ml	25	
21	Weight, g	4			V2, ml		=(F19*F20)/F18
22	Mol. weight	132.13					
23	# of mols/l		=B21/B22		**EXAMPLE 1-8**		
24	Molarity		=C23			INPUT	OUTPUT
25					g, solute	0.25	
26	**EXAMPLE 1-5**				ml solution	250	
27		INPUT	OUTPUT		% (w/v)		=(F25/F26)*100
28	**K phosphate:**						
29	Mol/liter	0.05			**EXAMPLE 1-9**		
30	Mol. weight	136.09				INPUT	OUTPUT
31	Wt., (g/liter)		=B29*B30		Wt. solution, g	100	
32	**Mannitol:**				Wt. solute, g	96	
33	Mol/liter	0.3			Wt. solvent, g		=F31-F32
34	Mol. weight	182.18			Mol. wt. acid	98.1	
35	Wt., (g/liter)		=B33*B34		Mol. acid/100g		=F32/F34
36	**Mg chloride:**				Molality		=(G35/G33)*1000
37	Mol/liter	0.025					
38	Mol. weight	203.31					
39	Wt., (g/liter)		=B37*B38				
40							

Spreadsheet solutions, Lotus 1-2-3

A	A	B	C	D	E	F	G
1	EXAMPLE 1-1				EXAMPLE 1-5		
2		INPUT	OUTPUT			INPUT	OUTPUT
3	Weight, g	0.0034			EDTA:		
4	Mol. weight	664.44			Mol/liter	0.0015	
5	No. of moles		5.12E-06		Mol. weight	202.24	
6					Wt., (g/liter)		0.30336
7	EXAMPLE 1-2						
8		INPUT	OUTPUT		EXAMPLE 1-6		
9	No. of mols	0.005				INPUT	OUTPUT
10	Mol. weight	230.11			Molarity, KOH	0.25	
11	Weight		1.15055		Mol. weight	56.11	
12					Wt., (g/liter)		14.0275
13	EXAMPLE 1-3				g/300 ml		4.20825
14		INPUT	OUTPUT		g/100 ml		1.40275
15	Normality	12					
16	Volume	0.015			EXAMPLE 1-7		
17	No. of equivs		0.18			INPUT	OUTPUT
18					C2, mg/ml	5	
19	EXAMPLE 1-4				V1, ml	0.5	
20		INPUT	OUTPUT		C1, mg/ml	25	
21	Weight, g	4			V2, ml		2.5
22	Mol. weight	132.13					
23	Mols/l		0.0302732		EXAMPLE 1-8		
24	Molarity		0.0302732			INPUT	OUTPUT
25					g. solute	0.25	
26	EXAMPLE 1-5				ml solution	250	
27		INPUT	OUTPUT		% (w/v)		0.1
28	K phosphate:						
29	Mol/liter	0.05			EXAMPLE 1-9		
30	Mol weight	136.09				INPUT	OUTPUT
31	Wt., (g/liter)		6.8045		Wt. solution, g	100	
32	Mannitol:				Wt. solute, g	96	
33	Mol/liter	0.3			Wt. solvent, g		4
34	Mol. weight	182.18			Mol. wt. acid	98.1	
35	Wt., (g/liter)		54.654		Mol acid/100g		0.9785933
36	Mg chloride:				Molality		244.64832
37	Mol/liter	0.025					
38	Mol. weight	203.31					
39	Wt., (g/liter)		5.08275				

Spreadsheet formulas, Lotus 1-2-3

A	A	B	C	D	E	F	G
1	EXAMPLE 1-1				EXAMPLE 1-5		
2		INPUT	OUTPUT			INPUT	OUTPUT
3	Weight, g	0.0034			EDTA:		
4	Mol. weight	664.44			Mol/liter	0.0015	
5	No. of moles		+B3/B4		Mol. weight	202.24	
6					Wt., (g/liter)		+F4*F5
7	EXAMPLE 1-2						
8		INPUT	OUTPUT		EXAMPLE 1-6		
9	No. of mols	0.005				INPUT	OUTPUT
10	Mol. weight	230.11			Molarity, KOH	0.25	
11	Weight		+B9*B10		Mol. weight	56.11	
12					Wt., (g/liter)		+F10*F11
13	EXAMPLE 1-3				g/300 ml		+G12*0.3
14		INPUT	OUTPUT		g/100 ml		+G13/3
15	Normality	12					
16	Volume	0.015			EXAMPLE 1-7		
17	No. of equivs		+B15*B16			INPUT	OUTPUT
18					C2, mg/ml	5	
19	EXAMPLE 1-4				V1, ml	0.5	
20		INPUT	OUTPUT		C1, mg/ml	25	
21	Weight, g	4			V2, ml		(F19*F20)/F18
22	Mol. weight	132.13					
23	Mols/l		+B21/B22		EXAMPLE 1-8		
24	Molarity		+C23			INPUT	OUTPUT
25					g. solute	0.25	
26	EXAMPLE 1-5				ml solution	250	
27		INPUT	OUTPUT		% (w/v)		(F25/F26)*100
28	K phosphate:						
29	Mol/liter	0.05			EXAMPLE 1-9		
30	Mol weight	136.09				INPUT	OUTPUT
31	Wt., (g/liter)		+B29*B30		Wt. solution, g	100	
32	Mannitol:				Wt. solute, g	96	
33	Mol/liter	0.3			Wt. solvent, g		+F31-F32
34	Mol. weight	182.18			Mol. wt. acid	98.1	
35	Wt., (g/liter)		+B33*B34		Mol acid/100g		+F32/F34
36	Mg chloride:				Molality		(G35/G33)*100
37	Mol/liter	0.025					
38	Mol. weight	203.31					
39	Wt., (g/liter)		+B37*B38				

1-12. The mathematics of measurement is at the core of data analysis

We started this chapter with a discussion of units of measurement. We conclude with a discussion of the mathematical nature of measurement. Our aim is to provide an introduction to some statistical methods for describing, and making inferences about, data obtained with laboratory procedures such as those of Chapters 2 to 9. The ideas introduced here fall into two categories: *descriptive statistics,* which deals with organizing and presenting data in a systematic way, and *inferential statistics,* or methods of making inferences about large populations from small samples. These ideas are discussed more fully in Chapter 10.

Measurements are subject to error. The term *error* in this discussion refers to *the difference between the true value and a measured value. Systematic errors* are those which yield incorrect results by a constant amount; they are often reproducible and may be eliminated by correcting the inaccuracies caused by instrumental, methodological or human factors. *Random errors,* however, are not reproducible or predictable; they are not eliminated by correcting experimental factors; but they may be minimized by the use of appropriate statistical procedures.

Because of random error laboratory workers learn early in their careers that even after the elimination of systematic errors, multiple measurements of the same item do not yield sets of identical values. Some reasons for this variability are that an investigator may be

- unable to control external factors,
- unable to recognize factors which cause errors, or
- unable to control statistical phenomena such as counts of radioactivity.

Experience has shown that most individual measurements in sets of data usually are concentrated or clustered around the means or averages of the respective sets; that is, measurements show *central tendency* so that individual values which differ from the mean by small amounts occur more frequently than values which differ from the mean by a large extent. The *arithmetic mean,* $\bar{x}$, of a sample is defined as

$$\bar{x} = \frac{x_1 + x_2 + x_3 \ldots + x_n}{n} = \frac{\sum\limits_{i=1}^{i=n} x_i}{n} \tag{38}$$

Random errors have a random disribution and follow the relevant laws of probability discussed in section 10-4. These laws are based on an exponential equation which describes the scatter of measurements around the mean. The exponential equation is known variously as the *normal curve,* the *normal law of error,* the *normal distribution* or the *Gaussian distribution:*

$$f = \frac{1}{\sigma\sqrt{(2\pi)}} e^{-(x-\mu)^2/2\sigma^2} \tag{39}$$

where   f   is the frequency of the variable x,

μ = the arithmetic mean of the population (section 10 –7),

σ^2 = the variance = Σ (x-μ)^2 / n (section 10 –7),

and σ = the standard deviation of the population (section 10 –7).

Two parameters of equation (39), μ and σ, fit the normal curve to a particular set of data.

The *standard deviation* of a sample, s, is a good measure of variability; it is the square root of the *variance*, (section 10 –7) so that

$$\text{the variance, } s^2 \; = \; \frac{\sum X_i^2 - \dfrac{(X_i)^2}{n}}{n-1} \, , \tag{40}$$

$$\text{and standard deviation, } s \; = \; \sqrt{\frac{\sum X_i^2 - \dfrac{(X_i)^2}{n}}{n-1}} \, , \tag{41}$$

In words, therefore, the *sample variance* is the *sum of squares* (the square of the deviations of each value from the mean), divided by n-1, the *degrees of freedom*. The latter, discussed in section 10 –7, varies according to the statistical test being used and is not always equal to the number of measurements minus one.

When reporting experimental results, one should provide a measure of the central tendency, or the opposite, the variability, of the results. The standard deviation, s, is such a measure. The standard deviation, moreover, yields several other values for evaluating the quality of data. These include (cf. section 10 –7) computing the *standard error* (s ÷ $\sqrt{n}$) for a variety of purposes such as

- estimating the precision of making a measurement;
- estimating the precision of computing an average;
- comparing the precision of making two sets of measurements; and
- evaluating the difference between two sets of means.

One therefore may use the standard deviation and the standard error to report variability of many types of results. We discuss these in greater detail in section 10 –7.

The *coefficient of variation*, section 10 –7, is designated by C (sometimes by CV or V). It is the standard deviation divided by the mean * 100. Since *s* and $\bar{x}$ have the same units, C is a dimensionless value which may be used to compare variability and central tendency in different samples:

$$C \; = \; \frac{\text{the standard deviation}}{\text{the mean}} \; * \; 100. \tag{42}$$

The coefficient of variation therefore expresses variability relative to the mean as a percentage.

After one calculates a result, it is important to ask "within what interval would the result be true?" This interval is the *confidence interval*, and it indicates how probable or likely it is that a mean lies between two values. Confidence intervals (section 10 –7) are usually calculated from the quantity known as Student's *t*, section 10 –8, defined by

$$t \quad = \quad \frac{(\bar{x} - \mu)}{s_{\bar{x}}} \tag{43}$$

Rearranging the equation for calculating Student's *t* yields

$$\text{Confidence interval} \quad = \quad \bar{x} \pm t. \frac{s}{\sqrt{n}} \tag{44}$$

The *t*-distribution is used for samples of small size, defined historically as samples with less than 30 items. Among the most important uses of the *t*-distribution is *testing for significance*. To explain the principle of hypothesis testing for significance, let us imagine comparing a standard bioanalytical method with a new one. And *to begin*, let's set up a *null hypothesis* which says that there is no difference between the two methods; i.e., we postulate that the two methods are identical. *Next*, we calculate *t* for the difference between the two means from

$$t \quad = \quad \frac{\bar{x}_1 - \bar{x}_2}{\sqrt{\dfrac{s_p^2}{n_1} + \dfrac{s_p^2}{n_2}}} \tag{45}$$

Then, we compare the calculated *t* with the theoretical distribution of *t* to estimate whether the calculated *t* provides an answer which supports the null hypothesis with a confidence level of, say, 95 or 99% (cf. sections 10 –5 and 10 –6).

Spreadsheet programs provide built-in functions for n (the number of items in a sample), as well as functions for the mean, standard deviation, variance and other values. From the built-in functions, it is easy to calculate *t* and other values with equations (38) to (45) as we show in section 1-14.

The fit of data for linear functions by the method of least-squares (section 10-14) is one of the most common procedures in biomedical research. In this method, the coefficients of the slope-intercept form of the equations of a straight line are chosen so that the sum of the deviations from the best fit line, squared, is a minimum. Thus, if we let Δ be the deviation, then

$$b.x + a - y \quad = \quad \Delta \tag{46}$$

and for the best-fit line, the correct equation becomes

$$b.x + a - y \quad = \quad 0 \tag{47}$$

When the sum of squares is expressed first as a function of b then as a function of a and the appropriate mathematical operations are done (section 10-14), one obtains equations for $\hat{b}$, the slope and $\hat{a}$, the y-intercept of the best-fit line:

Slope:
$$\hat{b} = \frac{n\Sigma xy - \Sigma x\Sigma y}{n\Sigma x^2 - (\Sigma x)^2} \qquad (48)$$

Intercept:
$$\hat{a} = \bar{y} - \hat{b}x. \qquad (49)$$

Substitution of the slope, the y-intercept and each value of the variables x and y into the slope-intercept equation will give Cartesian coordinates for the best-fit line. Least-squares calculations may be carried out with built-in spreadsheet functions or with the simple least-squares routines which we provide with the spreadsheet solutions of worked examples (cf. sections 1-10 and 1-14).

1-13. Three examples covering some common statistical calculations

In section 1-14, we will use Microsoft *Excel*® and *Lotus*®*1-2-3*®, to illustrate the use of spreadsheets in data analysis; specifically, we will use examples 1-10 to 1-12 of this section (a) to plot a graph of the best-fit line from linear least-squares data, and (b) to calculate *fifteen statistical measures*: the mean, range, variance, sum of squares, standard deviation, standard error, coefficient of variation, 95 percent confidence limits, correlation coefficient, slope, y-intercept, $s_{y.x}$ the standarderror of estimate, *F* the variance ratio, the *t* statisitc, and *chi-square*.

Example 1-10. In laboratory work, one often obtains several sets of data. The data may refer to items such as the activities of enzymes, the output from a chromatograph, the counts from a liquid scintillation spectrometer or the results of immunoassay. One of the first steps in evaluating such results is to determine their statistical reliability. Such evaluation usually involves computing measures of central tendency and dispersion. We evaluate four sets of data, *set a, set b, set c,* and *set d,* in the spreadsheet solutions and formulas which follow this section.

Example 1-11. As mentioned in section 10 −14, the term *simple regression* refers to the relation between two variables in which the magnitude of one variable (the *dependent* variable) is a function of the second variable (the *independent* variable). A relationship wherein one variable changes as the second changes but neither of the variables depends on the other is referred to as *correlation*. In simple regression, the independent variable usually is denoted by *x*, the dependent variable by *y*. In *linear regression*, the relation between the two variables is a straight line expressed by the *slope-intercept form of the equations of a straight line*. So, given the following values, use the equations of section 10 −15 to determine the slope, the y-intercept, the correlation coefficient, the standard error of estimate and the quality of the fit.

Independent variable, *x* :	1.0	2.0	3.0	4.0	5.0
Dependent variable, *y* :	2.0	4.0	5.9	8.0	10.1

Example 1-12. In biomedical research, investigators often compare two sets of data. The objective of the comparison usually is to learn whether there is a statistically significant difference between the samples. The t-test is used to compare the means of two small samples, the F-test is used to compare the variances or the standard deviations of two small samples and the *chi square* test is used to estimate whether the observed and expected frequencies of two sets of data differ. We show data and spreadsheet solutions illustrating all three of these tests in the spreadsheet results in section 1–14.

The reader may find it useful to review sections 10-7 to 10-10 as background for interpreting the spreadsheet solutions of examples 1-10 to 1-12.

1-14. Spreadsheet Solutions (Statistics), Microsoft Excel

	A	B	C	D	E	F	G
1	EX 1-10	Set a	Set b	Set c	Set d	EX 1-11, Plot	
2		111	330	200	195	X	Y-hat
3		110	322	183	185	0	-0.06
4		115	345	183	220	1	1.96
5		115	331	185	213	2	3.98
6		148	323	210	190	3	6
7		147	318	220	180	4	8.02
8	Sum	746	1969	1181	1183	5	10.04
9	Mean	124.33	328.17	196.83	197.17		
10	Range	38.00	27.00	37.00	40.00		
11	Variance	326.27	92.57	248.57	254.17		
12	SS	1957.60	555.40	1491.40	1525.00		
13	SD	18.06	9.62	15.77	15.94		
14	SE	7.37	3.93	6.44	6.51		
15	CV	14.53	2.93	8.01	8.09		
16	± 95% CL	18.96	10.10	16.55	16.73		
17							
18	EX 1-11	X	Y	X^2	XY	Y^2	Y-hat
19		1	2	1	2	4	1.96
20		2	4	4	8	16	3.98
21		3	5.9	9	17.7	34.81	6
22		4	8	16	32	64	8.02
23		5	10.1	25	50.5	102.01	10.04
24	Count	5					
25	Σ	15	30	55	110.2	220.82	
26	Σ^2	225	900	3025	12144.04	48761.47	
27	Mean	3	6				
28	Correl.	.9998	101	7.071068	14.286357		
29	Slope	2.02					
30	Intcp.	-0.06					
31	Sy.x	.073	40.82	40.804			
32							
33	EX 1-12	F-TEST		t-TEST		Chi Square	
34		7	3	111	320	Observed	Expected
35		18	8	110	213	1.9	5
36		5	5	115	227	15.8	14.8
37		8	7	115	293	7.1	8.6
38		3	5	148	211	2.7	4.3
39		10	10	147	330	62	64.8
40		3	3	124	322	9.2	8.1
41		5	15	195	345	1.1	0.3
42	Variance	24.2679	16.2857			0.2	0.1
43	F statistic	1.4901					
44	Mean			133.125	282.625		
45	t (calcns.)			685.5011	14		
46	t statistic			6.9975			
47							
48	Chi square statistic					5.350248	

Spreadsheet Formulas (Statistics), Microsoft Excel

	A	B	C
1	EX 1-10	Set a	Set b
2		111	330
3		110	322
4		115	345
5		115	331
6		148	323
7		147	318
8	Sum	=SUM(B2:B7)	=SUM(C2:C7)
9	Mean	=AVERAGE(B2:B7)	=AVERAGE(C2:C7)
10	Range	=MAX(B2:B7)-MIN(B2:B7)	=MAX(C2:C7)-MIN(C2:C7)
11	Variance	=VAR(B2:B7)	=VAR(C2:C7)
12	SS	=VAR(B2:B7)*COUNT(B2:B7)	=VAR(C2:C7)*COUNT(C2:C7)
13	SD	=STDEV(B2:B7)	=STDEV(C2:C7)
14	SE	=STDEV(B2:B7)/SQRT(COUNT(B2:B7))	=STDEV(C2:C7)/SQRT(COUNT(C2:C7))
15	CV	=B13/B9*100	=C13/C9*100
16	± 95% CL	=2.571*B14	=2.571*C14
17			
18	EX 1-11	X	Y
19		1	2
20		2	4
21		3	5.9
22		4	8
23		5	10.1
24	Count	=COUNT(B19:B23)	
25	Σ	=SUM(B19:B23)	=SUM(C19:C23)
26	Σ^2	=SUM(B19:B23)^2	=SUM(C19:C23)^2
27	Mean	=AVERAGE(B19:B23)	=AVERAGE(C19:C23)
28	Correl.	=C28/(D28*E28)	=B24*E25-B25*C25
29	Slope	=(B24*E25-B25*C25)/(B24*D25-B26)	
30	Intcp.	=+C27-(B29*B27)	
31	Sy.x	=SQRT((C31-D31)/(B24-2))	=F25-(C26/B24)
32			
33	EX 1-12	F-TEST	
34		7	3
35		18	8
36		5	5
37		8	7
38		3	5
39		10	10
40		3	3
41		5	15
42	Variance	=VAR(B34:B41)	=VAR(C34:C41)
43	F statistic	=B42/C42	
44	Mean		
45	t (calcns.)		
46	t statistic		
47			
48	Chi sq. statistic		

Spreadsheet Formulas (Statistics), Microsoft Excel

	D
1	Set c
2	200
3	183
4	183
5	185
6	210
7	220
8	=SUM(D2:D7)
9	=AVERAGE(D2:D7)
10	=MAX(D2:D7)-MIN(D2:D7)
11	=VAR(D2:D7)
12	=VAR(D2:D7)*COUNT(D2:D7)
13	=STDEV(D2:D7)
14	=STDEV(D2:D7)/SQRT(COUNT(D2:D7))
15	=D13/D9*100
16	=2.571*D14
17	
18	X^2
19	=B19*B19
20	=B20*B20
21	=B21*B21
22	=B22*B22
23	=B23*B23
24	
25	=SUM(D19:D23)
26	=SUM(D19:D23)^2
27	
28	=SQRT(B24*D25-B25^2)
29	
30	
31	=(E25-B27*C25)^2/(D25-B27*B25)
32	
33	t-TEST
34	111
35	110
36	115
37	115
38	148
39	147
40	124
41	195
42	
43	
44	=AVERAGE(D34:D41)
45	=STDEV(D34:D41)*COUNT(D34:D41)+STDEV(E34:E41)*COUNT(E34:E41)
46	=SQRT(D45/E45)
47	
48	

Spreadsheet Formulas (Statistics), Microsoft Excel

	E	F	G	H
	Set d	EX 1-11, Plot		
	195	X	Y-hat	
	185	=0	=F3*B29+B30	
	220	1	=B29*B19+B30	
	213	2	=B29*B20+B30	
	190	3	=B29*B21+B30	
	180	4	=B29*B22+B30	
	=SUM(E2:E7)	5	=B29*B23+B30	
	=AVERAGE(E2:E7)			
	=MAX(E2:E7)-MIN(E2:E7)			
	=VAR(E2:E7)			
	=VAR(E2:E7)*COUNT(E2:E7)			
	=STDEV(E2:E7)			
	=STDEV(E2:E7)/SQRT(COUNT(E2:E7))			
	=E13/E9*100			
	=2.571*E14			

	E	F	G	H
	XY	Y^2	Y-hat	
	=B19*C19	=C19*C19	=B19*B29+B30	
	=B20*C20	=C20*C20	=B20*B29+B30	
	=B21*C21	=C21*C21	=B21*B29+B30	
	=B22*C22	=C22*C22	=B22*B29+B30	
	=B23*C23	=C23*C23	=B23*B29+B30	
	=SUM(E19:E23)	=SUM(F19:F23)		
	=SUM(E19:E23)^2	=SUM(F19:F23)^2		
	=SQRT(B24*F25-C25^2)			

	E	F	G	H
			Chi Square	
		Observed	Expected	χ
	320	1.9	5	=(F35-G35)^2/G35
	213	15.8	14.8	=(F36-G36)^2/G36
	227	7.1	8.6	=(F37-G37)^2/G37
	293	2.7	4.3	=(F38-G38)^2/G38
	211	62	64.8	=(F39-G39)^2/G39
	330	9.2	8.1	=(F40-G40)^2/G40
	322	1.1	0.3	=(F41-G41)^2/G41
	345	0.2	0.1	=(F42-G42)^2/G42
	=AVERAGE(E34:E41)			
	=(COUNT(D34:D41)-1)+(COUNT(E34:E41)			
		=SUM(H34:H42)		

Spreadsheet Solutions (Statistics), Lotus 1-2-3

	A	B	C	D	E	F	G	H
1	EX 1-10	Set a	Set b	Set c	Set d	EX 1-11, Plot		
2		111	330	200	195	X	Y-hat	
3		110	322	183	185	0	-0.06	
4		115	345	183	220	1	1.96	
5		115	331	185	213	2	3.98	
6		148	323	210	190	3	6	
7		147	318	220	180	4	8.02	
8	Sum	746	1969	1181	1183	5	10.04	
9	Mean	124.3	328.17	196.8	197.17			
10	Range	38	27	37	40			
11	Variance	326.3	92.567	248.6	254.17			
12	SS	2284	555.4	1491	1525			
13	SD	18.06	9.6212	15.77	15.943			
14	SE	7.374	3.9278	6.436	6.5085			
15	CV	14.53	2.9318	8.01	8.0859			
16	± 95% CL	18.96	10.098	16.55	16.733			
17								
18	EX 1-11	X	Y	X^2	XY	Y^2	Y-hat	
19		1	2	1	2	4	1.96	
20		2	4	4	8	16	3.98	
21		3	5.9	9	17.7	34.81	6	
22		4	8	16	32	64	8.02	
23		5	10.1	25	50.5	102.01	10.04	
24	Count	5						
25	Σ	15	30	55	110.2	220.82		
26	Σ^2	225	900		12144			
27	Mean	3	6					
28	Correl.	1	101	7.071	14.286			
29	Slope	2.02						
30	Intcp.	-0.06						
31	Sy.x	0.073	40.82	40.8				
32								
33	EX-1-12	F-TEST		t-TEST		Chi-Square		
34		7	3	111	320	Observed	Expected	χ
35		18	8	110	213	1.9	5	1.92
36		5	5	115	227	15.8	14.8	0.07
37		8	7	115	293	7.1	8.6	0.26
38		3	5	148	211	2.7	4.3	0.60
39		10	10	147	330	62	64.8	0.12
40		3	3	124	322	9.2	8.1	0.15
41		5	15	195	345	1.1	0.3	2.13
42	Variance	24.27	16.286			0.2	0.1	0.10
43	F statistic	1.49						
44	Mean			133.1	282.63			
45	t (calcns.)			685.5	14			
46	t statistic			6.997				
47								
48	Chi square statistic					5.350248		

Plot (rows 9–16, columns F–H): Y-Axis scale 10, 6, 2, -2; X-Axis scale 0 1 2 3 4 5.

adsheet Formulas (Statistics), Lotus 1-2-3

A	B	C
EX 1-10	Set a	Set b
	111	330
	110	322
	115	345
	115	331
	148	323
	147	318
Sum	@SUM(B2..B7)	@SUM(C2..C7)
Mean	@AVG(B2..B7)	@AVG(C2..C7)
Range	@MAX(B2..B7)-@MIN(B2..B7)	@MAX(C2..C7)-@MIN(C2..C7)
Variance	@VARS(B2..B7)	@VARS(C2..C7)
SS	@VARS(B2..B7)*@COUNT(C1..C7)	@VARS(C2..C7)*@COUNT(C2..C7)
SD	@STDS(B2..B7)	@STDS(C2..C7)
SE	@STDS(B2..B7)/@SQRT(@COUNT(B2..B7))	@STDS(C2..C7)/@SQRT(@COUNT(C2..C7))
CV	+B13/B9*100	+C13/C9*100
± 95% CL	2.571*B14	2.571*C14
EX 1-11	X	Y
	1	2
	2	4
	3	5.9
	4	8
	5	10.1
Count	@COUNT(B19..B23)	
Σ	@SUM(B19..B23)	@SUM(C19..C23)
Σ^2	+B25*B25	+C25*C25
Mean	@AVG(B19..B23)	@AVG(C19..C23)
Correl.	+C28/(D28*E28)	+B24*E25-B25*C25
Slope	(B24*E25-B25*C25)/(B24*D25-B26)	
Intcp.	+C27-(B29*B27)	
Sy.x	@SQRT((C31-D31)/(B24-2))	+F25-(C26/B24)
EX 1-12 F-TEST		
	7	3
	18	8
	5	5
	8	7
	3	5
	10	10
	3	3
	5	15
Variance	@VARS(B34..B41)	@VARS(C34..C41)
F statistic	+B42/C42	
Mean		
(calcns.)		
t statistic		
Chi square statistic		

Spreadsheet Formulas (Statistics), Lotus 1-2-3

A	D	
1		Set c
2		200
3		183
4		183
5		185
6		210
7		220
8	@SUM(D2..D7)	
9	@AVG(D2..D7)	
10	@MAX(D2..D7)-@MIN(D2..D7)	
11	@VARS(D2..D7)	
12	@VARS(D2..D7)*@COUNT(D2..D7)	
13	@STDS(D2..D7)	
14	@STDS(D2..D7)/@SQRT(@COUNT(D2..D7))	
15	+D13/D9*100	
16	2.571*D14	
17		
18		X^2
19	+B19*B19	
20	+B20*B20	
21	+B21*B21	
22	+B22*B22	
23	+B23*B23	
24		
25	@SUM(D19..D23)	
26		
27		
28	@SQRT(B24*D25-B25^2)	
29		
30		
31	(E25-B27*C25)^2/(D25-B27*B25)	
32		
33	t-TEST	
34		111
35		110
36		115
37		115
38		148
39		147
40		124
41		195
42		
43		
44	@AVG(D34..D41)	
45	@STDS(D34..D41)*@COUNT(D34..D41)+@STDS(E34..E41)*@COUNT(E34..E41)	
46	@SQRT(D45/E45)	
47		
48		

adsheet Formulas (Statistics), Lotus 1-2-3

E	F	G	H
	Set d	EX 1-11, Plot	
	195	X	Y-hat
	185	0	+B29*F3+B30
	220	1	+B29*F4+B30
	213	2	+B29*F5+B30
	190	3	+B29*F6+B30
	180	4	+B29*F7+B30
@SUM(E2..E7)	5	+B29*F8+B30	
@AVG(E2..E7)			
@MAX(E2..E7)-@MIN(E2..E7)			
@VARS(E2..E7)			
@VARS(E2..E7)*@COUNT(E2..E7)			
@STDS(E2..E7)			
@STDS(E2..E7)/@SQRT(@COUNT(E2..E7))			
+E13/E9*100			
2.571*E14			

XY	Y^2	Y-hat	
+B19*C19	+C19*C19	+B29*B19+B30	
+B20*C20	+C20*C20	+B29*B20+B30	
+B21*C21	+C21*C21	+B29*B21+B30	
+B22*C22	+C22*C22	+B29*B22+B30	
+B23*C23	+C23*C23	+B29*B23+B30	
@SUM(E19..E23)	@SUM(F19..F23)		
+E25*E25			
@SQRT(B24*F25-C25^2)			

EST		Chi-Square		
	320	Observed	Expected	χ
	213	1.9	5	(F35-G35)^2/G35
	227	15.8	14.8	(F36-G36)^2/G36
	293	7.1	8.6	(F37-G37)^2/G37
	211	2.7	4.3	(F38-G38)^2/G38
	330	62	64.8	(F39-G39)^2/G39
	322	9.2	8.1	(F40-G40)^2/G40
	345	1.1	0.3	(F41-G41)^2/G41
		0.2	0.1	(F42-G42)^2/G42
@AVG(E34..E41)				
(@COUNT(D34..D41)-1)+(@COUNT(E34..E41)-1)				
	@SUM(H35..H42)			

1-15. Review questions

1-1. What are the SI base units of length, mass, and electric current? How are they related, respectively, to volume, mass concentration, and electric charge?

1-2. Define coherent and noncoherent derived quantities.

1-3. What is the product of normality and volume measured in milliliters? How is this product used in calculations which are required for preparing dilute solutions of acids and bases from concentrated solutions?

1-4. What is the product of molarity and volume measured in liters?

1-5. What result is obtained by dividing the first by the second of each of the following:
    weight, g/molecular weight;
    weight, g/number of moles;
    weight in g/(weight, g / number of moles);
    number of moles/molarity;
    number of milliequivalents/volume in milliliters?

1-6. What result is obtained from the following products:
    l*M;
    l*M*mol wt;
    number of moles*molecular weight;
    volume in ml*M?

1-7. How do moles, molarity and molality differ?

1-8. Explain without performing calculations how you would prepare 100 ml of an approximately 1.0 N solution of H_2SO_4.

1-9. Explain without calculations how you would prepare 250 ml of a 10 mM solution of glucose.

1-10. Distinguish clearly between solutions prepared on a weight/volume basis and those prepared on a weight/weight basis.

1-11. Explain the terms bar graph, scatter plot, line graph, random access memory, arithmetic operator and spreadsheet function.

1-12. How does a computer user tell Excel and Lotus 1-2-3 to expect a formula to be entered in a worksheet?

1-13. What, in Lotus 1-2-3, are the equivalent of the Excel statistical functions STDEV, STDEVP, AVERAGE, LINEST, MAX, MIN and TREND?

1-14. Explain the sequence of steps you would follow to plot linear regression data from Excel and Lotus 1-2-3.

1-15. If you were using both Excel and Lotus 1-2-3, how would you print out the results of section 1-14 for a manuscript such as that used for this book?

1-16. How would you adjust the height and width of a spreadsheet printout to fit the dimensions of a printed page in a textbook?

1-17. What is a spreadsheet template?

1-18. Explain how you would transfer graphics on Excel and Lotus 1-2-3 worksheets to other software applications.

1-19. How would you insert or delete columns and rows in Excel and 1-2-3?

1-20. Explain how you would add, insert, delete, or align text on spreadsheet charts prepared with Excel and Lotus 1-2-3?

1-16. Review problems

1-21.   How much phosphatidylcholine (in mg) would you have to weigh out to prepare 5 ml of a chloroform-methanol solution containing 2 μg per μl?

1-22.   If you added 18 ml of sulfuric acid very carefully to sufficient water to make 1.0 l of dilute solution, what would be the molarity and normality of the resulting solution?

1-23.   If you dissolved 100 mg of bovine serum albumin in glass distilled water and then made the solution up to 100 ml, what would be the concentration of the protein in terms of (a) % wt/vol and (b) % wt/wt?

1-24.   Given that the molecular weight of acetic acid is 60.1 and that its density is 1.0097 g ml $^{-1}$ at 20°C, calculate the concentration of a solution which contains 90.15 g of acetic acid per liter at 20°C. Express your answer in mol l $^{-1}$, molality, normality and % wt/wt.

1-25.   Show how you would calculate the amount of each component which is required for preparing 2.5 l of a solution containing: 0.02 M Hepes buffer; 0.25 M sucrose; 1.0 mM EDTA; 10 mM $MgCl_2$; 10 mM KH_2PO_4 and 1.0 % (w/v) bovine serum albumin. The relevant molecular weights are: Hepes(N-2-hydroxymethyl-piperazine-N-2-ethane-sulfonic acid) 238.31; sucrose, 342.3; EDTA, 292.24; $MgCl_2$. $6H_2O$, 203.31; KH_2PO_4,136.09.

1-26.   Given stock solutions of glucose (1M), asparagine (100 mM) and NaH_2PO_4 (50 mM), how much of each solution would you need to prepare 500 ml of a reagent which contains 0.05 M glucose, 10 mM asparagine and 2 mM NaH_2PO_4?

1-27.   A 750 ml bottle of California Cabernet Sauvignon wine contained 11.5% alcohol (ethanol) vol/vol. What is the total volume (in ml) of ethanol in the bottle of wine? Given that the density of ethanol is 0.789 g ml^{-1} and that its molecular wt. is 46.07, what is the molarity of the ethanol in the wine?

1-28.   (a) Calculate the number of millimoles in 500 mg of each of the following amino acids: alanine (molecular weight 89); leucine (molecular weight 131); tryptophan (molecular weight 204); cysteine (molecular weight 121) and glutamic acid (molecular weight 147).
(b) Calculate the weights in grams of each of the five amino acids listed in part (a) which are required for preparing five 250 ml solutions of 0.2 M concentrations.

1-29.   Calculate the amounts of each component which would be required for preparing 10 ml of an assay medium containing 0.25 mol l $^{-1}$ glucose; 1.0 mol l $^{-1}$ of EDTA; 10 mol $^{-1}$ of $MgCl_2$; 10 mol l $^{-1}$ of KH_2PO_4 and 0.02 mol l $^{-1}$ of Hepes buffer. The relevant molecular weights are Hepes (N-2-hydroxymethyl-piperazine-N-2-ethanesulfonic acid) 238.31; sucrose, 342.3; EDTA, 292.24; $MgCl_2$. $6H_2O$, 203.31; KH_2PO_4, 136.09.

1-30.   Calculate the amounts of each component which would be required for preparing 10 ml of an assay medium which contains 0.25 moles of glucose; 1.0 millimole of EDTA; 10 millimoles of $MgCl_2$; 10 m mol of KH_2PO_4 and 0.02 moles of Hepes buffer. (Molecular weights are given in Problem 1–29.)

1-17. Additional problems

A. Preparation of Solutions

1-31. How many grams of KOH are required for the preparation of 750 ml of a 0.05 M solution? Convert the answer to g l^{-1}, % (wt/vol) and normality. The molecular weight of KOH is 56.1.

1-32. Calculate the weight of sodium hydroxide (molecular weight 40) that is needed for preparing 1.5 liter of a 0.25 M solution. Convert your answer to g l^{-1}, percent (wt/vol) and normality. Recalculate the three values assuming that it was desired to prepare 2.5 liters of a 0.5 M solution.

B. Dilutions

1-33. What volume (milliliters) of 12.0 M HCl is required to prepare 300 ml of 0.2 M solution?

1-34. What molarity of HCl is needed so that 5 ml diluted to 300 ml will yield 0.2 M?

1-35. How much 0.2 M HCl can be made from 5.0 ml of 12.0 M HCl solution?

1-36. What is the molarity of 300 ml of HCl prepared from 5 ml of 12.0 M HCl?

1-37. You have been given 0.5 ml of a $CHCl_3$ solution containing 25 mg ml^{-1} each of cholesterol oleate, trilinolenin, oleic acid and ergosterol. To what volume should this $CHCl_3$ solution be diluted so that the dilute solution contains 5 µg of each component per µl of solution (i.e. 5 mg/ml)?

1-38. How many milliliters of a 0.0025 g ml^{-1} solution are needed to prepare 4 l of a 0.0005 g ml^{-1} solution?

1-39. What volume (milliliters) of a $1*10^{-2}$ g ml^{-1} solution of NaCl is needed to prepare 4 liters of a $5*10^{-5}$ g ml^{-1} solution?

1-40. What concentration of sodium potassium tartrate (in grams per milliliter or equivalent unit) should be prepared so that 5 ml of the stock solution diluted to 500 ml will yield a 0.001 g ml^{-1} solution?

1-41. What concentration of asparagine (in grams per milliliter or equivalent unit) should be prepared so that 50 ml of the stock asparagine solution diluted to 1000 ml will yield a $3*10^{-3}$ g ml^{-1} solution?

1-42. What concentration of glucose solution (in g ml^{-1}) should be prepared so that 15 ml of this solution diluted to 1 l will yield a $2*10^{-4}$ g ml^{-1} solution?

1-43. What weight (in grams) of 10% (w/w) ammonium hydroxide solution can be prepared from 1800 grams of 28% (wt/wt) ammonium hydroxide solution?

1-44. What volume (milliliters) of a $2*10^{-4}$ M solution of glycine can be prepared from 125 ml of a $2*10^{-3}$ M solution?

1-45. What is the concentration of 1500 ml of ethanol solution expressed as % (vol/vol) prepared from 500 ml of a 15% (vol/vol) ethanol solution?

1-46. What is the concentration (grams per milliliter of 1000 ml of potassium permanganate solution prepared from 50 ml of a 0.05 g ml^{-1} solution?

1-47. How many grams per milliliter of sodium chloride should be used to obtain a stock solution from which 50 ml diluted to 1000 ml will yield a 0.9% (wt/vol) solution? What is the total amount of NaCl that should be weighed out to make up 500 ml of the stock solution?

C. Percent Concentrations

Percent (Weight/Volume)

1-48. Calculate the weight of glucose required to prepare 2 l of a 5% (wt/vol) solution.

1-49. What weight of a protein (in grams) is needed to prepare 250 ml of a 0.1% (wt/vol) solution?

1-50. Calculate the concentration as % (wt/vol) of a solution of urea, used in preparing a protein for electrophoresis, if 50 ml contain 59.08 g of urea.

1-51. How many ml of an 8.56% solution can be prepared from 42.8 g of sucrose.

Percent (Weight/Weight)

1-52. How many grams of phenol are needed to prepare 300 g of a 5% (wt/wt) solution in water?

1-53. Calculate the concentration as % (wt/wt) of a solution which contains 0.75 g of a compound in 150 g of solution.

1-54. What weight of a 5% (wt/wt) solution of cupric sulfate can be prepared from 2 g of the compound?

Percent (Volume/Volume)

1-55. How many ml of $CHCl_3$ are needed to prepare a 2.5% (vol/vol) solution in 500 ml of CH_3OH.

1-56. If a 250 ml solution of ethanol in water were prepared with 4 ml of absolute ethanol, what was the concentration of ethanol as % (vol/vol)?

1-57. A chloroform-methanol solution contains 10% (vol/vol) of a polyunsaturated triacylglycerol. What volume of the solution will contain 7.5 ml of the triacylglycerol?

D. Molar Concentrations

1-58. Calculate each of the following:
(a) the volume required to make a 0.25 M solution from 21.4 g of sucrose, molecular weight 342.3;
(b) the molarity of a solution containing 8.56 g of sucrose in 250 ml;
(c) the weight (grams) of sucrose contained in 800 ml of a 0.3 M solution;
(d) the molecular weight of sucrose if a 0.1 M solution contains 8.5575 g of sucrose per 250 ml.

1-59.   The concentration of glucose (molecular weight 180.16) in capillary blood falls within the range 680 -1030 mg l $^{-1}$ from birth to 29 years of age. What is this range in mol l^{-1}.

1-60.   In diabetic patients, the concentration of glucose can be as high as 0.250 g ml $^{-1}$ of capillary blood depending on diet. Express this level of blood glucose in molarity.

1-61.   At 20°C, 53.1 g of ammonium sulfate per 100 ml of aqueous solution will yield a saturated solution. The molecular weight of $(NH_4)_2SO_4$ is 132.14. What is the concentration of this saturated solution of ammonium sulfate in molarity.

1-62.   What weight of each compound is required for preparing 500 ml of a solution for assay of mitochondrial respiration which contains 0.05 M KH_2PO_4; 0.03 M D-mannitol; 25 mM $MgCl_2$ and 115 mM EDTA, [(ethylenedinitrilo)tetraacetic acid]. The relevant molecular weights are KH_2PO_4, 136.09; D-mannitol, 182.18; $MgCl_2.6H_2O$, 203.31 and EDTA 292.24.

1-63.   A nutrient medium for growing *Neurospora crassa* contains 4 g of L-asparagine per liter. What is the molarity of asparagine in the medium? The molecular weight of L- asparagine is 132.13.

E.   PPM, Equivalents and Normality

1-64.   You have been given a solution which contains 10^{-4} M HPO_4^{2-} ion. Express this molar concentration in terms of ppm. The formula weight of HPO_4^{2-} is 95.97.

1-65.   What weight of KH_2PO_4 is needed for preparing 1 l of solution containing 50 ppm P? The molecular weight of KH_2PO_4 is 136.09 and the atomic weight of P is 30.97.

1-66.   Calculate the weight of $CuSO_4$ which is needed for the preparation of 500 ml of solution containing 1000 ppm Cu. The molecular weight of $CuSO_4$ is 159.60 and the atomic weight of Cu is 63.54.

1-67.   You have been given two solutions, one containing Na^+ ions and the other Hg^{2+} ions. Each solution contains 1.00 ppm of the respective ion. Convert the concentration, as ppm, to mol l^{-1}. The formula weights of Na and Hg are 22.99 and 200.59 respectively.

1-68.   The concentration of ammonia nitrogen in normal human blood is often reported as being in the range of 0.04 to 0.1 mg% or400 to 1000 ppm. Convert the range, expressed as ppm, to meq l $^{-1}$. The formula weight of NH_3 is 17.007.

1-69.   If you diluted 20 ml of concentrated sulfuric acid by adding the acid slowly to water and mixing to make a final volume of 1000 ml, you would obtain a solution of H_2SO_4 that is approximately 0.36 M. What is the normality of the diluted H_2SO_4 solution?

Suggested Reading

Causton, D. R. 1983. *A Biologist's Basic Mathematics,* Arnold, Baltimore, Maryland.

Causton, D. R. 1987. *A Biologist's Advanced Mathematics,* Allen & Unwin, Boston.

Christian, G.D. 1980. *Analytical Chemistry,* 3d ed., Wiley, New York.

Cornish-Bowden, A. 1981. *Basic Mathematics for Biochemists,* Chapman and Hall, London.

Dickerson, R. E., H. B. Gray, M. Y. Darensbourg and D. J. Darensbourg 1984. *Chemical Principles,* 4th ed., Benjamin/Cummings, Reading, Massachusetts.

Goldfish, D. M. 1983. *Basic Mathematics for Beginning Chemistry,* 3d ed., Macmillan, New York.

Kibby, M. R. 1985. The Electronic Spreadsheet as a General-Purpose Programming Tool, *Comput. Appl. Biosci.,* 1: 73-78.

Kibby, M. R. 1986. Four Spreadsheet Templates for the Laboratory, *Comput. Appl. Biosci.,* 2:1-4.

Montgomery, R., and C. A. Swenson 1976. *Quantitative Problems in the Biochemical Sciences,* 2d ed., Freeman, New York.

Morris, J. G. 1974. *A Biologist's Physical Chemistry,* 2d ed., Arnold, Baltimore, Maryland.

Stoklosa, M. J. and H. C. Ansel 1980. *Pharmaceutical Calculations,* 7th ed., Lea and Febiger, Philadelphia.

Note. References on statistics are listed at the end of Chapter 10.

Chapter 2
Acids, Bases and Buffers

2-1. Control of acid-base properties is important in biochemistry

---

The pH of solutions is important in the biomedical sciences for two main reasons. First, the proper functioning of biomolecules depends to an important degree on the control of pH. Second, changes as small as 0.1 or 0.2 pH unit can cause significant metabolic disturbances in certain cells, tissues, and organs. Because of the pH sensitivity of many biomolecules, control of pH also is important for the success of several procedures used in the biomedical laboratory. These include the separation, purification and assay for biological activity of several biomolecules. For example, separation and purification of cellular components by countercurrent distribution, chromatography, gel filtration, gel electrophoresis, and ultracentrifugation, discussed in Chapters 3 and 4, all require the use of buffered solutions. Then in Chapter 5, all aspects of enzyme activity depend on careful control of pH, as does equilibrium binding when studied with Scatchard and Hill plots. In Chapter 6, we will see that the ratios in which antigens and antibodies combine depend, to a large extent, on pH and other factors. Later, in Chapters 7 and 8, we may infer that virtually any biomolecule in aqueous solution which is investigated either by spectroscopic or radiochemical methods, would have been isolated by a pH-dependent method related to those discussed in Chapters 3 and 4. Then, in Chapter 9, we will see that the determination of midpoint potential is influenced by pH since many biological oxidations occur by dehydrogenations and these dehydrogenations yield protons which contribute directly to the concentration of hydrogen ions.

In the first part of this chapter, we will review briefly some fundamentals of acid-base chemistry. These include definitions of acids and bases; pH and pK; the ionization of water; the Henderson-Hasselbalch equation; the preparation of buffers and the correction of pK_a for ionic strength and temperature. We will examine the influence of pH on the ionization of biomolecules to conclude the chapter.

2-2.   Life scientists use the Lowry-Brønsted definition of acids

According to the 1923 definition of Thomas M. Lowry of England and Johannes N. Brønsted of Denmark, *acids release protons in their reactions whereas bases are substances which accept protons.* Other definitions of acids and bases such as those of Svante Arrhenius and Gilbert N. Lewis, are not as useful as the Lowry and Brønsted definition for biochemical purposes.

Strong acids release protons readily and almost completely in dilute aqueous solutions but *weak acids* do not so that, at equilibrum, in most cases, less than 1% of a weak acid is ionized to yield protons. Examples of strong acids are hydrochloric, nitric and sulfuric acids and examples of weak acids are acetic, malic, or boric acids. *Strong bases* have a great capacity for accepting protons whereas *weak bases* are poor acceptors of protons. Examples of strong and weak bases are sodium hydroxide and ethanolamine, respectively.

When an acid undergoes dissociation and gives up a proton, an anion is formed. The original acid which gave up the proton is known as a *conjugate acid*; the resulting anion which can act as a base, is known as a *conjugate base*. The two together are called a conjugate acid-conjugate base pair. Three examples of such dissociation follow:

$$CH_3CH_2COOH \longrightarrow CH_3CH_2COO^- + H^+ \tag{1}$$

$$\begin{array}{ccc} \text{Conjugate} & & \text{Conjugate} \\ \text{acid} & & \text{base} \end{array}$$

$$
\begin{array}{ccc}
\text{COOH} & & \text{COO}^- \\
| & & | \\
\text{H CH} & & \text{HOCH} \\
| & & | \\
\text{CH}_2 & \longrightarrow & \text{CH}_2 \ + \ 2\text{H}^+ \\
| & & | \\
\text{COOH} & & \text{COO}^- \\
\text{Conjugate} & & \text{Conjugate} \\
\text{acid} & & \text{base}
\end{array}
\tag{2}
$$

$$H_3N^+CH_2CH_2OH \longrightarrow H_2NCH_2CH_2OH + H^+ \tag{3}$$

$$\begin{array}{ccc} \text{Conjugate} & & \text{Conjugate} \\ \text{acid} & & \text{base} \end{array}$$

2-3.   pH and pK were developed for mathematical convenience

Since weak acids dissociate only to a small extent in dilute aqueous solution, the concentration of H^+ in dilute solutions of these acids is small. Frequently, the concentration of hydrogen ions in solutions of weak acids is less than 10^{-6} mol l $^{-1}$. It can be somewhat inconvenient mathematically to work with values of this small magnitude. To permit easier handling of such low values of [H+], the Danish chemist S. P. L. Sørensen proposed in 1909 that [H+] be expressed logarithmically as follows:

$$pH \quad = \quad -\log_{10} a_{H^+}, \tag{4}$$

where a_{H^+} is the activity of H+ ions. In biochemistry, measured molar concentrations are used instead of activities. Therefore equation (4) is rewritten as:

$$pH \quad = \quad -\log_{10}[H^+], \tag{5}$$

or
$$pH \quad = \quad \log_{10} \frac{1}{[H^+]}, \tag{6}$$

and
$$[H^+] \quad = \quad 10^{-pH} \ = \ \text{antilog -pH}. \tag{7}$$

Not only is this logarithmic equation useful for expressing low concentrations of H+ in aqueous solution, but, more generally, it is useful for expressing low concentrations of biomolecules and metal ions. Similarly, because the dissociation constants (K_a) of weak acids and bases are small, these constants also are expressed, for convenience, as pK_a' where p = the negative logarithm of and the prime shows that molar concentrations rather than activities are used. In mathematical terms:

$$pK_a' \quad = \quad -\log K_a' \ = \ \log \frac{1}{K_a'}. \tag{8}$$

2-4.   The ionization of water influences the behavior of biomolecules

Most biochemical reactions occur in aqueous environments. The ionization of water influences not only the behavior of acids and bases but that also of the major groups of biomolecules. Therefore it is important to consider the ionization of water to understand the behavior of biomolecules in cells and in the laboratory. Water has a slight tendency to ionize according to the equation:

$$H_2O \quad \longrightarrow \quad H^+ + OH^-. \tag{9}$$

To be chemically precise, when water ionizes, hydronium ions or hydrated hydrogen ions, H_3O^+, are formed rather than protons. In biochemistry, however, H+ is written instead of H_3O^+. The reason is that distinguishing between hydronium ions and protons is important when the behavior of H+ is considered both in aqueous and nonaqueous media. However, as mentioned above, most biochemical reactions occur exclusively in one environment i.e. in water. Therefore the need to distinguish between H_3O^+ and H^+ is diminished greatly. Because there often is no need to consider both aqueous and nonaqueous media in biological systems and, in addition, to achieve simplicity, biochemists generally write H^+ instead of H_3O^+. Nevertheless, it is understood that the chemical behavior of H^+ in most biochemical reactions is that of hydrated H^+ or H_3O^+, not that of H^+. For convenience then, we will follow biochemical convention and write H^+ instead of H_3O^+. The equilibrium constant for the ionization of water is thus

$$K' = \frac{[H^+][OH^-]}{[H_2O]} . \tag{10}$$

The concentration of H_2O in equation (10) is large since the ionization of water is small. Accordingly, $[H_2O]$ may be taken as being constant relative to $[H^+]$ and $[OH^-]$ and we can obtain a new constant, designated K_w', from multiplying K' by $[H_2O]$:

$$K_w' = [H^+][OH^-]. \tag{11}$$

Measurements of electrical conductivity have shown that

$$K_w' = 1*10^{-14} \text{ at } 25°C, \tag{12}$$

Thus,
$$K_w' = [H^+][OH^-] = 1*10^{-14}, \tag{13}$$

At equilibrium,
$$[H^+] = [OH^-] = \sqrt{1*10^{-14}} = 1*10^{-7}, \tag{14}$$
$$\therefore \text{ pH} = 7. \tag{15}$$

The contribution of H^+ or OH^- ions from this equilibrium must be considered in calculations involving low concentrations of acids or bases. In calculations involving large concentrations of acids or bases, the low concentrations of H^+ or OH^- due to the ionization of water may be ignored. In the laboratory, the ionization of water and the pH of aqueous solutions influence several phenomena. Examples are the extraction of biomolecules from cellular material, the separation of biomolecules by chromatography or electrophoresis and the measurement of enzyme activity.

2-5.   The Henderson-Hasselbalch equation shows how weak acids behave

If a weak acid is represented by HA then it will dissociate as follows:

$$HA \longrightarrow H^+ + A^-. \tag{16}$$

The equilibrium constant for this reaction is

$$K_a' = \frac{[H^+]\,[A^-]}{[HA]}, \tag{17}$$

Solving for $[H^+]$, $$[H^+] = K_a'\frac{[HA]}{[A^-]}, \tag{18}$$

Taking negative logarithms of both sides

$$-\log[H^+] = -\log K_a' - \log\frac{[HA]}{[A^-]}, \tag{19}$$

$$pH = pK_a' - \log\frac{[HA]}{[A^-]}, \tag{20}$$

$$pH = pK_a' + \log\frac{[A^-]}{[HA]}. \tag{21}$$

Equation (21) is known as the Henderson-Hasselbalch equation because it was developed independently by L. J. Henderson of the U.S.A and K. A. Hasselbalch of Germany in 1913 and 1916 respectively. In the biomedical sciences, the Henderson-Hasselbalch equation is used (with certain restrictions) for calculating values related to acid-base titration curves.

Figure 2-1 shows the titration curves of three weak acids of varying strengths. The three acids give curves of similar shapes when they are titrated against strong base; but each curve occupies a different position on the pH axis. The shapes of the three curves are determined by the logarithmic nature of the Henderson-Hasselbalch equation. The logarithmic nature of the Henderson-Hasselbalch equation, in turn, stems from the logarithmic derivation of pH. Given a particular ratio of proton donor to proton acceptor for such titration curves, the pH of the mixture which gives the ratio $[A^-]$ to $[HA]$ can be calculated from the Henderson-Hasselbalch equation. *At the midpoint of each titration curve in Figure 2-1, the ratio of [proton donor]/[proton acceptor] is equal to 1; and, since log 1 = 0, then pH = pK at the midpoint of each titration.* Thus, pK values indicate the relative strengths of acids, i.e., the lower the pK, the greater the relative strength. In addition, given the pH and the ratio $[A^-]$ to $[HA]$, the pK of the mixture which gave the ratio can be estimated. However, there is an important restriction on the use of such calculations. It is that calculated values and titrimetric data agree most closely when the ratio $[A^-]/[HA]$ is obtained within the range of one pH unit on either side of the pK or midpoint of the titration. This type of calculation applies both to the titration of weak acids against strong bases and to the titration of weak bases versus strong acids. Another use of the Henderson-Hasselbalch equation is to aid in the preparation of buffer solutions, and this application is described in the next section.

2-6. The preparation of buffers is based on two approaches

Buffers are solutions which contain relatively high concentrations of weak acids or bases and their conjugate partners. The conjugate partners generally are present as salts of the weak acid or base. A buffer solution which contains a weak acid and its salt is considered to be acidic. One which contains a weak base and its salt is considered to be a basic buffer. There are two approaches to the preparation of biochemical buffers:

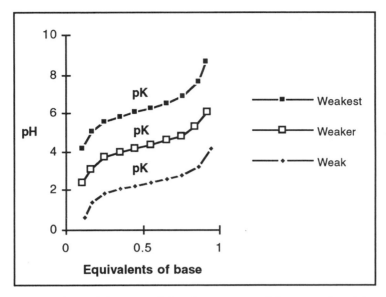

Figure 2-1. Diagram of titration curves of three weak acids

• Carefully calculated amounts of both conjugate acid and conjugate base are weighed out separately, dissolved, and made up to volume.

• One component of the conjugate pair, either the weak acid or base, is weighed out and dissolved in an appropriate volume of water. The conjugate partner then is obtained by adding to the solution of weak acid or base an amount of strong base or strong acid calculated to yield the desired pH value.

When the correct pK_a' values are available, two sets of calculations are needed for computing the amounts of conjugate acid-base pairs required for preparing buffer solutions: (1) the fractions of buffer present as dissociated and undissociated species and (2) the amounts of conjugate acid and conjugate base required.

2-7.   Step 1: Calculate the fraction of buffer dissociated

Computations of the fractions of buffer present as dissociated and undissociated species may be made readily from the equation:

$$f_d = \frac{\text{antilog } \Delta}{\text{antilog } \Delta + 1}, \tag{22}$$

where

$$f_d = \text{fraction of buffer present in dissociated form,}$$
$$1- f_d = \text{fraction of buffer in undissociated form, and}$$
$$\Delta = pH - pK_a'$$

Equation (22) may be derived from the Henderson-Hasselbalch equation as follows:

$$\Delta \ = \ \log\frac{[A^-]}{[HA]} \ = \ \log\frac{f_d}{1-f_d} \ ; \tag{23}$$

$$\text{Antilog} \ \ \Delta \ = \ \frac{f_d}{1-f_d} \ ; \tag{24}$$

$$f_d \ = \ \text{antilog } \Delta * (1-f_d); \tag{25}$$

$$\text{Antilog} \ \ \Delta \ = \ f_d + \text{antilog } \Delta * f_d \ ; \tag{26}$$

$$f_d \ = \ \frac{\text{antilog } \Delta}{\text{antilog } \Delta + 1} \ . \tag{27}$$

2-8. Step Calculate the conjugate acid and base required

Once the fractions of buffer present as dissociated and undissociated forms have been calculated, the amounts of conjugate acid and conjugate base needed can be calculated by applying principles of general chemistry. The calculations differ somewhat depending on whether the conjugate acid and conjugate base are weighed out separately, or whether both conjugate partners are obtained from one component. In both cases, however, "molarity of buffer" refers to the *total concentration* of buffer species: the sum of [A⁻] and [HA].

Case 1: If both conjugate partners are weighed out the following relationships should be used:

1. Vol. of buffer, L * Molarity of buffer = Number of mols of buffer;
2. Mols of buffer * f_d = mols of conjugate base, A;
3. Mols of buffer * $1-f_d$ = mols of conjugate acid, HA;
4. Mols of conjugate base * mol wt of conjugate base   = grams of A;
5. Mols of conjugate acid * mol wt of conjugate acid   = grams of HA.

If it is desired to prepare the buffer by mixing stock solutions of conjugate acid and conjugate base, the preparation may be guided by the next set of relationships:

6. Vol. of A, L = calcd. mols of conjugate base/molarity of buffer;
7. Vol. of HA , L = calcd. mols of conjugate acid/molarity of buffer.

Note regarding phosphate buffer. In *Basic Biochemical Laboratory Procedures and Computing*, we use the Henderson-Hasselbalch equation, *published in 1913 and 1916* to compute [A⁻] and [HA]. But handbooks usually cite 1909 *references by Sørensen* as their source for [A⁻] and [HA]. Also, handbooks often do not list the pK_2 of H_3PO_4. So, it is no surprise that, even with a good guess of the handbook pK_2, our amounts of [A⁻] and [HA] needed to prepare phosphate buffer of desired pH are not identical to those in handbooks.

Case 2: If both conjugate partners from one component, the following relationships should be used:

1. Volume of buffer, L * Molarity of buffer = Number of mols of buffer
2. Mols of buffer * f_d = mols of conjugate base, A⁻;
3. Mols of buffer * $1-f_d$ = mols of conjugate acid, HA;
4. Grams of buffer   = mols of buffer   *   mol weight of buffer;
5. Mols of OH⁻ required for acidic buffer   = mols of A⁻ (step 2);
6. Mols of H⁺ required for basic buffer   = mols of HA (step 3);
7. Volume of OH⁻ or H⁺ required = mols of OH⁻or H⁺ required/ molarity of base or acid available.

2-9. To use buffers effectively, certain rules should be followed

The following criteria should be adhered to if buffers are to be used effectively:

- The pH of the buffer should not be more than 1 pH unit above or below the pK_a'; i.e. effective buffering usually is obtained within the pH range of $pK_a' \pm 1$ pH unit as explained in sections 2-5 and 2-12.
- The buffer should not have any metabolic effects on the system being studied.
- The buffer should not react with any component of the assay medium.
- Prepare buffers in glassware instead of plasticware.
- Prepare buffers at temperatures close to the working temperatures.
- pH meters should be calibrated with at least two standard buffer solutions.
- Corrections should be made for the influence of ionic strength and temperature on pK_a.

2-10. Thermodynamic pK values should be corrected for ionic strength

Ideal or thermodynamic pK_a values are based on activities but concentrations are used in preparing buffer solutions in the biomedical laboratory. Concentration-based (or "practical") pK_a values, designated pK_a', therefore should be used for calculating the amounts of A⁻ and HA required in the biomedical laboratory. The pK_a' values may be obtained from pK_a by a modification of the Debye-Hückel equation:

$$pK_a' \quad = \quad pK_a - \left[\frac{0.509(2n + 1)\sqrt{\mu}}{1 + \sqrt{\mu}} \right] - 0.1*\mu , \tag{24}$$

where n =   the number of charges on the conjugate acid,
and μ =   ionic strength of the solution (which equals
                   concentration for weak acids or bases).

The differences between pK_a' and pK_a decrease as ionic strength decreases; that is, pK_a' approaches pK_a as ionic strength approaches zero. In addition, differences between pK_a' and pK_a depend on the buffer being considered and may vary considerably from buffer to buffer.

2-11. Thermodynamic pK values also should be corrected for temperature

The value of the equilibrium constant of a reaction (K_{eq}) depends on temperature. This was expressed quantitatively for the first time by Jacobus H. van't Hoff late in the nineteenth century. Van't Hoff showed that the rate at which K_{eq} changes with temperature is a function of the change in enthalpy. (Enthalpy is the heat released or absorbed in a process which occurs at constant pressure). Since K_{eq} changes with temperature, pK_a also will change with temperature. Concentration-based pK_a' values may be corrected for such temperature changes with equation (25):

$$pK_a' \text{ at } T_2 \quad = \quad pK_a' \text{ at } T_1 + \Delta pK_a' , \tag{25}$$

where T_1 =   temperature °C, at which pK_a' is known,
        T_2 =   temperature °C, for which pK_a' is to be calculated,
        $\Delta pK_a'$ = $(T_2 - T_1)*d(pK_a')/dt$,
and $d(pK_a')/dt$ =   change of pK_a' per °C (see Appendix 2).

If $d(pK_a')/dt$ is negative, as for many buffers used in biomedical research, (Appendix 2), then $\Delta pK_a'$ is positive when $T_2 < T_1$. Conversely, if $d(pKa')/dt$ is negative but $T_2 > T_1$, then $\Delta pK_a'$ is negative.

2-12. Values of pK_a may be estimated graphically

The Henderson-Hasselbalch equation can be rearranged to give

$$\log \frac{[A^-]}{[HA]} \quad = \quad pH - pK_a . \tag{26}$$

This is equivalent to the slope-intercept form of the equations of a straight line:

$$y \quad = \quad mx + b,$$

and a plot of log $[A^-]/[HA]$ versus pH should be a straight line with slope = 1 and y-intercept = $-pK_a$. By definition, pH = pK_a when the concentrations of proton donor and proton acceptor are equal so that log $[A^-]/[HA] = 0$. Therefore, equation (26) is satisfied by the coordinates x = pK_a and y = 0. In other words, the equation is satisfied when the curve intersects the pH-axis at pH = pK_a and the y-axis at $-pK_a$ as shown in Figure 2-2.

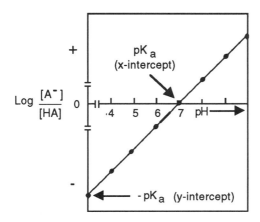

Figure 2-2. Graphical determination of pK_a

Experimentally, the determination of the pK_a of the acid-base indicator bromthymol blue (a weak acid) may be used as an example of the graphical approach to the estimation of pK_a. To conduct the experiment, one uses buffered solutions of the indicator. Then, the absorbances of these buffered solutions are read in a spectrophotometer at appropriate wavelengths – wavelengths at which the absorbance readings indicate the concentrations of A^- and HA, sometimes designated as In^- and HIn respectively. After the absorbances are obtained (at defined pH values), a curve of log $[A^-]/[HA]$ versus pH is plotted. Finally, the pK_a of the indicator is estimated from the x- and y-intercepts of the curve (cf. Figure 2-2). As we shall see in Chapter 8, such graphical estimation of pK_a is similar, in principle, to the estimation of the midpoint potentials of biological redox compounds, E_m, by potentiometric titration. The reason for this similarity is that many biochemical reactions, other than those of acids and bases, may properly be classified as dissociation reactions. Examples are oxidation-reduction reactions, the dissociation of enzyme-substrate complexes, the dissociation of drug-receptor complexes and the dissociation of ligands from metal coordination compounds such as hemoglobin. Thus, in addition to the dissociation of the weak acid HA, we may consider a general example, depicted by dissociation of the compound QX:

$$QX \longrightarrow Q + X \tag{27}$$

so that the equilibrium constant, K' is

$$K' = [Q][X]/[QX] \tag{28}$$

$$[Q] = K' \frac{[QX]}{[X]} \tag{29}$$

and, by analogy to the Henderson-Hasselbalch equation,

$$-\log [Q] = -\log K' - \log \frac{[QX]}{[X]} \tag{30}$$

$$pQ = pK' + \log \frac{[X]}{[QX]} \tag{31}$$

and
$$\log \frac{[X]}{[QX]} = pQ - pK' \tag{32}$$

Thus, the dissociation of the compound QX in equation (27) may be evaluated either from a linear plot of log [X]/[QX] against pQ as in Figure 2-3, or from a logarithmic plot of pQ versus [X] as in Figure 2-4.

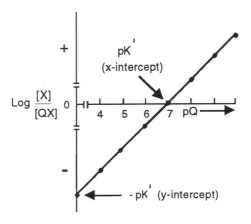

Figure 2-3.   Relation of pK' to log [X]/[QX] for the dissociation reaction QX ⟶ Q + X

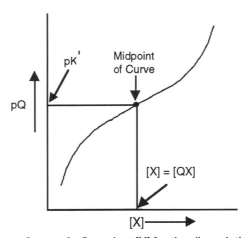

Figure 2-4. Curve of pQ against [X] for the dissociation reaction

In the linear plot, Figure 2-3, the y-intercept is -pK' and the x-intercept, where log [X]/[QX] = 0, is equal to pK'. In the logarithmic curve, Figure 2-4, pK' is the midpoint of the curve, i.e. the point of inflection where QX is 50% dissociated and the proportions of X and QX are equal. The logarithmic curve of Figure 2-4 therefore is analogous both to acid-base and to oxidation-reduction titration curves.

2-13. Two- or three-buffer systems are needed for a wide pH range

Biomedical experiments often call for measurements of biological activity over a range of pH values much greater than ± 1 pH unit above or below the pK_a' of a buffer. It may seem that this problem may be solved by using individual buffers which span the desired pH range. Examination of this approach shows, however, that it introduces some new problems. For example, use of different, single buffers to cover a wide pH range may lead to chemical or biological effects caused by each buffer rather than by the change in pH. Moreover, different buffers may yield different ionic strengths or may react differently with the bio-molecule being assayed, thereby changing its kinetic and other properties. Another single buffer approach, conducting the experiment with higher concentrations of a buffer to maintain the pH value, also may cause undesirable side effects such as inhibition of biological activity, or a change of biological activity due to a change of ionic strength. A more desirable solution would be to use a mixture of two or three buffers, each with overlapping pH ranges, to cover the desired range of pH values. But, improper use of such mixtures may introduce new problems such as inhibition, changes in ionic strength and changes in the ionization of ionizable groups on proteins and other biomolecules. Therefore, it is desirable to devise mixtures of buffers in which ionic strength is kept constant and chemical and biological effects on the assay system are eliminated. Since there is no single buffer mixture that is suitable for all possible experiments, it is desirable to test a proposed mixture before embarking on any experiment. As a starting point, previously determined mixtures of buffers may be consulted and tried, especially when these mixtures have been devised according to chemically sound principles. Such a list is given in the reference by Ellis and Morrison which is cited in the references at the end of this chapter. The following are some of the points which should be considered in the use of buffer mixtures to measure biological activity over a wide pH range:

- The pK_a' values of each buffer component may be ± 2 pH units apart and still provide good buffering capacity.
- If metal ions are essential for activity, the buffer should not have chelating activity.
- It is desirable to use relatively large organic acids or bases for adjusting the pH of a buffer mixture so as to minimize interaction with ionic groups on the biomolecules being studied.
- To adjust the pH of a buffer mixture add acid or base slowly, with rapid stirring, to avoid local zones of high or low pH.
- The components of a buffer system should not absorb light if activity of the biological system is to be measured by spectrophotometry.

In the rest of this chapter, we will examine the effect of pH on the titration curves of amino acids and proteins. Our purpose will be to show how the net charges of amino acids (and other biomolecules) depend on pH.

2-14.   Titration curves show the influence of pH on ionization

The titration curves of the monoamino monocarboxylic amino acids are analogous to curves from the stepwise titration of diprotic acids. The reason is that the monoamino monocarboxylic amino acids, e.g. alanine (Table 2-1), have two titratable functional groups. These functional groups, an amino and a carboxyl group, are *alpha* to each other; and at physiological pH, i.e. approximately pH 6 to 7, internal proton transfer from the carboxyl to the amino group occurs thereby leading to the formation of a *dipolar ion or zwitterion*:

$$NH_3^+$$
$$|$$
$$R\text{-}CH\text{-}COO^-$$
zwitterion

The stepwise formation and ionization of the dipolar ion yield three ionic species, a cation (or fully protonated species), a dipolar ion and an anion:

$$
\begin{array}{ccccc}
NH_3^+ & & NH_3^+ & & NH_2 \\
| & -H^+ & | & -H^+ & | \\
R\text{-}CH\text{-}COOH & \rightleftharpoons & R\text{-}CH\text{-}COO^- & \rightleftharpoons & R\text{-}CH\text{-}COO^- \\
cation & & dipolar\ ion & & anion
\end{array}
\quad (33)
$$

The curve of Figure 2-5 shows how the proportions of these species vary with pH during titration of a fully protonated, monoamino monocarboxylic amino acid. The titration curve has two legs, each of which may be divided into three major regions: a relatively flat middle region in which pH changes very little and two other regions (one at either end of the flat region) in which the changes in pH are somewhat greater than in the middle region. The first leg is obtained from the titration of the α-carboxyl group. Because the α-NH_3^+ group exerts an electron-withdrawing effect on the α-COOH group, the latter is a relatively strong acid (it has a pK_a of ≈ 2 compared with the COOH of acetic acid, pK_a 4.76). Therefore, in solution, the α-carboxyl group is markedly dissociated *before* the addition of base. At the start of the titration, as NaOH is added, more carboxylate anion, α-COO$^-$, is formed from the neutralization of α-COOH; consequently, buffering action is enhanced and the slope of the curve decreases slightly. As more NaOH is added, the pH increases slowly because the concentration of α-COO$^-$ continues to increase and to repress further ionization of α-COOH to α-COO$^-$. A point of inflection occurs at the midpoint where the ratio [α-COO$^-$]/ [α-COOH] = 1; this is pK_1, the pK_a of the α-COOH group. Thus, the rate of change in pH during the titration depends on the extent to which [α-COO$^-$]/[α-COOH] differs from 1. That is, as mentioned in section 2-5, buffering ability is greatest at $\pm$ 1 pH unit around the pK_a where [α-COO$^-$]/ [α-COOH] = 1; conversely, buffering ability is least when [α-COO$^-$]/[α-COOH] is much greater or less than 1.

A second point of inflection occurs between the two legs of the titration. This is the equivalence point for the titration of the α-COOH group. Here, a stoichiometric amount of base has neutralized the α-COOH group. This equivalence point is known as the *isoelectric point,* designated pI or pH_I, because it is the point at which the amino acid does not have a net charge. The isoelectric point of a monoamino monocarboxylic acid is the arithmetic mean of the pK_a values of the carboxyl and amino groups (cf. section 2-16). Above the isoelectric point, amino acids have a net negative charge; below pI, amino acids have a net positive charge. The ionic species of monoamino monoarboxylic acids at the major stages of both legs of the titration are shown on Figure 2-5 as structures I, II, and III.

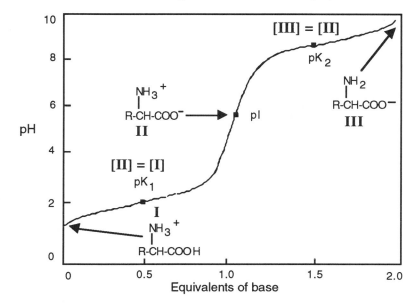

Figure 2-5. Titration curve of a fully protonated monoamino monocarboxylic acid

We see also from Figure 2-5 that the shapes of the titration curves produced by the α-COOH and α-NH_3^+ groups are similar. This indicates that the sequence of events which takes place during the second leg of the titration is analogous to the sequence which takes place during the first leg. Therefore, one has, essentially, only to substitute α-NH_3^+ for α-COOH, $[NH_2]/[NH_3^+]$ for $[\alpha\text{-}COO^-]/[\alpha\text{-}COOH]$ and pK_2 (the pK_a of the NH_3^+ group) for pK_1 to obtain a description of the course of titration of the NH_3^+ group. The article "Amino acid titration curves - misshapen or mislabeled?", by Darvey and Ralston, *Trends Biochem. Sci.* 18: 69-71, 1993, has additional information on, and an analysis of, amino acid titration curves. The ionic charge of amino acids and other biomolecules at given pH values is exploited in the laboratory in several important techniques two of which are ion-exchange chromatography and electrophoresis. We will discuss ion-exchange chromatography (a form of chromatography based on electrostatic attractions between the polyelectrolyte surfaces of biomolecules and the ion-exchange surfaces of ion-exchange resins) in Chapter 3. A brief discussion of electrophoresis, (defined as *the transport of electrically charged particles through a buffered electrolyte under the influence of an applied electric field*), follows.

There are three general types of methods for carrying out electrophoresis: *moving boundary electrophoresis, zone electrophorseis* and *capillary electrophoresis*. Moving boundary electrophoresis dates back to 1937 when it was introduced by Arne Tiselius, the Swedish physical chemist. *Moving boundary and capillary electrophoresis* both may be *classified as free-zone methods* because both are carried out without supports such as paper, cellulose acetate strips, agarose, or polyacrylamide gels. Zone electrophoresis, although known before moving boundary electrophoresis, only became an important biomedical technique after paper electrophoresis was developed in the 1950s. But, until recently, zone electrophoretic techniques have been the methods of choice and free-zone electrophoretic procedures had fallen into disfavor. Currently, however, free-zone methods are receiving renewed interest and are experiencing explosive growth. The reasons for this are that capillary electrophoresis is rapid and that it yields separations of remarkable resolution and sensitivity. Nevertheless, it is still informative to consider how moving boundary electrophoresis (the original free-zone electrophoretic method) works since many of the principles underlying its operation also may be applied to capillary and to zone electrophoresis.

In moving boundary electrophoresis one uses an apparatus which generally resembles a U-tube (Fig2-6).

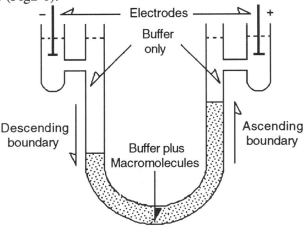

Figure 2-6. Diagram of Tiselius moving boundary apparatus

To conduct a run in this apparatus, the sample consisting of buffer plus charged macromolecules is placed in the U-tube and buffer alone is layered carefully over the solution of macromolecules. Next, the apparatus is placed in a water bath at constant temperature, then electrodes are placed in the buffer and the electric field is applied. In Figure 2-6, it is assumed that the macromolecules are acidic, i.e., they are negatively charged in the buffered electrolyte. These anions therefore migrate toward the anode (positively charged). If the macromolecules were cationic when placed in the buffered electrolyte they would, alternatively, move toward the cathode (negatively charged). In moving boundary electrophoresis, migration of biomolecules is observed by optical methods and is recorded photographically. A key feature of the moving boundary method, therefore, is that *charged molecules move freely* in buffered solution.

Table 2-1.   Structures and Some Properties of Amino Acids

Name and Abbreviation	R-group	$pK_a'\alpha\text{-COOH}$	$pK_a'\alpha\text{-NH}_3^+$	$pK_a'\,R$	pH_I
Nonpolar Amino Acids:					
Alanine, Ala	CH_3-	2.34	9.69		6.00
Valine, Val	$\begin{array}{c}CH_3\\CH_3\end{array}\!\!>\!CH-$	2.32	9.62		5.96
Leucine, Leu	$\begin{array}{c}CH_3\\CH_3\end{array}\!\!>\!CH\,CH_2-$	2.36	9.60		5.98
Isoleucine, Ileu	$CH_3CH_2\underset{\underset{CH_3}{\mid}}{CH}-$	2.36	9.60		6.02
Proline, Pro	(R)—COOH (with N–H)	1.99	10.60		6.30
Methionine, Met	$CH_3SCH_2CH_2^-$	2.28	9.21		5.74
Phenylalanine, Phe	$\bigcirc-CH_2^-$	1.83	9.13		5.48
Tryptophan, Trp	indole$-CH_2-$	2.83	9.39		5.89
Polar Amino Acids:					
Glycine, Gly	$H-$	2.34	9.60		5.97
Serine, Ser	CH_2OH-	2.21	9.15		5.68
Threonine, Thr	CH_3CHOH-	2.63	10.43		5.60
Cysteine, Cys	CH_2SH-	1.71	10.78	8.33	5.07
Tyrosine, Tyr	$HO-\bigcirc-CH_2^-$	2.20	9.11	10.07	5.66
Asparagine, Asn	H_2NCOCH_2-	2.02	8.80		5.41
Glutamine, Gln	$H_2NCOCH_2CH_2-$	2.17	9.13		5.65

Table 2-1.   Structures and Some Properties of Amino Acids Continued

Name and Abbreviation	R-group	$pK_a' \alpha\text{-COOH}$	$pK_a' \alpha\text{-NH}_3^+$	$pK_a' R$	pH_1
Acidic Amino Acids:					
Aspartic Acid, Asp	$^-OOCCH_2-$	2.09	9.82	3.86	2.77
Glutamic acid, Glu	$^-OOCCH_2CH_2-$	2.19	9.67	4.25	3.22
Basic Amino Acids:					
Lysine, Lys	$H_3N^+CH_2CH_2CH_2CH_2-$	2.18	8.95	10.53	9.74
Arginine, Arg	$\overset{\overset{+NH_2}{\|\|}}{H_2NCNH}CH_2CH_2CH_2-$	2.17	9.04	12.48	10.76
Histidine, His	$-CH_2-$ (imidazole)	1.82	9.17	6.00	7.58

In *zone electrophoresis, charged molecules do not move freely in solution.* Instead, the samples to be analyzed are applied in small spots or thin bands to inert supports such as filter paper, polyacrylamide gels, agarose gels or strips of cellulose acetate. After application of the sample, the inert support is immersed in a buffered solution in an appropriate vessel and an electric field is applied. Components of the mixture then are separated from each other on the basis of differences in net charge and, in some cases, differences in molecular size. A variation of zone electrophoresis that has become an important tool in modern biochemistry is *isoelectric focusing* which usually is carried out in polyacrylamide gels. In this method, a stable pH gradient is produced in the gel by the inclusion of a mixture of ampholytes. *Ampholytes* are defined here as mixtures of low molecular weight polyelectrolytes which can yield H^+ or OH^- and thus act as proton donors or proton acceptors. As the electric field is applied through the gel, the ampholytes migrate to positions corresponding to their isoelectric points, thus giving a pH gradient. Then, as a mixture of proteins is placed on the top of the gel, individual proteins migrate to positions in the gels corresponding to their isoelectric points and form sharp, narrow bands. These sharp bands of proteins are said to be *focused.*

Capillary electrophoresis (see sections 4-19 to 4-22) usually is carried out in tubes made of fused silica. The tubes range in length from 10 to 100 cm and have internal diameters of 20 to 200 μm. To date, capillary electrophoresis has been conducted mostly in uncoated and unpacked tubes, but the capillaries also may be coated with inert polymers or packed with an appropriate chromatographic or gel electrophoretic support. Generally, the capillaries are filled with a buffer, and the ends of the tubes are placed in reservoirs containing electrodes, as shown in Figure 2-7.

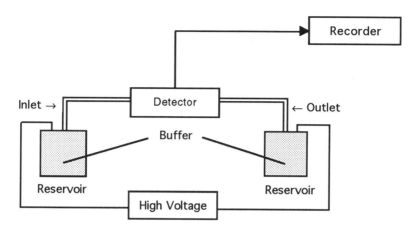

Figure 2-7. Diagram of a capillary electrophoresis system

Samples for capillary electrophoresis are often in the picoliter range and are introduced into the capillaries by a variety of methods such as gravity (raising the sample and achieving entry by siphoning), electromigration (entry under voltage), or injection under positive or negative pressure. Additional information on this and other electrophoretic methods is given in sections 4-6 to 4-7 and 4-19 to 4-22.

2-15.   The peptide bond influences the properties of proteins

All naturally occurring amino acids, except glycine (Table 2-1) are optically active and have the L-configuration. Proteins are formed from head to tail reactions between the α-carboxyl and α-amino groups of these amino acids. Such head to tail reactions between amino acids occur with the elimination of water and lead to the formation of the -CO-NH- group; that is, a substituted amide link known as a *peptide* bond. Dipeptides, tripeptides and tetrapeptides contain two, three, and four amino residues respectively; but peptides up to molecular weight 10, 000 are known as *polypeptides* and those of molecular weight greater than 10, 000 as *proteins*. The amino acid residues at the NH_2 and COOH ends of a polypeptide chain are known as the amino-terminal *(N-terminal)* and carboxyl-terminal *(C-terminal)* residues respectively. We may, therefore, represent peptide bond formation between two L-amino acids as

$$\overset{\overset{\textstyle R_1}{\textstyle |}}{NH_2\text{-}CH\text{-}COOH} + \overset{\overset{\textstyle R_2}{\textstyle |}}{NH_2\text{-}CH\text{-}COOH} \rightarrow \overset{\overset{\textstyle R_1 \qquad R_2}{\textstyle | \qquad |}}{NH_2\text{-}CH\text{-}CONHCH\text{-}COOH} + H_2O \quad (34)$$

A protein thus may have many acidic and basic R groups and may be *oligomeric,* consisting of two or more polypeptide chains, each of which has a unique sequence of amino acids, a particular molecular weight, and a specific or *native* three-dimensional structure or conformation. Usually, the native conformation of proteins is the only biologically active one among the many conformations that are theoretically possible. These properties raise a question which we can approach only parenthetically in this chapter on biochemical aspects of acids and bases. The question is: how do cells transmit and decode information on unique amino acid sequences and specific three-dimensional conformations? We provide brief answers and references to additional information for the interested reader.

The *sequence of amino acids in a polypeptide* is determined from the *transcription* and *translation* of information stored in the DNA of an organism. These processes, by which the amino acid sequences of proteins are transcribed and translated from genetic information stored in DNA, are now understood and are well-described in textbooks by Bohinski; Conn, Stumpf, Bruening, Doi; Lehninger, Nelson, Cox; Mathews, van Holde; Stryer; Zubay; and others.

The course by which *three-dimensional structure is formed from polypeptide sequences* (an aspect of the *decoding* of genetic information) is not yet well-understood. But, it appears that the *correct folding, assembly, transport and three-dimensional structure of proteins* all depend on the participation of a family of oligomeric proteins known as *molecular chaperones*. The best known of the latter are the *hsp60/groEL* proteins or *chaperonins* which belong to the group of *heat shock* or *stress proteins* found in plastids, mitochondria and prokaryotes. More information on molecular chaperones may be found in references by Debbie Ang et al., by Burston et al., by Ellis and Hemmingsen, by Ellis and van der Vies, by Georgopoulos, and by Shinde and Inouye, all listed at the end of this chapter.

To return to the acid-base properties of proteins, we note that the reaction for peptide formation, equation (34) above, indicates that the *acid-base properties of polypeptides* at physiological pH *depend on their many projecting R groups* also referred to as side chains; stated another way, the acid-base properties of polypeptides do not depend on the carboxyl and amino groups involved in peptide linkages. Other biological, chemical and physical properties of peptides do, however, depend on the nature of the peptide bond. Among them are the following:

- The C-N single bond contains approximately 40 % double bond character. Consequently, the C-N bond of polypeptides is partly rigid and does not rotate freely.
- The C=O double bond of peptides contains approximately 40 % single bond character but still does not rotate freely.
- The peptide bond, as a consequence of the previous two points, is a flat, rigid structure. This rigidity limits the ways in which polypeptide chains may fold. These properties were first described by Linus Pauling and colleagues from x-ray spectroscopic studies of peptides. We discuss x-ray spectroscopy in Chapter 7.

2-16. Titrations complement chromatography and electrophoresis

As we indicated above, proteins generally have several side chains which usually have both acidic and basic groups. Therefore the titration of proteins may be viewed in the same light as the stepwise titration of a polyprotic acid. Glutamic acid and lysine, each with more than two titratable groups (three each) may be regarded as simple polyprotic acids. Accordingly, when compared with monoamino monocarboxylic acids, the titration curves of glutamic acid and lysine should provide some insight into the complexities introduced by the presence of several titratable groups in polypeptides. Figures 2-8 and 2-9 show titration curves of glutamic acid and lysine. The main point of interest in these curves is that it is not possible to distinguish different legs between the two carboxyl groups of glutamic acid or between the two amino groups of lysine. As a result, pK_a' values cannot be assigned to either of the two carboxyls of glutamic acid or to either of the two amino groups of lysine from titration curves - as can be done for the pK_a' values of the α-carboxyl and α-amino groups of alanine. Not surprisingly, it is even more difficult (than with glutamic acid and lysine) to assign pK_a' values to the large numbers of acidic and basic groups of proteins. But, since during the titration of proteins, one molecule of acid adds a proton and each molecule of base removes a proton, titration curves of proteins can provide information on the number of ionizable protons that are present in a protein. Such information is useful for checking the amino acid composition of proteins obtained by ion exchange chromatography or other methods. Titrimetric information also can provide clues to the three-dimensional structures of proteins.

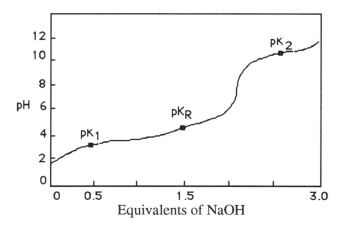

Figure 2-8.   Titration curve of fully protonated glutamic acid

Consideration of the titration curves of Figures 2-5, 2-8 and 2-9 will indicate that, not only the amino acids but proteins, as well, are negatively charged at pH values above their isoelectric points and positively charged at pH values below their pI values. Moreover, the further the pH is above the isoelectric points of these amphoteric biomolecules, the greater is their negative charge; conversely, the further the pH is below the isoelectric point, the greater is the positive charge.   We

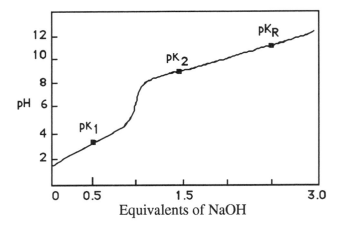

Figure 2-9. Titration of fully protonated curve of lysine

can view the process of calculating the pI values of amino acids as one of converting the fully protonated forms of the acids to their dipolar ions as indicated in the following summary:

If the fully protonated form contains	*Then calculate pI from these pK values*		
1 COOH and 1 NH_3^+	(α-COOH + α-NH_3^+)	÷	2
2 COOH and 1 NH_3^+	(α-COOH + R-COOH)	÷	2
1 COOH and 2 NH_3^+	(α-NH_3^+ + R-NH_3)	÷	2

Instead of using titration curves for deriving information on isoelectric points of proteins, estimation of pI can be made from the movement of the biomolecules during electrophoresis. Thus, estimation of the isoelectric point of a protein may be made from a plot of electrophoretic mobility versus pH. In such plots, as shown in Figure 2-10, the pH value at the x-intercept is the isoelectric point, indicating that, at the isoelectric point, the protein is stationary and mobility is zero. Frequently, however, there are significant differences between such estimates of pI and values obtained by careful experimentation. These differences may be due, in part, to cations and anions from the buffer solution being bound by the protein. Nevertheless, since each protein has a unique sequence of amino acids and a characteristic titration curve, then theoretically, in a mixture of proteins, it should be possible to locate a pH value at which the difference in charge between a given protein and all others in the mixture is at a maximum. This pH value could be a useful starting point for developing procedures for the separation and purification

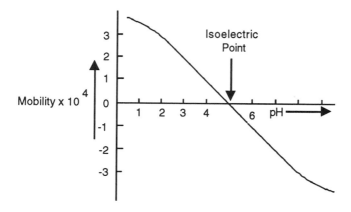

Figure 2-10. Electrophoretic estimation of the isoelectric point

of the desired protein by ion-exchange chromatography or by electrophoresis if two points are borne in mind:

- that in ion-exchange chromatography, the separation of proteins and other amphoteric biomolecules depends both on the amount, and on the surface distribution or symmetry of charge, on oppositely charged polyelectrolytes (i.e., the protein and the ion-exchange resin); and
- that in electrophoresis, the separation of proteins is independent of the symmetry or asymmetry of charge distribution, and depends mainly on the net charge and the isoelectric points of the biomolecules.

2-17.   Worked examples

EXAMPLE 2-1. Calculate the pH of a solution in which $[H^+]$ is 0.0005 M.

From equation (5): pH = $- \log [H^+]$,
                     pH = $- (\log 0.0005)$,
                           = $- (-3.3010)$,
                     pH = 3.3.

Alternative 1: pH = $- \log [H^+]$,
                     pH = $- \log (5 \times 10^{-4})$ $= - (-4 + \log 5)$,
                           = $- (-4 + 0.699)$ $= - (- 3.301)$,
                     pH = 3.3.

Alternative	pH	=	$\log_{10} 1/[H^+]$,
	pH	=	$\log_{10} 1/0.0005$,
	pH	=	$\log_{10} 2000$,
	pH	=	3.3.

EXAMPLE 2-2. Calculate $[H^+]$ of an aqueous solution of pH 6.8.

From equation (7):	$[H^+]$	=	10^{-pH} = antilog -pH,
	$[H^+]$	=	$10^{-6.8}$,
	$[H^+]$	=	1.59×10^{-7} M.

EXAMPLE 2-3. (a) Calculate the pK_a' of KH_2PO_4 if $K_a' = 1.38 \times 10^{-7}$ M. (b) Calculate K_a' of acetic acid given its $pK_a' = 4.76$.

(a) From equation (8) pK_a'	=	- log $K_a' = \log_{10}, 1/K_a'$,
pK_a'	=	- $\log_{10}$, 1.38×10^{-7} ,
	=	$\log_{10}$, $(1/1.38 \times 10^{-7})$,
pK_a'	=	6.86,

(b)	K_a'	=	$10^{-K_a'}$ = antilog - K_a' ,
		=	$10^{-4.76}$,
	K_a'	=	1.74×10^{-5} .

EXAMPLE 2-4. Calculate the fraction of a buffer, pK_a' 6.15, that is protonated at p 6.4.

From equation (22):	f_d	=	$\dfrac{\text{antilog } \Delta}{(\text{antilog } \Delta) + 1}$,		
	Δ	=	pH - pK_a'	= 6.4 - 6.15 ,	= 0.25,
	f_d	=	$10^{0.45}/(10^{0.45})$ +	1,	
	f_d	=	1.78/2.78	= 0.64,	
	1 - f_d	=	1-0.64 = 0.36 =	fraction which is protonated.	

EXAMPLE 2-5. What is the ratio of A- to HA in 300 ml of 0.2 M TES buffer, pH 7.95? The pK_a' of TES is 7.50 at 20°C.

From equation (22): $f_d = \dfrac{\text{antilog } \Delta}{(\text{antilog } \Delta) + 1}$,

$$\Delta = pH - pK_a' = 7.95 - 7.50 = 0.45,$$
$$f_d = 10^{0.25}/(10^{0.25}) + 1,$$
$$f_d = 2.82/3.82 = 0.74.$$

Total # mmoles TES		=	300 * 0.2	=	60 mmol,
	A⁻	=	60 * 0.74	=	44.4 mmol,
	HA	=	60 * 0.26	=	15.6 mmol,
	A-/HA	=	44.4/15.6	=	2.85.

EXAMPLE 2-6. A reaction produced 0.25 mmol of H^+ in a buffered solution which contained 100 ml of 0.02 M phosphate buffer, pH 7.2. What was the pH of the medium after the reaction was completed? The pK_a' of phosphate buffer is 6.85 at 20°C.

From equation (22): $f_d = \dfrac{\text{antilog } \Delta}{(\text{antilog } \Delta) + 1}$,

$$\Delta = pH - pK_a' = 7.2 - 6.85 = 0.35,$$
$$f_d = 10^{0.35}/(10^{0.35}) + 1,$$
$$f_d = 2.24/3.24 = 0.691.$$

At the start, # mmol A⁻	=	0.691 x (100 ml * 0.02M)	=	1.38 mmol /100r
At the start, # mmol HA	=	(1-0.691) * (100 ml * 0.02M)	=	0.62 mmol /100ı
Moles of H⁺ produced in reaction			=	0.25 mole.

After reaction, # mmol A⁻	=	1.38 - 0.25	=	1.13 mmol,
After reaction, # mmol HA	=	0.62 + 0.25	=	0.87 mmol,
After reaction, A-/HA	=	1.13/0.87	=	1.30.

The Henderson-Hasselbalch equation, equation (19), states that

	pH	=	$pK_a'+ \log [A^-]/[HA,$
∴ after reaction:	pH	=	$6.85 + \log_{10}1.30 = 6.85 + 0.11,$
	pH	=	6.96.

EXAMPLE 2-7. Calculate the proportions of KH_2PO_4 (molecular weight 136) and K_2HPO_4 (molecular weight 174) required for the preparation of 1.5 liters of 0.25 M potassium phosphate buffer, pH 7.4. The pK_a' of 0.25 M phosphate buffer is 6.72 at 20°C. Express your answer in terms of (a) moles of conjugate acid and conjugate base required (b) weights of conjugate acid and conjugate base required (c) volume (ml) of 0.25 M conjugate acid and conjugate base required for obtaining the total of 1.5 liters of buffer.

From equation (22): $f_d = \dfrac{\text{antilog } \Delta}{(\text{antilog } \Delta) + 1}$,

$$\begin{aligned}
\Delta &= pH - pK_a' &= 7.4 - 6.72 = 0.68, \\
f_d &= 10^{0.68}/(10^{0.68}) + 1, \\
f_d &= 4.79/5.79 &= 0.827, \\
1 - f_d &= 1 - 0.827 &= 0.173,
\end{aligned}$$

Total # mols buffer in 1.5 l	=	1.5 liter x 0.25 mol l^{-1}	= 0.375 m
mols of A-	=	f_d * 0.375 = 0.827 * 0.375 mol	= 0.310 n
mols of HA	=	1-f_d * 0.375 = 0.173 * 0.375 mol	= 0.065 n
grams of A-	=	0.310 mols * 174 g mol^{-1}	= 53.94 g
grams HA	=	0.065 mols * 136 g mol^{-1}	= 8.84 g
Vol (ml) of 0.25 M A-	=	0.310 mols/0.25 mols l^{-1}	= 1240 m
Vol (ml) of 0.25 M HA	=	0.065 mols/0.25 mols l^{-1}	= 260 m

EXAMPLE 2-8. Given a bottle of HEPES crystals (N-2-hydroxyethyl-piperazine-. ethanesulfonic acid), and a 0.2 N solution of sodium hydroxide, how much HE should you weigh out and what volume of the 0.2 N NaOH would you need to pro 250 ml of 0.1 M buffer at a pH of 7.8? The pK$_a'$ of HEPES is 7.55 and its mole weight is 238.3.

From equation (22): $f_d = \dfrac{antilog\ \Delta}{(antilog\ \Delta) + 1}$,

$$\begin{aligned}
\Delta &= pH - pK_a' &= 7.8 - 7.55 &= 0.25, \\
f_d &= 10^{0.25} / (10^{0.25}) + 1 &= 1.78/2.78 &= 0.64, \\
1 - f_d &= 1.0 - 0.64 &= 0.36,
\end{aligned}$$

Total # mols buffer in 0.25 liter	=	0.25 liter * 0.1 mol l^{-1}	= 0.025
mols of A$^-$	=	f_d * 0.025 = 0.64 * 0.025 mol =	0.016
mols of HA	=	1-f_d * 0.025 = 0.36 * 0.025 mol =	0.009
grams of HEPES required	=	0.025 mol * 238.3 g mol^{-1}	= 5.957
mols NaOH required	≡	mol A$^-$ = 0.016 mol,	
Vol of 0.2 N NaOH required	=	(0.016 mol/0.2 mol l^{-1}) * 10^3 ml l^{-1} =	80 ml

EXAMPLE 2-9. The thermodynamic pK$_a$ of KH$_2$PO$_4$ is 7.21 at 25°C. What is the practical pK$_a'$ of a 0.5 M solution of this compound?

From equation (24): $pK_a' = pK_a - \left[\dfrac{0.509\ (2n + 1)\ \sqrt{\mu}}{1 + \sqrt{\mu}} \right] - 0.1*\mu,$

$$= 7.21 - \left[\dfrac{0.509\ (2 + 1)\ \sqrt{0.5}}{1 + \sqrt{0.5}} \right] - 0.05,$$

$$\begin{aligned}
&= 7.21 - (0.632 - 0.05), \\
&= 6.61.
\end{aligned}$$

EXAMPLE 2-10. The temperature coefficient of TRIS buffer, $\Delta pK_a/°C$ is -0.031. Calculate the pK_a' of this buffer at 30°C given that the pK_a' at 20°C is 8.3.

From equation (25):

$$pK_a' \text{ at } T_2 \quad = \quad pK_a' \text{ at } T_1 + \Delta pK_a',$$

$$\Delta pK_a' \quad = \quad (30 - 20) * -0.031,$$

$$= \quad -0.31.$$

$$pK_a' \text{ at } 30°C \quad = \quad 8.3 - 0.31,$$

$$= \quad 7.99.$$

2-18. Spreadsheet solutions

	A	B	C	D	E	F	G
1	EXAMPLE 2	INPUT	OUTPUT		EXAMPLE 2-7	INPUT	OUTPUT
2	[H+]	0.0005			pH	7.4	
3	pH		3.30103		pKa'	6.72	
4	pH		3.30103		Δ		0.68
5					fd		0.827178
6	EXAMPLE 2-2				1-fd		0.172822
7		INPUT	OUTPUT		liters buffer	1.5	0.345644
8	pH	6.8			molarity of buffer	0.25	
9	[H+]		1.585E-07		mol buffer		0.375
10					mol A-		0.310192
11	EXAMPLE 2-3				mol HA		0.064808
12		INPUT	OUTPUT		Mol wt A-	174	
13	Ka'	1.4E-07			Mol wt HA	136	
14	pK'		6.8601209		grams of A-		53.97337
15	pK'		6.8601209		grams of HA		8.813921
16	pK'	4.76			ml of 0.25 M A-		1241
17	Ka'		1.738E-05		ml of 0.25 M HA		259
18	EXAMPLE 2-4				EXAMPLE 2-8		
19		INPUT	OUTPUT			INPUT	OUTPUT
20	pH	6.4			pH	7.8	
21	pKa'	6.15			pKa'	7.55	
22	Δ		0.25		Δ		0.25
23	fd		0.640065		fd		0.640065
24	1-fd		0.359935		1-fd		0.359935
25					liters buffer	0.25	0.71987
26	EXAMPLE 2-5				molarity of buffer	0.1	
27		INPUT	OUTPUT		mol buffer		0.025
28	pH	7.95			mol A-		0.016002
29	pKa'	7.5			mol HA		0.008998
30	Δ		0.45		Mol wt of HEPES	238.3	
31	fd		0.738109		grams of HEPES		5.9575
32	ml of TES	300	0.261891		Mol NaOH reqd.		0.016002
33	Molarity TES	0.2			ml 0.2 N NaOH reqd.		80
34	mmol TES		60				
35	A-		44.286542		EXAMPLE 2-9		
36	HA		15.713458			INPUT	OUTPUT
37	A-/HA		2.8183829		pKa	7.21	
38					Molarity	0.5	
39	EXAMPLE 2-6				n	1	
40		INPUT	OUTPUT		pKa'		6.627496
41	pH	7.2					
42	pKa'	6.85					
43	Δ		0.35				
44	fd		0.6912362		EXAMPLE 2-10		
45	mmol A-, start		1.3824723			INPUT	OUTPUT
46	mmol HA, start		0.6175277		T1 (20°C)	20	
47	mol H+ produce	0.25			T2 (30°C)	30	
48	mmol A-, after		1.1324723		pKa' at T1	8.3	
49	mmol HA after		0.8675277		d(pKa)/°C	-0.03	
50	A-/HA after rxn.		1.3054019		ΔpKa'		-0.31
51	pH after rxn.		6.9657442		pKa' at 30°C		7.99

Spreadsheet formulas

	A	B	C
1	EXAMPLE 2-1	INPUT	OUTPUT
2	[H+]	0.0005	
3	pH		=-LOG10(B2)
4	pH		=LOG10(1/(B2))
5			
6	EXAMPLE 2-2		
7		INPUT	OUTPUT
8	pH	6.8	
9	[H+]		=10^-B8
10			
11	EXAMPLE 2-3		
12		INPUT	OUTPUT
13	Ka'	0.000000138	
14	pK'		=LOG10(1/0.000000138)
15	pK'		=-LOG10(B13)
16	pK'	4.76	
17	Ka'		=10^-B16
18			
19	EXAMPLE 2-4		
20		INPUT	OUTPUT
21	pH	6.4	
22	pKa'	6.15	
23	Δ		=B21-B22
24	fd		=10^C23/((10^C23)+1)
25	1-fd		=1-C24
26			
27	EXAMPLE 2-5		
28		INPUT	OUTPUT
29	pH	7.95	
30	pKa'	7.5	
31	Δ		=B29-B30
32	fd		=10^C31/((10^C31)+1)
33	ml of TES	300	=1-C32
34	Molarity TES	0.2	
35	mmol TES		=B33*B34
36	A-		=C35*C32
37	HA		=C35*(1-C32)
38	A-/HA		=C36/C37
39			
40	EXAMPLE 2-6		
41		INPUT	OUTPUT
42	pH	7.2	
43	pKa'	6.85	
44	Δ		=B42-B43
45	fd		=10^C44/((10^C44)+1)
46	mmol A-, start		=C45*(100*0.02)
47	mmol HA, start		=(1-C45)*(100*0.02)
48	mol H+ produced	0.25	
49	mmol A-, after		=C46-B48
50	mmol HA after		=C47+B48
51	A-/HA after rxn.		=C49/C50
52	pH after rxn.		=B43+LOG10(C51)

Spreadsheet formulas

D	E	F	G
1	EXAMPLE 2-7	INPUT	OUTPUT
2	pH	7.4	
3	pKa'	6.72	
4	Δ		=F2-F3
5	fd		=10^G4/((10^G4)+1)
6	1-fd		=1-G5
7	liters buffer	1.5	=(1-G5)*(100*0.02)
8	molarity of buffer	0.25	
9	mol buffer		=F7*F8
10	mol A-		=G5*G9
11	mol HA		=G6*G9
12	Mol wt A-	174	
13	Mol wt HA	136	
14	grams of A-		=G10*F12
15	grams of HA		=G11*F13
16	ml of 0.25 M A-		=G10/F8*1000
17	ml of 0.25 M HA		=G11/F8*1000
18			
19	EXAMPLE 2-8		
20		INPUT	OUTPUT
21	pH	7.8	
22	pKa'	7.55	
23	Δ		=F21-F22
24	fd		=10^G23/((10^G23)+1)
25	1-fd		=1-G24
26	liters buffer	0.25	=(1-G24)*(100*0.02)
27	molarity of buffer	0.1	
28	mol buffer		=F26*F27
29	mol A-		=G24*G28
30	mol HA		=G25*G28
31	Mol wt of HEPES	238.3	
32	grams of HEPES		=G28*F31
33	Mol NaOH reqd.		=G29
34	ml 0.2 N NaOH reqd		=(G33/0.2)*1000
35			
36	EXAMPLE 2-9		
37		INPUT	OUTPUT
38	pKa	7.21	
39	Molarity	0.5	
40	n	1	
41	pKa'		=F38-((0.509*(2*F40+1)*SQRT(0.5))/(1+SQRT(0.5))-0.1*F39)
42			
43			
44			
45	EXAMPLE 2-10		
46		INPUT	OUTPUT
47	T1 (20°C)	20	
48	T2 (30°C)	30	
49	pKa' at T1	8.3	
50	d(pKa)/°C	-0.031	
51	ΔpKa'		=(F48-F47)*F50
52	pKa' at 30°C		=F49+G51

2-19.   Review questions

Indicate whether the amino acids of questions 2-1 to 2-10 have net charges of -2, -1, 0, +1, or +2 in the aqueous solutions of specified acidity or basicity:

2-1.   Lysine at pH 2.0.
2-2.   Histidine at pH 2.0.
2-3.   Alanine at pH 2.0.
2-4.   Glutamic acid at pH 2.0.
2-5.   Asparagine at pH 2.0.
2-6.   Lysine at pH 12.0.
2-7.   Histidine at pH 12.0.
2-8.   Alanine at pH 12.0.
2-9.   Glutamic acid at pH 12.0.
2-10.   Asparagine at pH 12.0.

2-11.   Why do biochemists think it unnecessary to distinguish H^+ from H_3O^+?
2-12.   Why is it considered essential to use corrected rather than thermodynamic pK_a values in biochemical work?
2-13.   What corrections to pK_a are required for precise biochemical work? Describe them.
2-14.   What criteria should one follow for effective use of biological buffers?
2-15.   Outline the two main approaches to the preparation of biochemical buffers.
2-16.   By what factor does $[H^+]$ change when pH is decreased by two units?
2-17.   By what factor does $[OH^-]$ change when pH changes (increase or decrease) by two units?
2-18.   How is pH related to pK_a?
2-19.   How is the pH of pure water related to K'_w?
2-20.   Define *isoelectric point,* and explain how it is used in the separation of proteins.
2-21.   Define *ampholyte, moving boundary electrophoresis,* and *zone electrophoresis.*
2-22.   What is the practical pK_a of a buffer and how is it obtained?
2-23.   Explain the effectiveness of isoelectric focusing.
2-24.   Write equations showing the stepwise formation and ionization of the three ionic species (cation, dipolar ion and anion) of serine, aspartic acid and lysine.
2-25.   Outline the major points to be considered in the preparation of a buffer to measure enzyme activity over a wide range of pH.
2-26.   Compare the titration curves of proteins with those of monoamino monocarboxylic acids.
2-27.   What is the Henderson-Hasselbalch equation and why is it important in the biomedical sciences?
2-28.   List the major acidic and basic R groups of proteins. and explain how the ionization of these groups is influenced by pH.

2-20. Review problems

2-29. Calculate the pH of an aqueous solution in which $[H^+] = 0.0005$ mol l^{-1}.

2-30. Calculate pH if $[H^+]$ is $4 * 10^{-2}$ mol l^{-1}.

2-31. Calculate $[H^+]$ of a solution of pH 7.2.

2-32. Calculate the pK_a of KH_2PO_4 if $K_a' = 1.38 * 10^{-7}$ mol l^{-1}.

2-33. Calculate K_a' of acetic acid given $pK_a = 4.76$.

2-34. Calculate the fraction of MES buffer, pK_a 6.15, which is protonated at pH 6.60.

2-35. Calculate, in increments of 0.1 pH unit, f_d and $1-f_d$ for phosphate buffer at 0.1 mol l^{-1} and pH 6.6 to 7.4. Assume that the pK_a' of phosphate buffer is 6.85 at 20°C.

2-36. What are the concentrations in grams l^{-1} of A^- and HA in 300 ml of 0.25 mol l^{-1} TES buffer pH 7.6? The pK_a' of TES is 7.50 at 20°C and its molecular weight is 229.3. The formula of TES is: $(HOCH_2)_3-N^+HCH_2CH_2SO_3H$. Write the structures of A^- and HA.

2-37. How much 5 M NaOH is needed to bring the pH of 1 l of a 0.25 mol l^{-1} Hepes buffer solution from pH 7.0 to pH 8.0? The pK_a' of Hepes is 7.55 at 20°C.

2-38. You are planning to carry out an enzyme assay. The stoichiometry of the reaction indicates that the reaction will produce 0.025 mmol of hydrogen ions, H^+, in an assay medium which will contain 10 ml of 0.02 mol l^{-1} phosphate buffer, pH 7.2. With this information, please answer the following questions:

(a) what will be the amounts of A^- and HA in mmol before and after H^+ is produced?

(b) what will be the ratio of $[A^-]/[HA]$ before and after H^+ is produced?

(c) what will be the pH at the end of the assay?

The pK_a' of phosphoric acid is 6.85; the molecular weight of Na_2HPO_4 is 141.98, and the molecular weight of NaH_2PO_4 is 138.01.

2-21. Additional problems

A.   pH

2-39. Calculate each of the following:

(a) the molarity of a solution of HNO_3, pH 3.5.

(b) the pH of a 0.005 mol l^{-1} solution of HCl.

2-40.   Given the pH values of the biological fluids in (a) to (h), calculate the corresponding H_3O^+ concentrations (mol 1^{-1}): (a) saliva (children), 7.32; (b) saliva (adults), 6.4; (c) gastric juice (men), 1.92; (d) gastric juice (women); 2.59; (e) bile, 7.50; (f) semen, 7.19; (g) cow's milk, 6.61; and (h) lemon juice, 2.34.

2-41.   What is the difference in hydrogen ion concentration (mol 1^{-1}) between a sample of urine of pH 3.8 and another of pH 7.9?

2-42.   If the hydroxide ion concentration of an assay medium is 10^{-8} mol l^{-1}, what is the pH of the medium?

B.   Buffers

2-43.   The pK_2' of phosphoric acid is 6.86, the molecular weight of KH_2PO_4 is 136.09, and the molecular weight of K_2HPO_4 is 174.18, calculate the amounts of conjugate acid and conjugate base which should be weighed out to prepare 500 ml of a 0.2 mol 1^{-1} buffer solution, pH 7.2.

2-44.   Assuming that ionic strength is equivalent to concentration for weak acids, calculate the extent to which pK_2' of phosphate buffer at 0.1 mol 1^{-1}, 0.25 mol 1^{-1} and 1.0 mol 1^{-1} concentrations will differ from the thermodynamic pK_a of 7.21 (at 20°C). Express your answer as percent change from the thermodynamic pK_a.

2-45.   At 20°C, the pK_a of TRIS buffer [tris(hydroxymethyl)aminomethane] is 8.30. The dissociation constant of TRIS changes with temperature to the extent of $\Delta pK_a/°C = - 0.031$. If the pH of a 0.05 mol 1^{-1} TRIS buffer were initially 7.5 when prepared in the laboratory at 20°C, calculate the pH of the same buffer if used at 4°C and 37°C. Correct the thermodynamic pK_a both for ionic strength and temperature and express your answer as percent change in pH.

2-46.   The pK_a' of Hepes is 7.55 at 20°C and its molecular weight is 238.31. Calculate the amounts of Hepes (in grams) and of 1.0 N sodium hydroxide (in ml) which are required for preparing 250 ml of 0.2 mol 1^{-1} Hepes buffer at pH 7.2. Report the following values: fd,1-fd, moles of A, moles of HA, moles of OH⁻, grams of Hepes and ml of OH⁻. Hepes is N-2-hydroxyethylpiperazine-N'-ethanesulfonic acid.

2-47.   Calculate the amounts of 1.0 N NaOH (in ml) required for preparing 100 ml of 0.05 mol 1^{-1} Hepes buffer at pH 6.6, 6.8, 7.0, 7.2, and 7.4. Use data on Hepes from question 2-46.

2-48.   How much BICINE ($pK_a = 8.35$ at 20°C; molecular weight $= 163.17$) and how much 6 N HCl are required for making 1.5 l of 0.02 mol 1^{-1} buffer at pH 8.5? BICINE is [N,N-bis-(2-hydroxyethyl)glycine].

2-49. Compute the amounts of $NaH_2PO_4 \cdot H_2O$ (Molecular weight 138.01) and of Na_2HPO_4 (molecular weight 141.98) required for preparing 1 l quantities of 0.1 mol l $^{-1}$ buffer at intervals of 0.1 pH unit between pH 7.0 and 7.8. Express your answer in terms of mols of conjugate base, mols of conjugate acid, grams of conjugate base, grams of conjugate acid, ml of 0.1 mol l^{-1} conjugate base and ml of 0.1 mol l $^{-1}$ conjugate acid. Assume that the pKa' of $H_2PO_4^{2-}$ is 6.85 at 20°C.

2-50. Compute the amounts of TES, 2-{[tris-(hydroxymethyl)methyl]amino} ethane sulfonic acid (molecular weight 229.3) and 1N HCl required for preparing 1 l volumes of 0.1 mol l $^{-1}$ buffer at intervals of 0.1 pH unit between pH 7.7 and 8.0. Express your answer in terms of fd, 1-fd, mols A, mols HA, grams of buffer, and ml of 1N acid. The pK_a' of TES is 7.50 at 20°C.

Suggested Reading

Alberty, R.A. and A. Cornish-Bowden 1993. The pH Dependence of the Apparent Equilibrium Constant, K', of a Biochemical Reaction. *Trends in Biochem. Sci.*, 18: 288-291.

Ang, Debbie, K. Liberek, D. Skowyra, M. Zylicz, and C. Georgopoulos 1991. Biological Role and Regulation of the Universally Conserved Heat Shock Proteins. *J. Biol. Chem.* 266: 24233-24236.

Barker, G. R. 1981. The Concept of the Activity of Ions in Relation to Buffers, *Trends in Biochem. Sci.,* 6 No. 12: XI.

Barrante, J. R. 1977. *Physical Chemistry for the Life Sciences,* Prentice-Hall, New Jersey.

Blanchard, J. S. 1984. *Buffers for Enzymes,* Methods in Enzymology, Academic Press, New York, 104: 404-414.

Bohinski, R.C. 1987. *Modern Concepts in Biochemistry, 5th ed.,* Allyn and Bacon, Boston.

Bray, H. G. and K. White 1966. *Kinetics and Thermodynamics in Biochemistry,* 2d. ed., Academic Press, New York.

Burston, S.G., R. Sleigh, D.J. Halsall, C.J. Smith, J.J. Holbrook, and A.R. Clarke 1992. The Influence of Chaperonins on Protein Folding. *Ann. New York Acad. Sci.,* 67: 2-9.

Christensen, H. N. 1964. *pH and Dissociation, 2d ed.,* Saunders, Philadelphia, Pennsylvania.

Conn, E.E., P.K. Stumpf, G. Bruening and R.H. Doi 1987. *Outlines of Biochemistry, 5th ed.,* John Wiley & Sons, New York.

Darvey, I.G. and G. B. Ralston 1993. Amino Acid Titration Curves- Misshapen or Mislabeled, *Trends in Biochem. Sci.,* 18: 69-71.

Day, R. A. and A. L. Underwood 1980. *Quantitative Analysis, 4th ed.,* Prentice-Hall, New Jersey.

Dickerson, R. E., H. B. Gray, M. Y. Darensbourg and D. J. Darensbourg 1984. *Chemical Principles, 4th ed.,* Benjamin/Cummings, Reading, Massachusetts.

Douzou, P. and P. Maurel 1977. Ionic Control of Biochemical Reactions, *Trends in Biochem. Sci.,* 2: 14-17.

Ellis, K. J. and J. F. Morrison 1982. *Buffers of Constant Ionic Strength for Studying pH-Dependent Processes.* Methods in Enzymology Academic Press, New York, 87: 405-426. Note. This reference contains computer programs for calculating the ionic strengths of two- and three-buffer mixtures.

Ellis, R. John and S. M. Hemmingsen 1989. Molecular Chaperones: Proteins Essential for the Biogenesis of Some Macromolecular Structures. *Trends in Biochem. Sci., 14:* 339-342.

Ellis, R. John and S. M. van der Vies 1991. Molecular Chaperones. *Annu. Rev. Biochem.* 60: 321-347.

Frisell, W. R. 1982. *Human Biochemistry,* MacMillan, New York.

Georgopoulos, C. 1992. The Emergence of the Chaperone Machines. *Trends in Biochem. Sci.,* 17: 295 - 299.

Gomori, G. 1955. *Preparation of Buffers for Use in Enzyme Studies,* Methods in Enzymology, Academic Press, New York, 1: 138-146.

Good, N. E., G. D. Winget, W. Winter, T. N. Connolly, S. Izawa and R. M. M. Singh 1966. Hydrogen Ion Buffers for Biological Research, *Biochemistry 5:* 467-477.

Good, N. E. and S. Izawa 1972. *Hydrogen Ion Buffers,* Methods in Enzymology, Academic Press, New York, 24: 53-68.

Hendrick, J.P. and F-U. Hartl 1993. *Molecular Chaperone Functions of Heat-Shock Proteins.* Annu. Rev. Biochem. 62: 349-384.

Hengen, P.N. 1993. Methods and Reagents. *Trends Biochem. Sci.,* 18: 446-448.
This article recounts how the author obtained help in selecting buffers for electrophoresis over the Internet Computer Network

Johnson, R. J. and D. E. Metzler 1971. *Buffer Preparation,* Methods in Enzymology, Academic Press, New York, 22: 3-5.

Katz, M. A. 1976. *Calculus for the Life Sciences.* Marcel Dekker, New York.

Kull, F. J., G. T. Vellekamp, E. E. Button 1982. Ionic Strengths and Enzyme Activities, *Trends in Biochem. Sci.,* 7: 317-318.

Landers, J.P. 1993. Capillary Electrophoresis: Pioneering New Approaches for Biomolecular Analysis. *Trends in Biochem. Sci.* 18: 409-414.

Lehninger, A. L. 1975. *Biochemistry,* 2d ed., Worth, New York.

Lehninger, A. L., D. J. Nelson and M. M. Cox 1992. *Principles of Biochemistry,* 2d. ed., Worth, New York.

Mathews, C.K. and K.E. van Holde 1990. *Biochemistry.* The Benjamin/Cummimgs Co., NY.

Matthews, C.R. 1993. *Pathways of Protein Folding.* Annu. Rev. Biochem. 6653-684.

Montgomery, R., and C. A. Swenson 1976. *Quantitative Problems in the Biochemical Sciences,* 2d ed., Freeman, New York.

Morris, J. G. 1974. *A Biologist's Physical Chemistry, 2d ed.,* Arnold, Baltimore, Maryland.

Morimoto, R.I., K.D. Sarge and K.Abravaya 1992. Transcriptional Regulation of Heat Shock Genes. A Paradigm for Inducible Genomic Responses. *J. Biol. Chem.* 267: 21987-21990.

Schlesinger, M.J. 1990. Heat Shock Proteins. *J. Biol. Chem.* 265: 12111-12114.

Shinde, U. and M. Inouye 1993. Intramolecular Chaperones and Protein Folding. *Trends in Biochem Sci.,* 18: 442-446.

Stoll, V.S. and J.S. Blanchard 1990. *Buffers: Principles and Practice,* Methods in Enzymology, Academic Press, New York, 182: 24-38.

Strang, R. 1981. The Theoretical and Practical pK_2 of Phosphoric Acid, *Trends in Biochem Sci.,* 6 # 5: VII -VIII

Stryer, L. 1987. *Biochemistry.* W. H. Freeman and Co., San Francisco..

Tipton, K. F. and H. B. F. Dixon 1979. *Effects of pH on Enzymes,* Methods in Enzymology, Academic Press, New York, 63: 183-234.

Tiselius, A. 1957. *Electrophoresis,* Methods in Enzymology, Academic Press, NY, 4: 3-20.

Chapter 3

Countercurrent Distribution and Chromatography

3-1. Effective chromatography requires a grasp of basic theory

---

Countercurrent distribution and chromatography are physical methods which separate compounds from each other by repetitive partitions of the compounds between two immiscible phases: a *mobile phase and a stationary phase.* Separation is achieved because each compound associates to a differing degree with the stationary phase and therefore migrates at a differing rate in the mobile phase. In countercurrent distribution, both phases are liquid. In chromatography, the mobile phase may be a gas or liquid whereas the stationary phase may be a liquid, a gel or a solid of large surface area. Thus, in chromatography, repetitive distribution of solutes may occur between liquid and liquid, between liquid and gel, between gas and liquid or between a liquid and a solid which may have adsorptive, gel-filtration, ion exchange, complex-forming, or ligand-binding properties. In countercurrent distribution, the series of partitions is carried out in discrete steps. In chromatography, the partitions take place continuously. The extent to which a solute partitions between the two phases depends on its *partition coefficient,* K, defined as the ratio of the concentrations of solute in the two immiscible phases after mixing and equilibration of the phases.

In this chapter, we will discuss some essentials of five aspects of chromatographic theory:

- *the partition coefficient,* an index of solute mobility;
- *the capacity factor,* an index of peak position relative to void volume;
- *the theoretical plate,* an index of column efficiency;
- *the van Deemter equation,* a tool for optimizing chromatographic separations;
- *the resolution* of closely related peaks, an index of good separations.

Following the presentation of these five aspects of chromatographic theory, we will describe some major chromatographic methods used in biomedical laboratories. Our purpose in presenting theory and practice in this manner is twofold, first:

• to identify the mobile and stationary phases of each method so as to emphasize in a nonmathematical way that the underlying principle of every chromatographic process is repetitive partitioning between two phases. Our second goal is:

• to indicate that chromatographic procedures should not be followed as if they are precise recipes.

Although the cookbook approach to performing chromatographic separations sometimes works adequately, there is strong likelihood, also, that this approach may yield barely satisfactory results. Indeed, the recent spectacular advances in gas-liquid, high-performance and other forms of chromatography have been due chiefly to systematic application of theoretical principles. Therefore, investigators-in-training should have a sufficiently good understanding of the fundamentals of the chromatographic process to permit modification of chromatographic procedures to obtain a desired type of separation. We will discuss the five aspects of chromatographic theory mentioned above (the partition coefficient, the capacity factor, the theoretical plate, the van Deemter equation and resolution of peaks) in sections 3-2 to 3-6. Then we will present major chromatographic methods in the following sequence and sections of the text:

SECTION 3-7:   *High-performance liquid chromatography*

SECTION 3-8:   *Liquid-liquid chromatography*
        1.   Traditional liquid chromatography
        2.   Paper chromatography

SECTION 3-9:   *Liquid-solid chromatography*
        1.   Ion-exchange chromatography
        2.   Affinity chromatography
        3.   Thin-layer chromatography

SECTION 3-10:   *Gas-liquid chromatography*
        1.   Packed columns
        2.   Capillary columns

SECTION 3-11:   *Countercurrent chromatography*

We will discuss the separation of macromolecules by *gel filtration* in Chapter 4.

3-2. The partition coefficient is an indicator of solute mobility

We will start our discussion of chromatographic theory by using countercurrent distribution as a model of chromatography. *Our goal is to show that the partition coefficient determines solute mobility.* To simplify the discussion of solute mobility in countercurrent distribution, we will make two assumptions: (a) that the two immiscible solvents are present in equal volumes; (b) that the solute has a partition coefficient of 1.0 in the solvent pair. *A countercurrent distribution apparatus consists of pairs of upper and lower tubes.* The mobile phase of the solvent pair is placed in the upper tubes (labeled U in Fig. 3-1), the stationary phase is placed in the lower tubes (labeled L). The pairs of tubes are connected to each other in a series numbered 0, 1, 2, ... r.

In the five-tube apparatus of Figure 3-1, the process is started by dissolving 80 mg of solute in the stationary phase of tube L_0; then tube U_4 containing mobile phase is moved into contact with L_0 and the two phases are shaken, equilibrated and allowed to separate. At this stage, (transfer 0, Figure 3-1), tubes L_0 and U_4 each contains 40 mg of solute since the partition coefficient is 1.0 and the volumes of the two immiscible solvents are equal. For the next stage, the upper tubes containing mobile phase are shifted one step forward so that the contents of U_4 are mixed and   equilibrated with the contents of L_1 and those of U_3 mixed and equilibrated with L_0. One transfer now has taken place. The total amount of solute in both phases of each of the two pairs of tubes (i.e. $U_4 + L_1$ and $U_3 + L_0$) is thus 40 mg since the partition coefficient is 1.0 and the volumes of the two immiscible solvents are equal. After the 2nd, 3rd and 4th transfers, the contents of tube U_0 are mixed and equilibrated with L_0; U_1 with L_1, U_2 with L_2, U_3 with L_3 and U_4 with L_4. The total amounts of solute in both phases after four transfers are: 5 mg (tube 0); 20 mg (tube 1); 30 mg (tube 2); 20 mg (tube 3) and 5 mg (tube 4). Instead of calculating the amounts of solute per tube arithmetically as just described, the fraction of solute, $T_{n,r}$ in the r th pair of tubes for n transfers, may be computed from the equation:

$$T_{n,r} \quad = \quad \frac{n!}{r!(n-r)!} \left(\frac{1}{K+1} \right)^n K^r , \qquad (1)$$

where n =   the total number of transfers (this value does not vary in the calculation),

r =   tube number (this varies in the calculation),

K =   the partition coefficient.

Once $T_{n,r}$ has been calculated, the total amount of solute present in both layers of a pair of tubes may be obtained by multiplying $T_{n,r}$ by the amount of solute that was dissolved at the start of the experiment.

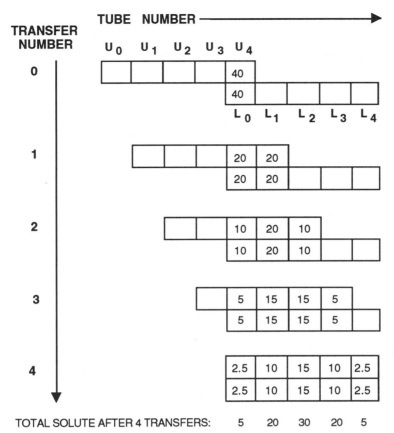

Figure 3-1. Diagrammatic representation of countercurrent distribution

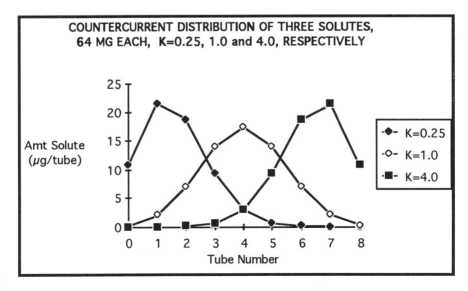

Figure 3-2. Countercurrent distribution of three solutes in a nine-tube apparatus

To demonstrate the dependence of rate of solute migration, or mobility, on partition coefficient, we will devise an experiment in which a mixture of three solutes is separated by countercurrent distribution in a nine-tube apparatus. We will assume that 64 mg of each solute is present and that the three solutes have partition coefficients of 0.25, 1.0 and 4.0 respectively. When the amount of solute in both layers (calculated as in equation 1) is plotted as ordinate against tube number as abscissa, the curve of Figure 3-2 is obtained. Figure 3-2 shows that the larger the partition coefficient, the faster the rate of migration. Figure 3-2 shows also that the three solutes, with K values of 0.25, 1.0 and 4.0, were not separated completely from each other after eight transfers. To obtain better separations, more than eight partitions or transfers would have been required. And for the moment, in somewhat over-simplified terms, chromatography yields better separation of mixtures of solutes than countercurrent distribution because chromatographic methods achieve larger numbers of partitions than countercurrent distribution. In contrast to the stepwise partitioning of solute between mobile phase and stationary phase occurring in countercurrent distribution, continuous partitioning of solutes between the mobile and stationary phases occurs in all chromatographic methods. A major reason for the popularity of chromatographic procedures is that the good separations they produce are often achieved with simpler and less expensive equipment than that used in countercurrent distribution. However, when relatively expensive and sophisticated chromatographic equipment is used, such as gas-liquid or high-performance liquid chromatographs, the principle of separation is the same as that underlying use of the simple equipment of thin-layer or paper chromatography. The principle: *solutes are held back to differing degrees by the stationary phase and therefore migrate at different rates in the mobile phase.* This principle is adequate for explaining why solutes having strong association with the stationary phase are eluted more slowly (have a lower mobility) than solutes which favor the mobile phase. But the principle, based mainly on partition coefficients, says little about parameters which determine the positions of solute peaks. Toward this end, we will next introduce the quantity known as the capacity factor.

3-3.   The capacity factor is an index of peak position

K, the partition coefficient, is the thermodynamic equilibrium constant for the partitioning of solute between two immiscible phases. Moreover, during chromatography, the distribution of solute between stationary and mobile phases is a dynamic process and the concentration of solute in the two phases varies with the distance traveled by the solute. Therefore, the partitioning of solute between stationary and mobile phases may be written as

$$K \quad = \quad \frac{C_S}{C_M} \tag{2}$$

where K = the partition coefficient; C_S = concentration of solute in the stationary phase, and C_M = concentration of solute in the mobile phase. Related to the partition coefficient is the *capacity factor*, k', variously known as the partition ratio or the column capacity ratio. Because k' is related to the partition coefficient, an expression for the capacity factor may be derived from distribution equilibria involving the partitioning of solute between stationary and mobile phases. This expression is

$$k' = \frac{C_S V_S}{C_M V_M},\tag{3}$$

Substituting (2) in (3), $k' = K(V_S/V_M).\tag{4}$

The capacity factor thus is the ratio of the total amount of solute in the stationary phase to the total amount of solute in the mobile phase at equilibrium; k' also may be defined as the product of the thermodynamic equilibrium constant and the ratio of the volume of stationary phase (V_S) to the volume of mobile phase (V_M).

Because the distance traveled by a solute during chromatography is related to the retention volume of the solute and to the time required for elution, the capacity factor may be expressed, alternatively, in terms of volume or time

$$k' = \frac{V_R - V_M}{V_M}\,; \qquad k' = \frac{t_R - t_M}{t_M}\tag{5}$$

V_R = retention volume of solute peak (volume required to elute the solute);
V_M = void volume (volume in the column unoccupied by stationary phase);
t_R = retention time of solute peak (time needed for the solute to exit the column);
t_M = time required for the solvent peak to appear.

According to equation (5) then, k', *the capacity factor may be defined either as*

- the difference between retention volume and void volume divided by the void volume; or as
- the retention time of a solute peak relative to the retention time of the solvent peak.

The capacity factor therefore can be used to relate the distribution of solute in a chromatographic column to the properties of that column.

Although partition coefficients can provide indices of the rate of solute mobility during chromatography, and although the capacity factor relates peak position thermodynamically to properties of chromatographic columns, these important values say little about the factors which influence the efficiency of chromatographic separations, or how such efficiency may be described or attained. To describe separation efficiency, we will next introduce the idea of the theoretical plate.

3-4.   The theoretical plate is a gauge of column efficiency

The concept of the theoretical plate was first used to refer to a hypothetical section of fractional distillation columns in which equilibration took place between vapor and circulating liquid. Later, the American biochemist Lyman Craig applied the idea of the theoretical plate to countercurrent distribution. Craig considered each tube in a countercurrent distribution apparatus to be a theoretical plate. Then in the 1940s, the British biochemists A. J. P. Martin and R. L. M. Synge recognized that numerous equilibrations occur in chromatography, and they further expanded the idea of theoretical plates, defining one theoretical plate as a segment of a chromatographic column in which one equilibration takes place. Since numerous equilibrations take place on chromatographic columns, it follows that each column usually consists of a very large number of theoretical plates. In column chromatography, the height equivalent to a theoretical plate is defined as the length of the column divided by the number of theoretical plates, n. The value of n is derived from experimental data and is given by the relationship:

$$n = \left(\frac{4t_R}{W}\right)^2 \tag{6}$$

where t_R = retention and W = peak width.

Because t_R and W on chromatograms usually are measured in the same units, (e.g. cm), n is a dimensionless number. The number of theoretical plates, n, for a given column depends on the solute, on flow rate, on how the column was packed, and on other variables. The reason for introducing n at this time is that it provides an index of column efficiency - *the larger the value of n, the greater is the efficiency of the column and the narrower the peaks produced.* Because large values of n are desirable, it is of interest to learn how such large values may be obtained. There are two general approaches: (a) increase the length of the column and (b) increase the number of equilibrations which occur within the column. Scientists who attempt to achieve the second objective find that use of the *height equivalent to a theoretical plate,* abbreviated HETP (or H), is a useful value. Mathematically, HETP is defined as the length of a column, L, divided by the number of theoretical plates:

$$\text{HETP} = \frac{L}{n} = \frac{L}{16}\left(\frac{W}{4t_R}\right)^2. \tag{7}$$

From equation (7), it is evident that the dimensions of L/n are cm plate^{-1}, and it is clear also that as n increases, HETP becomes smaller; therefore, small values of HETP are equivalent to superior column performance. Consequently, it is desirable, in the laboratory, to achieve operating conditions which make HETP as small as possible. Achieving such conditions in a systematic manner has been made possible from the work of the Dutch chemist J. J. van Deemter and colleagues who, in 1956, introduced (originally for optimizing separations in gas-liquid chromatography) the mathematical relationship that we will discuss next.

3-5.   The van Deemter equation measures column performance

Van Deemter and colleagues proposed that HETP is dependent on, and may be expressed as the sum of, three interdependent terms. In a simplified form, this relation usually is written as:

$$\text{HETP} = A + B/\mu + C\mu \tag{8}$$

where   A   =   eddy diffusion (defined below),
B   =   diffusion of solute in the mobile phase along the column,
C   =   a factor related to the time required for equilibration of solute, between mobile and stationary phases,
μ =   flow rate of the mobile phase (or carrier gas), cm sec $^{-1}$.

Eddy diffusion (term A) refers to the fact that, molecules of a solute placed on a chromatographic column at the same time, exit the column over a range of different times. A major reason for this variation of exit time is that the solute molecules travel across several different paths, each of which differs in length. Hence, a range of exit times occurs which is observed experimentally as zone (or peak) broadening. Term B in the van Deemter equation depends on the fact that concentration gradients of solute occur in the mobile phase, and these gradients cause diffusion of the solute from regions of high to regions of low concentration. But, since diffusion is much slower in liquids than in gases, B exerts much less influence with liquid mobile phases than with gases. As with eddy diffusion, molecular diffusion (due to term B) also is observed experimentally as peak broadening. Term C expresses the fact that finite time periods are required for equilibration of solute between mobile and stationary phases so that, if the flow rate of the mobile phase is too fast, equilibrium is not attained. Term C therefore expresses peak broadening as a function of nonequilibrium.

The foregoing discussion indicates that, in order to optimize column performance, the three terms on the right hand side of equation (8) should be kept as small as possible. In the biomedical laboratory, efforts to minimize these terms have resulted in the enunciation of several useful general rules for optimizing chromatographic procedures. Among these rules are the following:

•   The particle size of the chromatographic support, and the diameter of the column, should be as small as possible to minimize eddy diffusion.
•   To minimize molecular diffusion along the length of the column, the flow rate of the carrier gas should not be too slow. The reason is that the slower the flow rate, the greater the time available for molecular diffusion to occur. For liquid mobile phases, however, molecular diffusion is not as important as in gases (and frequently can be ignored), since diffusion in liquids is of the order of 10^5 times slower than in gases.
•   To minimize nonequilibration of solute, the flow rate of the mobile phase should not be too fast, i.e. terms B and C of the van Deemter equation both depend on flow rates, but in opposite directions.

• To minimize the time required for equilibration of solute between mobile and stationary phases, coatings of stationary phase on solid supports should not be too thick or deep.

So far, we have seen that solute mobility (expressed in terms of partition coefficients) and column efficiency (expressed by HETP) determine, to a large extent, the success of chromatographic procedures. Equally, or more important to successful chromatography is the resolution, or degree of separation, of closely related biomolecules, our next topic.

3-6. Resolution depends on retention times and peak widths

Successful resolution of two closely related compounds depends both on partition coefficients and column efficiency. Thus, to improve resolution, one may use three approaches:

• increase the difference between partition coefficients of similar compounds (change the mobile or stationary phase) to increase separation between the peaks;
• decrease peak width by increasing column efficiency (decrease HETP); or
• increase the differences between partition coefficients and change the mobile or stationary phase simultaneously. Mathematically, the resolution of two peaks is defined by the relationship:

$$\text{resolution} \quad = \frac{2(t_{2,R} - t_{1,R})}{W_2 + W_1} \tag{9}$$

where $t_{2,R}$ and $t_{1,R}$ =   retention times of peaks 2 and 1 respectively, and
      W_2 and W_1 =   corresponding widths of peaks 2 and 1.

In general, values of resolution higher than 1.0 are desirable. Values greater than 1.5 are equivalent to complete resolution of similar compounds. Values of 1.0 indicate good separations of two peaks with a small amount of overlap. Recently the theoretical principles outlined above have been applied brilliantly to the development of high-performance liquid chromatography, discussed next.

3-7. HPLC exemplifies the successful application of theory to practice

The *mobile phase in high-performance liquid chromatography* is always a liquid. The *stationary phase* may be a solvent bound to an inert solid, or the stationary phase may be a solid of small particle size endowed with any of a wide range of possible chemical or physical properties. In other words, HPLC may, for example, encompass chromatographic methods otherwise classified as liquid-liquid, adsorption, ion-exchange or gel-filtration. In traditional liquid chromatography, columns with wide diameters, packed with large diameter particles are operated at low flow rates. The result is that, because of the low diffusion rate of solutes in liquids, the separation of biomolecules may take several hours. However, in high-performance liquid chromatography, separation takes only minutes because the intrinsically low rate of diffusion (eddy diffusion) in liquids is circumvented by two approaches. The first is to use small diameter, regularly shaped particles of solid support in narrow columns. This decreases the volume of mobile phase within the column by allowing maximal packing density and by decreasing, simultaneously, dead space between the particles of solid support. A second approach is to coat the liquid stationary phase as a thin film on specially constructed particles of solid support. This avoids the occurrence of deep layers (i.e., large volumes) of the liquid stationary phase i.e., this minimizes the time required for equilibration. By these two strategies, the distance through which the solutes have to diffuse is reduced considerably, both in the mobile and stationary phases. This leads to increased speed and resolving power and is a reflection of the fact that the van Deemter equation can be modified empirically for application to high-performance liquid chromatography in the following way:

$$\text{HETP} \quad = \quad A + C\mu^m, \tag{10}$$

where A and C are as defined previously (in section 3-5), and m is an experimentally determined constant with values between 0.3 and 0.6.

Term B has been left out of the version of the van Deemter equation numbered (10) above since B is approximately equal to 0 under most operating conditions - except at very low flow rates. (At these very low flow rates, molecular diffusion can become an important factor and HETP, consequently, will increase). Placing small diameter, densely packed particles in narrow columns results in slower flow rates than in traditional liquid-liquid chromatography which uses large diameter particles in wide columns. Therefore to obtain acceptable flow rates in HPLC, the use of high pressures, commonly 1 000 to 3 000 psi, is called for (the SI equivalent of 1 psi is 703.070 kg m^{-2}). In some HPLC methods, pressures of 1 200 psi may yield flow rates of 1 to 2 ml min^{-1} in 2 to 4 mm diameter columns of 10 to 50 cm in length, packed with 50 μm diameter particles. In turn, high pressures require that the particles of solid support can withstand the additional pressure. Three broad types of pressure-resistant particles (which will not be discussed here) have been developed to meet the demands placed on HPLC solid supports. With the introduction of these improvements in materials and technique, classical liquid-liquid chromatography has been transformed into a method which is setting new standards for the separation and purification of biomolecules. Indeed, this new version of an old technique is rivaling gas-liquid chromatography in speed and resolving power. There are five major components in a high-performance liquid chromatography apparatus. They are (1) a pump assembly which supplies mobile phase under pressure (2) an injector system (3) the chromatographic column (4) a detector and (5) a recorder. Since 80 % to 90 % of all compounds of biomedical interest are nonvolatile and therefore cannot be separated by gas-liquid chromatography, the availability of a method for nonvolatile compounds which rivals gas-liquid chromatography in speed and resolving power is an important development for the biomedical scientist.

3-8. Column and paper are forms of liquid-liquid chromatography

3-8-1. Traditional liquid-liquid, column chromatography

In *liquid chromatography* (our abbreviation for liquid-liquid, column chromatography), water, immobilized to a solid support is the *stationary phase*, and an organic solvent, running through a glass column, is the *mobile phase*. Liquid chromatography, which is the oldest form of chromatography, was developed at the University of Warsaw by the Russian botanist Mikhail Tswett in 1906, and it avoids some of the shortcomings of countercurrent distribution such as relatively poor separation and the need for relatively large amounts of solute. In liquid chromatography, an inert solid support is placed in a cylindrical glass column. The solid support which was used in early work was silica gel, and water, immobilized to the silica gel by hydrogen bonding, was the *stationary phase*. An organic solvent running through the column was the mobile phase. Although silica gel and other solid supports are supposed to be "inert" when used for liquid chromatography, complete inertness is not always achieved. For example, some interaction does occur frequently between silica gel and compounds which are being separated; under these circumstances, the two phases could either be classified as liquid-liquid because solutes are retarded in the liquid stationary phase, or as liquid-solid because solutes also are retarded by the solid support. In other words, classification of chromatographic methods on the basis of the physical nature of stationary and mobile phases is not perfect. Traditional liquid chromatography is carried out by dissolving or suspending the mixture to be separated in a small volume of mobile phase and applying the mixture to the column containing adsorbent plus bound water. The compounds in the mixture then are eluted from the column, and small fractions of eluate are collected and analyzed. There are two general types of elution: *stepwise elution* and *gradient elution*. Stepwise elution is carried out with a succession of distinct and separate mobile phases, each of which has increasing elution power. In gradient elution, on the other hand, eluting power is increased gradually and smoothly by continuous change of the proportions of individual solvents in the mobile phase. A disadvantage of column chromatography is that the separation of biomolecules may take several hours, the reason for this long separation time being the low rates of diffusion of solutes in liquids. However, as we have seen in section 3-7, by systematic application of principles embodied in the van Deemter equation, low diffusion rates are avoided in high-performance liquid chromatography.

3-8-2. Paper chromatography

Paper chromatography, another form of liquid-liquid chromatography, is frequently referred to as paper partition chromatography to emphasize the partitioning of solutes between a mobile phase and water adhering to the filter paper during chromatography. Paper chromatography is carried out by placing samples to be analyzed on strips or sheets of filter paper and allowing an organic solvent, saturated with water, to develop the chromatogram i.e. to flow slowly up or down over the paper by capillary action so as to separate components of the mixture. The organic solvent is the *mobile phase*, the water which is held to the cellulose fibers of the paper by hydrogen bonding, is the *stationary phase*. Separation of individual biomolecules of the mixture occurs on the basis of differential partitioning of components of the sample between mobile and stationary phases. After separation, the spots of the separated components may be made visible by spraying the paper with reagents which form colored derivatives.

Sometimes, the individual compounds in mixtures of biomolecules exhibit closely similar physicochemical properties, e.g. closely similar partition coefficients in a given solvent system. Under these conditions, the compounds are not separated by paper chromatography in the single solvent system, and two-dimensional paper chromatography should be tried in order to resolve the mixture with a greater degree of success. Generally, solvent systems with markedly different characteristics (e.g., one acidic the other basic) should be used in two-dimensional paper chromatography. Two-dimensional paper chromatography is carried out by placing a single spot of the mixture in the lower corner of a sheet of filter paper and subjecting the mixture to sucessive development in the two solvent systems. Thus, after development in the first direction, the sheet of paper is removed, allowed to dry and then run in the second solvent system in a direction at right angles to the first. Separated compounds are visualized by spraying the completed chromatogram with appropriate reagents.

3-9. Liquid-solid chromatography includes ion-exchange, TLC & affinity

3-9-1. Ion-exchange chromatography

In *ion-exchange chromatography*, the *mobile phase* is a buffered electrolyte and the *stationary phase* is a finely divided solid which contains covalently linked anionic or cationic groups. Ion-exchange separations are used in traditional liquid chromatography, in high-performance liquid chromatography and in thin-layer chromatography. Separations depend mainly on differences in the net ionic charges of the solutes at specified pH values. However, the separations depend also on the extent to which hydrophobic or hydrophilic moieties of the biomolecules interact with the hydrophobic or hydrophilic ion-exchange resins. Among the various types of ion-exchange resins commercially available are hydrophobic beads made from polystyrene cross-linked with divinylbenzene, and hydrophilic (polar) supports made from cellulose, dextran, or agarose. Acidic or basic groups of varying strengths are introduced into these polar or nonpolar polymers to yield ion-exchange resins. *Cation exchangers* are negatively charged (acidic) resins which can bind cations. *Anion exchangers* are positively charged (basic) nitrogenous resins which bind anions.

We will use the separation of a mixture of amino acids on a sulfonated polystyrene (a cation exchange resin) to illustrate how ion exchange chromatography works. Sulfonated polystyrenes contain SO_3^- (sulfonic acid) groups covalently linked to polystyrene beads:

$$CH_2-CH-CH_2-CH-CH_2-CH-$$

Sulfonated Polystyrene

The SO_3^- form of the resin may be converted to the sodium form by washing a solution containing sodium ions through the column, thereby linking SO_3^- and Na^+ electrostatically according to the following equation:

$$nResin\text{-}SO_3^-H^+ + Na^+ \longrightarrow (Resin\text{-}SO_3)_nNa + nH^+$$

One starts the ion exchange separation of amino acids by placing a solution of the amino acids at pH 2.2 on the column in the sodium form. Since at pH 2.2 amino acids generally are present as cations, they will displace some Na^+ ions from the resin and will be bound to the column at sites vacated by Na^+. Elution is started at pH 3.25 and continued to pH 5.28 with the Na^+ concentration being increased concurrently with pH. Since charge and hydrophobic/hydrophilic interactions are involved in the separations, the order of elution can be explained on the basis of one or both properties. For example, in the elution of Moore and Stein reproduced in Figure 3-3, aspartic acid is eluted first and arginine last.

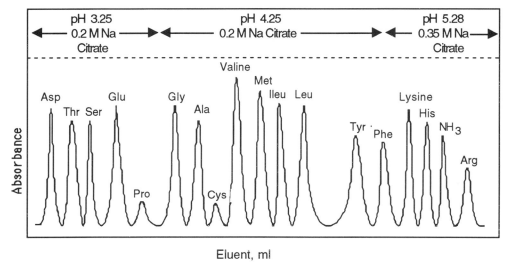

Figure 3-3. Representation of the order of separation of amino acids on a sulfonated polystyrene ion-exchange column.

We note also that threonine and serine are eluted at the beginning, immediately after aspartic acid and *before* glutamic acid, whereas the aromatic amino acids tyrosine, phenylalanine and histidine (which have hydrophobic rings as does styrene) are among the last eluted. This elution pattern is entirely consistent with the principle of separation outlined above. Thus, aspartic acid, the first eluted, is the most negatively charged (least positive) amino acid at pH 3.25 and is eluted first. Arginine is the most positively charged amino acid, is held most tightly to the SO_3^- groups, and is eluted last. On the other hand, threonine and serine are hydrophilic and do not associate readily with the hydrophobic polystyrene beads; conversely, the aromatic amino acids are rather hydrophobic and do interact hydrophobically with the polystyrene beads. With this background, the reader is encouraged to explain why the other amino acids are eluted in the order shown in Figure 3-3.

3-9-2. Affinity chromatography

In *affinity chromatography*, macromolecules (generally proteins) are purified by exploiting their abilities to bind ligands. (In biochemistry, a ligand is often defined as a substance which binds with high specificity to a protein; cf. section 5-10). The chosen ligand is linked covalently to an insoluble chromatographic support such as agarose, dextran, polyacrylamide or silica gel, with pore sizes that are of the appropriate range for achieving separation of macromolecules. A chromatographic column then is prepared from the support containing bound ligand. This ligand-linked, solid support becomes the stationary phase. After the column is packed, the sample containing protein to be purified is placed on the column and elution is started. The only macromolecule retained on the column is the one which binds specifically to the covalently bound ligand. All other biomolecules are eluted. The required biomolecule can be eluted with a solution which contains a second, soluble ligand to which the desired protein binds more tightly than it does to the first ligand. This solution containing the second ligand is the *mobile phase*. For the method to be successful, binding and elution of the desired biomolecule should occur without denaturation and the support should not, as a general rule bind proteins.

Choice of ligand depends on the protein or other biomolecule to be isolated. For example, if the protein is an enzyme, the ligand selected for linkage to the support may be related chemically to the substrate, to the product or to a competitive inhibitor. If the desired protein is a receptor, the ligand selected for binding to the support may be structurally related to a specific ligand (see section 5-10). If an antibody is to be isolated, the ligand selected for binding to the support may be a compound which is chemically similar to the antigen (see sections 6-2, 6-3 and 6-5). Affinity chromatography thus presents a unique example of the principle that, during chromatography, compounds are held back to differing degrees by the stationary phase and therefore migrate at different rates in the mobile phase.

3-9-3. Thin-layer chromatography

Thin-layer chromatography (TLC) is performed on thin layers of adsorbent coated on glass, aluminum, or sheets of plastic. Silicic acid, aluminum oxide and cellulose are the most commonly used *solid (stationary) phases*. Organic solvents saturated with water, or with aqueous solutions of acids or bases, constitute the *mobile phase*. Separation of compounds usually occurs because the solutes interact differentially with the solid phase by adsorption, hydrogen bonding, or ion-exchange. In addition, differential partitioning of solutes between the mobile phase and water, or aqueous solutions associated with the solid phase, may take place. Because of such differential interactions, the solutes migrate at different rates in the mobile phase. Spotting and development of the chromatograms is carried out in essentially the same way as in paper chromatography. Solutes in thin-layer chromatography may be held back either, or both, by partitioning between the mobile and stationary phases, and by interaction with the solid phase. In the case of silicic acid as adsorbent, solutes may interact with the adsorbent through hydrogen bonding or even through ion exchange with the surface silanol groups (-SiOH). The separated compounds are visualized by spraying the chromatograms with reagents which form colored derivatives. A major advantage of thin-layer chromatography is that it is more rapid, more sensitive and more reproducible than paper chromatography. Another advantage of thin-layer chromatography is that corrosive spray reagents can be used with silicic acid and other adsorbents. Thin-layer chromatography can be carried out quantitatively by scraping the spots off the plates, eluting the solute and analyzing it by an appropriate method. Figure 3-4 shows a thin-layer chromatogram of known lipids and microbial lipids. The chromatogram was run in a solvent system which separates neutral lipids.

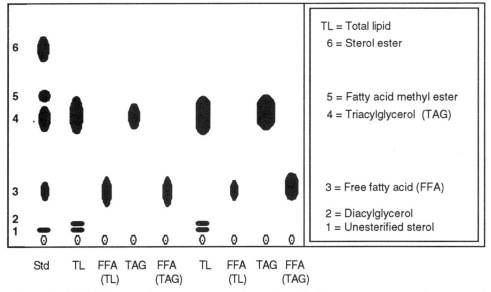

Figure 3-4. Thin-layer chromatogram of standard lipid compounds, and of samples of total lipid, triacylglycerols, and free fatty acids from hydrolyzed samples of the total lipids and triacylglycerols

Recently, some of the principles that were applied earlier to the development of high-performance liquid chromatography from column chromatography, have been applied to the development of *high performance thin-layer chromatography* (HPTLC). The successful application of a single set of principles to the development of more than one chromatographic method underscores the underlying similarity of various forms of chromatography. For the development of HPTLC, small diameter, regularly shaped particles of adsorbent have been introduced to replace the traditional thin-layer chromatographic adsorbents. Concurrently, methods of spreading the adsorbent on the plates, of spotting (applying) the samples to the coated adsorbent and of quantitating the compounds after separation have been automated and computerized. The result, according to some, is that high-performance thin-layer chromatography has become a technique that can equal, or surpass, high-performance liquid chromatography in certain respects.

3-10.   Gas chromatography is run on packed and capillary columns

3-10-1. Gas-liquid chromatography on packed columns

In *gas-liquid chromatography* (abbreviated GLC or GC) on packed columns, the *mobile phase* is an inert carrier gas and the *stationary phase* is a nonvolatile liquid, coated on a finely divided (i.e., small diameter) solid support such as diatomaceous earth. Most often, packed gas-liquid chromatographic columns are made of 1 to 2 m long tubing of stainless steel (or glass) with a diameter of 3 to 10 mm. To pack the columns, the coated diatomaceous earth is poured into the empty column with gentle vibration. After packing, both ends of the column are plugged with glass wool.

Samples usually are injected into gas chromatographs through a heated (approximately 200°C) injection port. The heated port causes the sample to become volatile (if it is not a gas) and the volatilized sample is carried onto the column by the carrier gas. Separation of the compounds in the sample occurs on the basis of partitioning between the mobile phase (carrier gas) and the stationary phase (nonvolatile liquid) coated on the solid support. Those compounds which are most similar in polarity to the stationary phase interact most efficiently with that phase and are retarded in their progress out of the column. Conversely, compounds in the sample which do not associate with the stationary phase are eluted before compounds which associate with the stationary phase. As compounds leave the column, they are detected automatically with one of several types of detectors currently available. The response of the detector is registered on a recorder in the form of peaks which, today, are conveniently quantitated with microprocessors (computer microchips).

Gas-liquid chromatography separates (resolves) closely related compounds very well because its relatively short (1 to 2 m) columns contain very large numbers of theoretical plates. In addition, gas-liquid chromatography is a rapid technique (takes only minutes) because the flow rate of gases is much greater than that of liquids. Equally important, gas-liquid chromatography is a sensitive technique (separates and detects µg to ng quantities easily) because the detectors which are used are among the most sensitive known. Figure 3-5 shows a gas-liquid chromatogram of fatty acid methyl esters. The separation of 18-carbon fatty acids which is obtained by this method would be extremely difficult or impossible by other methods.

3-10-2. Gas-liquid chromatography on capillary columns

The efficiency of separation of biomolecules in gas-liquid chromatography with packed columns (3 to10 mm i.d,1 to 2 m long) is high but is limited by the rate of diffusion of the samples between gas and liquid phases. However, when gas-liquid chromatography is carried out with capillary columns (0.10 and 0.25 mm i.d., 5 to 50 m long), the efficiency of separation is improved greatly because the much smaller internal diameters of the capillary columns lead to increased rates of diffusion, to shortened times of analysis and to improved sensitivity. In capillary columns, the liquid phase is coated onto the wall of the capillary. The wall of the tubing thus serves as the support for the liquid phase. Uncoated capillary tubing is commercially available for investigators who desire to coat columns for themselves rather than purchase commercially, precoated ones.

If the gas chromatograph were purchased initially to be used with packed columns, it may have to be modified in order to accept capillary columns. Once the modifications have been made, it is important to ensure that leaks and dead spaces do not occur. Dead spaces are most likely to occur toward the ends of the column - the regions nearest the injector and detector. Since the efficiency of separation on capillary columns results in sharp, narrow peaks, the areas of the peaks should be determined by digital integrators rather than by triangulation or other manual methods.

Until 1979, capillary columns were made only from tubing of soda-lime or borosilicate glass since heated metal is unsuitable for the analysis of many biomolecules. Since 1979, however, fused silica (an inert, flexible material) has been used for capillary gas-chromatographic columns. Capillary columns of fused silica can be made in a variety of sizes: short   (5 to 15 m) and small (i.d. 0.10 to 0.25 mm); long (30 to 50 m) and small (0.10 to 0.25 m i.d.) and megabore columns (0.53 mm i.d.) of varying lengths.

3-11. CC chromatography has the strong points of CCD and LC

Countercurrent chromatography (CCC) is a relatively new form of partition chromatography in which solid supports such as silica gel or filter paper are eliminated.

Countercurrent chromatography is carried out in coiled tubes ranging from 0.5 mm to 2.6 mm i.d. and from 5 m to 100 m long. In one type of CCC system, which has been referred to as the *hydrostatic equilibrium system*, the coiled tube is kept stationary during chromatography. In a second system, described as the *hydro-*

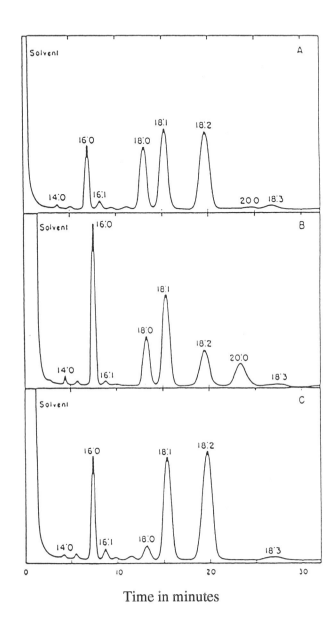

Figure 3-5. Separation of methyl esters of fatty acids by GC on a packed column

dynamic equilibrium system, the coiled tube is rotated about its axis during chromatography. In both systems the coiled tube initially is filled by pumping into it the stationary phase of a two-phase solvent system, making sure to avoid air bubbles. Then the sample, dissolved in equal proportions of the two-phase solvent system, is injected into the column (coiled tube). The mobile phase is now pumped through the column at an appropriate rate. The mobile phase, as a result, percolates through (but does not displace) the stationary phase and the sample is partitioned between the two phases leading to separation of the components in the sample according to partition coefficients. It has been reported that analytical columns in CCC yield partition efficiencies of "several thousand" theoretical plates, whereas separations on preparative columns have been obtained at partition efficiencies of 1 000 theoretical plates. The eluted components may be monitored continuously, for example with UV monitors.

Some stated advantages of CCC compared with traditional liquid-liquid chromatography are that there is diminished denaturation of samples in CCC, that CCC yields improved recovery and higher purity of separated components than liquid-liquid chromatography, and also that reproducibility and predictability are higher in CCC than in liquid chromatography. Moreover, equipment for countercurrent chromatography is relatively inexpensive and compact.

To date, countercurrent chromatography has been used for such tasks as the separation and purification of peptides, prostaglandins, and certain drugs, as well as for the partitioning of cells and macromolecules.

3-12. Worked examples

Example 3-1. A given amount of solute S has a partition coefficient of 1.0 in an immiscible solvent pair. Calculate the fractions of the total solute S in each tube of a five-tube apparatus after countercurrent distribution.

$$T_{n,r} \quad = \quad \frac{n!}{r!(n-r)!} \left(\frac{1}{K+1}\right)^n K^r$$

Tube 0: $\qquad T_{n,r} \quad = \quad \dfrac{4!}{0!\,(4-0)!} \left(\dfrac{1}{1+1}\right)^4 1^0 \quad = \quad (24/24)\,(1/2)^4\,1^0$

$\qquad\qquad\qquad = \quad 1 \times 0.0625 \times 1 \qquad\qquad = \quad 0.0625$

Tube 1: $\qquad T_{n,r} \quad = \quad \dfrac{4!}{1!\,(4-1)!} \left(\dfrac{1}{1+1}\right)^4 1^1 \quad = \quad (24/6)\,(1/2)^4\,1$

$\qquad\qquad\qquad = \quad 4 \times 0.0625 \times 1 \qquad\qquad = \quad 0.25$

Tube 2: $T_{n,r}$ $= \dfrac{4!}{2!\,(4-2)!}\left(\dfrac{1}{1+1}\right)^4 1^2$ $= (24/4)\,(1/2)^4\,1$

$= 6 \times 0.0625 \times 1$ $= 0.375$

Tube 3: $T_{n,r}$ $= \dfrac{4!}{3!\,(4-3)!}\left(\dfrac{1}{1+1}\right)^4 1^3$ $= (24/6)\,(1/2)^4\,1$

$= 4 \times 0.0625 \times 1$ $= 0.25$

Tube 4: $T_{n,r}$ $= \dfrac{4!}{4!\,(4-4)!}\left(\dfrac{1}{1+1}\right)^4 1^4$ $= (24/24)\,(1/2)^4\,1$

$= 1 \times 0.0625 \times 1$ $= 0.0625$

Example 3-2. If 48 mg of S were present at the start of the countercurrent distribution experiment of Example 1, how many mg of S were present in both layers of the five tubes.

Tube 0: $T_{n,r}$ $= 0.0625$
Amount of S $= 0.0625 \times 48$ $= 3$ mg

Tube 1: $T_{n,r}$ $= 0.25$
Amount of S $= 0.25 \times 48$ $= 12$ mg

Tube 2: $T_{n,r}$ $= 0.375$
Amount of S $= 0.375 \times 48$ $= 18$ mg

Tube 3: $T_{n,r}$ $= 0.25$
Amount of S $= 0.25 \times 48$ $= 12$ mg

Tube 4: $T_{n,r}$ $= 0.0625$
Amount of S $= 0.0625 \times 48$ $= 3$ mg

3-13. Spreadsheet solutions

	A	B	C	D	E	F
1	EXAMPLE 3-1				FACTORIALS:	
2		INPUT	OUTPUT			
3	n	4			0	1
4	K	1			1	1
5					2	2
6	Tube #, r	0			3	6
7	Tn,r		0.0625		4	24
8					5	120
9	Tube #, r	1			6	720
10	Tn,r		0.25		7	5040
11					8	40320
12	Tube #, r	2			9	362880
13	Tn,r		0.375		10	3628800
14					11	39916800
15	Tube #, r	3			12	479001600
16	Tn,r		0.25		13	6227020800
17					14	87178291200
18	Tube #, r	4			15	1.30767E+12
19	Tn,r		0.0625		16	2.09228E+13
20					17	3.55687E+14
21					18	6.40237E+15
22					19	1.21645E+17
23					20	2.4329E+18
24	EXAMPLE 3-2				21	5.10909E+19
25		INPUT	OUTPUT		22	1.124E+21
26					23	2.5852E+22
27					24	6.20448E+23
28	S at start	48			25	1.55112E+25
29					26	4.03291E+26
30	Tn,r	0.0625			27	1.08889E+28
31	Amt. of S		3		28	3.04888E+29
32					29	8.84176E+30
33	Tn,r	0.25			30	2.65253E+32
34	Amt. of S		12		31	8.22284E+33
35					32	2.63131E+35
36	Tn,r	0.375			33	8.68332E+36
37	Amt. of S		18		34	2.95233E+38
38					35	1.03331E+40
39	Tn,r	0.25			36	3.71993E+41
40	Amt. of S		12		37	1.37638E+43
41					38	5.23023E+44
42	Tn,r	0.0625			39	2.03979E+46
43	Amt. of S		3		40	8.15915E+47
44						

SPREADSHEET FORMULAS

	A	B	C	D	E	F
1	EXAMPLE 3-				FACTORIALS:	
2		INPUT	OUTPUT			
3	n	4			0	1
4	K	1			1	=E4*F3
5					2	=E5*F4
6	Tube #, r	0			3	=E6*F5
7	Tn,r		=F7/(F3*F7)*1/2^4*1		4	=E7*F6
8					5	=E8*F7
9	Tube #, r	1			6	=E9*F8
10	Tn,r		=F7/(F4*F6)*1/2^4*1		7	=E10*F9
11					8	=E11*F10
12	Tube #, r	2			9	=E12*F11
13	Tn,r		=F7/(F5*F5)*1/2^4*1		10	=E13*F12
14					11	=E14*F13
15	Tube #, r	3			12	=E15*F14
16	Tn,r		=F7/(F6*F4)*1/2^4*1		13	=E16*F15
17					14	=E17*F16
18	Tube #, r	4			15	=E18*F17
19	Tn,r		=F7/(F7*F3)*1/2^4*1		16	=E19*F18
20					17	=E20*F19
21					18	=E21*F20
22					19	=E22*F21
23					20	=E23*F22
24	EXAMPLE 3-				21	=E24*F23
25		INPUT	OUTPUT		22	=E25*F24
26					23	=E26*F25
27					24	=E27*F26
28	S at start	48			25	=E28*F27
29					26	=E29*F28
30	Tn,r	=C7			27	=E30*F29
31	Amt. of S		=B30*B28		28	=E31*F30
32					29	=E32*F31
33	Tn,r	=C10			30	=E33*F32
34	Amt. of S		=B33*B28		31	=E34*F33
35					32	=E35*F34
36	Tn,r	=C13			33	=E36*F35
37	Amt. of S		=B36*B28		34	=E37*F36
38					35	=E38*F37
39	Tn,r	=C16			36	=E39*F38
40	Amt. of S		=B39*B28		37	=E40*F39
41					38	=E41*F40
42	Tn,r	=C19			39	=E42*F41
43	Amt. of S		=B42*B28		40	=E43*F42
44						

3-14. Review questions

3-1.　State the principle which underlies all chromatographic separations.
3-2.　State the principle which underlies countercurrent distribution.
3-3.　Why is countercurrent distribution a model for partition chromatography?
3-4.　Define the term partition coefficient.
3-5.　How is the partition coefficient related to the mobilities of solutes in countercurrent distribution and in partition chromatography?
3-6　How does HPLC differ from traditional liquid chromatography?
3-7.　Why is it easier to predict the results of countercurrent distribution than the results of partition chromatography?
3-8.　Explain the term "liquid-liquid" chromatography.
3-9.　Define the terms "mobile phase" and "stationary phase."
3-10.　List five different types of mobile and stationary phases between which chromatographic separation occurs.

Predict the relative mobilities of the members of each pair of amino acids assuming that paper chromatography was conducted in butanol : acetic acid : water:

3-11.　The faster of leucine and threonine.
3-12.　The slower of glutamic acid and threonine.
3-13.　The slower of glutamic acid and aspartic acid.
3-14.　The faster of tyrosine and arginine.
3-15.　The slower of tyrosine and threonine.
3-16.　The slower of lysine and alanine.
3-17.　The faster of tyrosine and phenylalanine.
3-18.　The slower of proline and glycine.
3-19.　The faster of valine and serine.
3-20.　The faster of tryptophan and threonine.

3-15. Review problems

3-21.　If the partition coefficient of an amino acid in an immiscible solvent pair is 4.00, how much of the amino acid would be dissolved in each layer after mixing and separation in a separatory funnel. Assume that 75 mg of amino acid and equal volumes of the two solvents were used.

3-22.　The partition coefficient of a nucleoside of molecular weight 267.2 between an organic solvent and water was 10.00. If 50 ml of a 0.1 M aqueous solution of the nucleoside were mixed with 100 ml of the organic solvent, what weight (in grams) of the nucleoside was there in each of the two layers after mixing and equilibration?

3-23.　You were given 100 ml of an aqueous solution which contained 80 μg of a cytosine derivative. After mixing and equilibration with an immiscible solvent, spectrophotometric analysis showed that 16 μg remained in the aqueous phase. What was the partition coefficient of the compound in the solvent pair?

3-24.   Given 200 mg of a solute with a partition coefficient of 20.00 in a solvent pair, show which is the more effective: extraction of the solute once with 100 ml volumes of each of the two solvents, or extraction of the solute twice with 50 ml volumes of each of the solvents?

3-25.   Calculate the partition coefficient of a dinitrophenyl amino acid if two-thirds of it is removed from 100 ml of an aqueous solution with two 50 ml volumes of an organic solvent.

3-16.   Additional problems

3-26.   The partition coefficients of two dinitrophenyl amino acids, A and B, in a two-phase solvent system, are 10.0 and 1.0 respectively. Calculate the fractions and total amounts of each amino acid in each tube of an eight-tube (0 to 7) apparatus after countercurrent distribution if the initial amounts of A and B were 80 and 64 mg respectively.

3-27.   Repeat problem 3-26 with $K = 100$ and $K = 5$ for A and B respectively.

Suggested Reading

Bush, M.T. 1981. *Design of Solvent Extraction Methods.* Methods in Enzymology, Academic Press, New York, 77: 353-372.

Christian, G.D. 1980. *Analytical Chemistry, 3d ed.,* Wiley, New York.

Craig, L. C., D. Craig and E. G. Scheibel 1956. *Laboratory Extraction and Countercurrent Distribution,* In Technique of Organic Chemistry, Vol. III, ed. A. Weissberger, Interscience, New York.

Day, R. A. and A. L. Underwood 1980. *Quantitative Analysis, 4th ed.,* Prentice-Hall, New Jersey.

Gaucher, G. M. 1969 An Introduction to Chromatography, *J. Chem. Educ.* 46: 729-733.

Gordon, H. 1977. The Beginnings of Chromatography, *Trends Biochem. Sci.* 2: 243-246.

Horvath, C. March 1987. About Chromatography and Language, *Chromatography pp 5 - 8.*

Ito, Y. 1982. Countercurrent Chromatography, *Trends Biochem. Sci.* 7: 47-50.

Ito, Y. 1983. *Countercurrent Chromatography,* Methods in Enzymology, Academic Press, New York, 91: 335-351.

Lehninger, A. L. 1975. *Biochemistry, 2d ed.,* Worth, New York.

Lehninger, A. L. D. J. Nelson and M. M. Cox 1992. *Principles of Biochemistry,* 2d. ed., Worth, New York.

Maclouf, J. and M. Rigaud 1982. *Open Tubular Capillary Gas Chromatography Columns,* Methods in Enzymology, Academic Press, New York, 86: 612-631.

McNair, H. M. 1985. Chromatography. HPLC, GC, HPTLC, *Science* 232: 27.

Patton, G. M., S. Cann, H. Brunengraber and J. M. Lowenstein 1981. *Separation of Methyl Esters of Fatty Acids by Gas Chromatography on Capillary Columns, Including the Separation of Deuterated from Nondeuterated Fatty Acids,* Methods in Enzymology, Academic Press, New York, 72: 8-20.

Pecsok, R. L., L. D. Shields, T. Cairns and I. G. McWilliam 1976. *Modern Methods of Chemical Analysis, 2d ed.,* Wiley, New York.

Regnier, F. E. 1984. *High-Performance Ion-Exchange Chromatography*, Methods in Enzymology, Academic Press, New York, 104: 170-189.

Svasti, J. 1980. Automated Amino Acid Analysis Comes of Age: but Textbook Errors Persist, *Trends Biochem. Sci.*, 5: VIII-IX.

Willard, H. H., L. L. Merritt, J. A. Dean and F. A. Settle, Jr. 1981. *Instrumental Methods of Analysis, 6th ed.*, Wadsworth, Belmont, California.

PART II.

MACROMOLECULAR STRUCTURE AND ASSAY

The Separation and Characterization of Macromolecules

4-1.   Transport processes are pivotal in studying macromolecules

A simple bacterial cell can contain 3 000 or more proteins. Many of these proteins occur in small quantities and may differ only slightly from each other in physical and chemical properties; yet, to carry out an assay for the physical, chemical, or biological properties of a given protein or other macromolecule, scientists often have to isolate that single macromolecule from all others of the same type. Isolation of individual proteins or other macromolecules thus requires methods which have high selectivity, high reproducibility, and high resolving power. The methods which meet these criteria today are all *transport processes*. A *transport process is one in which solutes are moved through a solvent or membrane at a measurable rate*.

In transport processes, as in kinetics (section 5-14), time is a variable and the system is not at equilibrium. But, whereas the driving force in chemical kinetics is a decrease in free energy, transport processes are driven by other kinds of forces which cause solute molecules to be moved as a function of time. For example, in gel filtration, hydrostatic pressure causes macromolecules to be eluted; in electrophoresis an electric field forces macromolecules toward one electrode; and in centrifugation, centrifugal force causes macromolecules to be sedimented. Transported (separated) molecules usually are recovered in the same form in which they were introduced at the start of the transport (separation) process. Thus, whereas kinetics deals with the rate of chemical change, equivalent chemical changes usually do not accompany the physical movement of solutes during transport.

After one isolates a macromolecule, the next step is to characterize it physicochemically because such characterization makes it possible to identify the macromolecule and investigate its biological behavior. One of the identifying characteristics of any macromolecule is its molecular weight, and three of the most powerful physicochemical methods for determinations of molecular weights are gel filtration, sodium dodecyl sulfate (NaDodSO4) polyacrylamide gel electrophoresis, and sedimentation in the ultracentrifuge. These three methods are all transport processes.

In this chapter, we will describe some transport procedures for the determination of molecular weight, for the separation and identification of proteins and nucleic acids, for the determination of sedimentation coefficient, and for the fractionation of cellular material by differential centrifugation. Because the separation and analysis of polynucleotides have become major tools in molecular biology and biotechnology, we will undertake in sections 4-2 and 4-3 a brief survey of some aspects of recombinant DNA research and sequence methodology. Our purpose will be to place the various physicochemical methods in a context which reflects the increasingly informational nature of results currently being produced, or evaluated with, transport tools in laboratories throughout the world.

4-2. Recombinant DNA technology uses transport and biochemical tools

The term *recombinant DNA technology* refers to procedures in which unrelated segments of DNA from the same or different organisms are joined covalently *in vitro* and then inserted into appropriate living cells to be replicated or expressed as proteins. Such transfer of genetic material between unrelated species has opened up entirely new areas of research and is now carried out routinely in numerous laboratories; but these advances had to await three key developments:

- discovery of restriction endonucleases (restriction enzymes)
- development of methods for joining segments of DNA, and
- use of plasmids or bacteriophages to insert unrelated DNA into host cells.

Restriction enzymes recognize and cleave specific nucleotide sequences in double-stranded DNA. Restriction enzymes usually are associated, in function, with *modification methylases.* The latter methylate the same nucleotide sequences recognized by the matching restriction enzyme and serve to prevent restriction enzymes from attacking their own cellular DNA. But, foreign DNA has different patterns of methylation, is recognized as foreign and is degraded. The major steps in expressing recombinant DNA molecules are:

- Digest the DNA molecules to be joined (a segment of DNA and derivatives of plasmids or λ phage) with a restriction endonuclease. Then anneal the two complementary ends of the digested DNAs (allow them to associate spontaneously by base-pairing at an appropriate temperature and ionic strength) and *join them covalently with DNA ligase.*

• Step two is to mix the recombined DNA with cells of the bacterial host to permit some recombined DNA to enter the host. The frequency of entry is generally low, approximately one entry per one million recombined DNA molecules. But transformed cells may contain more than one vector plus recombined DNA.

• The third step is to select cells which have replicated the desired sequences of DNA. This selection may be made from a property of the vector (e.g., antibiotic resistance) or from a property of the foreign DNA. For example, the foreign DNA may be hybridized with pure labeled DNA. Or, the foreign DNA may be hybridized with a sample of RNA containing mRNA; in this case, selection is possible because translation of the (hybridized) mRNA *in vitro* is prevented. Sometimes, also, the recombined DNA may be constructed to have a desirable buoyant density so as to facilitate its identification.

Generally, strains of *Escherichia coli* which lack restriction endonucleases (nonrestricting strains) are used as host cells because these strains will not cleave the inserted (foreign) DNA. In addition, strains of *E. coli* usually are selected which cannot carry out recombination of DNA, thus ensuring that the inserted DNA will be replicated but not modified through recombination. Recombinant DNA technology may be used for at least five types of applications:

• to produce proteins or peptides of biomedical or commercial interest.
• to clone (replicate) desired sequences of foreign DNA.
• to alter DNA sequences to understand their functions.
• to study gene expression, and
• to construct genes which confer properties such as resistance to plant diseases.

Two transport methods, electrophoresis and blot techniques, are important adjuncts of recombinant DNA research. Electrophoresis is used to separate fragments from digestion with restriction enzymes according to size; blot techniques are used to identify the separated fragments (sections 4-8 to 4-10). More specifically,

• Low molecular weight fragments may be separated by polyacrylamide gel electrophoresis and large fragments by agarose gel electrophoresis.

• Intermediate-sized fragments may be separated on composite gels containing both polyacrylamide and agarose, but gels containing a gradient of polyacrylamide concentrations may be needed for separating fragments with a wide range of molecular weights.

• For the estimation of molecular weights (or number of base pairs, bp), it is standard practice to base the estimates on standard curves in which the logarithm of molecular weight is inversely proportional to electrophoretic mobility.

Thus, recombinant DNA technology uses as major tools:

- cutting enzymes (restriction endonucleases);
- joining enzymes (DNA ligases);
- vehicles for insertion of DNA (plasmids or bacteriophages);
- transport methods for separating fragments produced by restriction endo-nucleases; and
- methods for selecting cells which have replicated DNA sequences.

We discuss briefly in section 4-23 software for storing, analyzing, and comparing the enormous volume of data produced with recombinant DNA technology.

4-3. Cleavage maps and DNA sequencing also use transport tools

A goal of many studies with restriction endonucleases is the mapping of genomes. The first step in such investigations is to obtain a preparation of the desired DNA. This may be achieved by one of two approaches: by using RNA (mRNA, rRNA, or tRNA) as a template for the enzymatic synthesis of *complementary DNA*, abbreviated *cDNA*, or by extracting and purifying intact cellular DNA from an appropriate source, making sure to remove degradative enzymes (nucleases) as well as other contaminating biomolecules and subcellular fractions. The specific RNA required for synthesizing a desired cDNA may be obtained by combinations of biochemical, immunological, or chromatographic methods. Once obtained, the RNA is used as a template in an assay system containing *RNA-dependent DNA polymerase (reverse transcriptase)*, plus other necessary assay components, to synthesize the cDNA corresponding to the genome under study. Instead of following the cDNA route to obtain the desired genome, intact cellular DNA may be extracted and purified from an appropriate source; but, given the necessity of ensuring intactness when DNA is obtained by extraction, it is easiest to work with small DNA molecules. In a combination of the two approaches for obtaining DNA, radioactively labeled cDNA is sometimes used as a probe to form hybrids with cellular DNA by complementary base pairing. The desired genome can then be located and isolated ("fished-out") from the entire set of genes in the extract. Once the DNA has been obtained, the next step in cleavage mapping is to subject the purified DNA to digestion with a series of restriction endonucleases to obtain small overlapping fragments. These fragments are separated by agarose or polyacrylamide gel electrophoresis. The goal of the electrophoresis is to determine the sizes of the separated fragments and their order in the DNA. Size is determined from electrophoretic mobility or other appropriate method; order is deduced from the overlaps of specific restriction cleavage sites. Ordering the restriction fragments permits the construction of cleavage maps, defined as maps which show specific resriction cleavage sites. Cleavage maps are also referred to as *fragment maps* or *physical maps*. They permit one to compare mutant and wild-type DNA and to study the biological activities of various restriction fragments in relation to order on the genome. In addition to being used for mapping, specific cDNA probes are also being used for the identification of pathogenic microorganisms in clinical laboratories by complementary base pairing of the probes with RNA released from the organism under study.

More detailed than cleavage mapping of genomes is nucleotide sequencing of DNA by the chain terminator procedure of Fred Sanger and colleagues and the chemical degradation method of Allan Maxam and Walter Gilbert. Both methods generate mixtures of fragments of varying lengths but all terminating at the same 5'-end, and both separate restriction fragments by electrophoresis. Details follow.

Sanger's method. This procedure uses dideoxyribonucleoside triphosphates ddNTPs. These 2',3'-dideoxy analogs of the deoxyribonucleoside triphosphates, (dNTPs) are incorporated into growing oligonucleotide chains; but since such analogs do not have 3'-hydroxyl groups, the oligonucleotide chains into which they are incorporated cannot be extended further; i.e. termination occurs where the dideoxy analogs are located. To carry out Sanger's method, restriction fragments are prepared and annealed with complementary, single-stranded DNA. The unpaired end of the complementary strand serves as the template, the end which is base-paired with the restriction fragment serves as the primer. The single-stranded template DNA and the double-stranded primer are incubated with the Klenow enzyme (that is, the larger of two fragments from mild proteolysis of DNA polymerase I. The *Klenow fragment,* named after H. Klenow, has polymerase and 3' → 5' activity but lacks the 5' → 3' exonuclease activity of the intact enzyme). For the incubations, four tubes are prepared so that each tube contains buffered assay medium; modified polymerase I; one of the four ddNTP inhibitors; a labeled dNTP, usually labeled dAT^{32}P; and the three other unlabeled dNTPs. The various incubations are subjected to NaDodSO$_4$ gel electrophoresis and, after electrophoresis, the bands are visualized by autoradiography. The resulting bands show the distribution of labeled dNTPs in newly synthesized DNA. The sequence of nucleotides in the copied DNA can be read off from the pattern of bands on the gels.

The Maxam-Gilbert method. This procedure is based on ^{32}P labeling of one end of DNA. Labeling, however, is followed by chemical cleavage at selected linkages to generate differing sets (lengths) of end-labeled fragments, which, subjected to electrophoresis in polyacrylamide gels, yield bands corresponding to the various chain lengths. The procedure requires that four separate chemical reactions be run, one for cleaving each base. These reactions result in the selective breakage of DNA chains at A, G, C or T. After cleavage, the products of the four reactions are subjected to electrophoresis and the bands are visualized by autoradiography. The sequence of nucleotides can be read off from the patterns on the exposed films for each of the four sets of products, each set having one common (labeled) end, but differing in length at the unlabeled end. To make sequencing of large DNA molecules manageable, the DNAs are first fragmented with restriction endonucleases to yield smaller fragments. Use of more than one restriction enzyme each with different specificities, will give overlapping fragments. When these restriction fragments are end-labeled and chemically degraded, the results make it possible to deduce the nucleotide sequence of the original DNA.

In both the Sanger and Maxam-Gilbert procedures, one obtains sizes of the nucleotide fragments from plots of log molecular weight versus electrophoretic mobility. A similar linear relation between molecular weight and elution volume exists for the separation of biomolecules by gel filtration, the next topic.

4-4. Gel filtration separates macromolecules according to size

Gel filtration is a form of liquid chromatography (section 3-8-1). In this method, molecules are separated from each other on the basis of differences in their molecular sizes. Three major types of gels are available: dextran gels (Sephadex), polyacrylamide gels (Bio-Gel P), and agarose gels (Sepharose or Bio-Gel A). Dextran and polyacrylamide gels are marketed as small, spherical beads; agarose gels are marketed in a hydrated form. All three types of gel have a three-dimensional, porous matrix of controlled dimensions. The sizes of the pores in dextran and polyacrylamide gels are controlled by the degrees of cross-linking. Pore size in agarose gels is controlled by the concentration of agarose.

In *gel filtration*, small molecules diffuse into pores of the gel matrix and are delayed in their passage through the columns. Large molecules, however, are too big to diffuse into the pores and are eluted rapidly from the gel chromatographic columns. Both theoretically and experimentally, it has been demonstrated that, in gel filtration, elution volume (defined in section 4-5 below) is related to molecular weight. To calibrate a gel column for estimations of molecular weights, the elution volumes of standard compounds (or one of the many functions related to elution volume), are plotted against the logarithms of the molecular weights of the standards. Since elution volume also is related to molecular shape, however, the estimates of molecular weight do not depend exclusively on size but, depend also on the shapes of the eluted molecules. Nevertheless, gel filtration may be used with some success to estimate the molecular weights of biomolecules by interpolation from calibration curves.

4-5. The physicochemical basis of gel filtration is in dispute

What occurs during gel filtration is generally agreed upon. *How* it occurs is not. Thus, while it is agreed that in gel filtration molecules are sorted according to size, there is far less agreement on how sorting occurs. Some authors assume that the sieving of molecules in gel filtration depends on restricted diffusion. These authors suggest that the diffusion of molecules into and out of pores takes a long time relative to the time which the molecules spend in the liquid around the pores. Other authors suggest that molecular sieving in gel filtration depends on steric exclusion. These authors assume that because of steric factors, the time which solute molecules spend in the liquid around the pores is considerably longer than the time required for diffusion into and out of the pores.

If we define the volume of solvent which elutes the solute from gel filtration columns (the elution volume) as V_e; the liquid within the pores (the *inner volume*) as V_i; the *volume occupied by the gel* or polymer as V_p; and the volume between the gel packing and the column (the void volume) as V_o, then, the total volume of the packed column of gel is given by the equation

$$V_t \quad = \quad V_o + V_i + V_p \tag{1}$$

where V_t is the *total volume* of the packed colum. Rearranging equation (1), we obtain the volume of packed gel plus inner volume as

$$V_p + V_i \quad = \quad V_t - V_o \qquad (2)$$

Since gel filtration is a form of liquid chromatography, the rate of elution of a solute is related to its partition coefficient (see section 3-1). Because the mobile phase in gel filtration is considered to be the elution volume minus the void volume, and because the stationary phase is equated with the inner volume, V_i, then the equation for the partition coefficient is:

$$K \qquad = \quad (V_e - V_o)/V_i \qquad (3)$$

It is difficult to measure V_i but easy to measure the combined volume of the packed gel plus inner volume. So if we substitute the value of $V_p + V_i$ from equation (2) for V_i in equation (3), the equation for the partition coefficient becomes

$$K \qquad = \quad (V_e - V_o)/(V_t - V_o) \qquad (4)$$

Not only is the estimate of K from equation (4) readily available, but, it is independent of the geometry of packed gel filtration columns. The estimate of K from equation (4) is therefore useful for calculating the amounts of gel, and the volumes of sample which should be applied, when investigators switch from columns of one size to columns of different sizes.

In addition to this relationship between elution volume and partition coefficients, other experimentally derived relationships have been developed relat-ing elution volume to the molecular weights of macromolecules. One of these is:

$$V_e/V_o \qquad = \quad - b * \log \text{ mol wt } + a \qquad (5)$$

where V_e/V_o is the relative elution volume. Equation (5) indicates that if V_e/V_o for a set of standard macromolecules is plotted against the logarithms of their molecular weights, a straight line should be obtained. In practice, however, this type of plot yields a smooth sigmoid curve such as that of Figure 4-1. The region of the curve which is approximately linear corresponds to the fractionation range of the gel. When the linear portions of these calibration curves have steep slopes (i.e., large values of b in equation (5)), the columns from which they were derived will separate compounds efficiently within the linear range. Conversely, columns which produce calibration curves with small slopes will not separate compounds as efficiently as columns which produce curves with steep slopes. However, columns which produce calibration curves with small slopes will fractionate molecules over a larger range of molecular weights than columns which yield calibration curves with steep slopes. Interpolation within calibration curves yields molecular weights corresponding to V_e/V_o of samples run under the same conditions as the standards. Extrapolation of calibration curves to $V_e/V_o = 1$ on the ordinate gives a point known as the exclusion limit. In terms of pore size, *the exclusion limit of a gel filtration column is the maximum molecular size which can pass through its pores.* In terms of equation (5), the intercept, a, is equal to 1 at the exclusion limit.

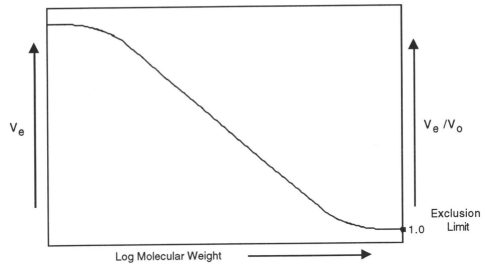

Figure 4-1.   Relation of elution volume to molecular weight

4-6. NaDodSO$_4$ electrophoresis can yield molecular weights of peptides

At present, there are many uncertainties concerning the physicochemical basis for the electrophoretic mobility of macromolecules in solution. Indeed, the view has been expressed that none of the physicochemical models which have attempted to relate electrophoretic mobility to molecular size is theoretically sound. Nevertheless, laboratory experience has demonstrated that electrophoretic mobility in polyacrylamide gels can, in practice, be related successfully to molecular weight - even though there is no rigorous theoretical explanation for this relationship. In biomedical laboratories, electrophoresis in polyacrylamide gels is carried out in the presence or absence of denaturing agents. When polyacrylamide gels are used for the electrophoresis of proteins in the absence of a denaturing agent, the electrophoretic mobilities of the proteins depend both on the ionic charges of the separated proteins and on the molecular sieving properties of the gel. On the other hand, when polyacrylamide gel electrophoresis is carried out in the presence of the amphiphilic, denaturing agent *sodium dodecyl sulfate (NaDodSO$_4$)*, electrophoretic mobility is independent of charge. NaDodSO$_4$ gel electrophoresis thus functions as a method of molecular sieving and becomes a powerful tool for determining the molecular weights of polypeptides. Let us now examine the basis of this method.

Oligomericproteins (section 2-15) consist of more than one polypeptide chain. NaDodSO$_4$ denatures oligomeric proteins and causes them to be dissociated into their component polypeptide chains. In the presence of NaDodSO$_4$, polypeptide chain-amphiphile complexes are formed in the proportion of approximately 1.4 g of amphiphile to each gram of polypeptide. This ratio is equivalent, approximately, to 1 molecule of amphiphile to 2 amino acid residues. Usually, the amphiphile binds to the polypeptides along their long axes and causes them to behave as if they were rods of constant diameter thus eliminating the effect of shape. Usually, also, polypeptide chain-NaDodSO$_4$ complexes are negatively charged to an equivalent degree. There are two reasons for this equivalence. First, the net negative charge of the amphiphile is considerably greater than the net charge of the protein. Second, the degree to which NaDodSO$_4$ can be bound, and the maximum amount which is bound, are the same for many proteins. Usually, therefore, the electrophoretic mobility of the strong, negatively charged NaDodSO$_4$-polypeptide complexes on NaDodSO$_4$ gels depends primarily on the molecular weights of the polypeptide chains. Frequently, 7 cm gels are prepared in 10 cm long tubes of 6 mm inside diameter and the apparatus and power supply required can be built for a few hundred dollars. To calculate the molecular weights of unknown polypeptides, calibration curves are prepared by plotting the logarithm of molecular weight against electrophoretic mobilities for standard proteins over an appropriate range of molecular weights. The molecular weights of unknown polypeptide chains then can be obtained from interpolation from the calibration curve. Equation (6) shows how one may calculate the mobilities of polypeptides after gel electrophoresis:

$$\text{mobility} = \frac{\text{distance migrated by protein}}{\text{distance migrated by tracking dye}} * \frac{\text{length of gel before staining}}{\text{length of gel after staining}} \cdot (6)$$

Use of this relative instead of absolute method of measuring mobilities makes it possible to compare results obtained by different investigators.

When used under appropriate conditions, NaDodSO$_4$ gel electrophoresis can provide estimates of molecular weights of polypeptides with an accuracy of $\pm$ 10% or better. But despite the power of this technique, it and other one-dimensional polyacrylamide electrophoresis techniques that are carried out in tubes can resolve a maximum of only 100 polypeptides from cellular homogenates. Such homogenates, as we have seen, may contain 3 000 or more proteins in simple cells. This number may rise to as many as 10 000 to 15 000 proteins in more advanced cells. Clearly, therefore, to gain more detailed knowledge of cellular processes, it is desirable to have electrophoretic techniques of sufficient resolving power to detect many more of the proteins in cells than is possible with one-dimensional tube or cylindrical gels. Such a technique is two-dimensional polacrylamide gel electrophoresis of proteins.

4-7.   Two-dimensional gel electrophoresis resolves many proteins

For several years, many investigators were aware that, as is the case with paper chromatography and thin-layer chromatography, a two-dimensional version of gel electrophoresis should separate many more cellular proteins than one-dimensional tube or slab techniques. Success eluded most who tried to develop such a technique until Patrick O'Farrell, in 1975, developed a two-dimensional method based on isoelectric focusing in tube gels in the first dimension followed by NaDodSO$_4$ electrophoresis in slab gels in the second dimension. A crucial element in O'Farrell's success was finding suitable experimental conditions for preparing samples for each electrophoretic stage. These stages are detailed in suggested reading cited at the end of this chapter. A brief description of the method follows.

An isofocusing gel solution (which includes a mixture of ampholytes, see section 2-14) is prepared in glass tubes and allowed to polymerize. The polymerized gels are prerun without sample under conditions of constant current. This prerun focuses the ampholytes which form a stable pH gradient. The sample then is applied and subjected to electrophoresis. Individual proteins move to positions in the gel corresponding to their isoelectric points. At the end of the isofocusing run, the gel is extruded and prepared for $NaDodSO_4$ electrophoresis. The treated, extruded tube gel is attached to the top edge of a slab gel and electrophoresis with $NaDodSO_4$ is carried out in the second direction. The completed slab gel is visualized by staining with dyes or silver ions, or by autoradiography when the proteins are radioactive. As indicated above, thousands of proteins may be separated by the O'Farrell method. But the physical nature of the gels plus the large number of polypeptides embedded in the gels makes identification of individual components difficult. Some software has been written to assist in the quantitation of separated proteins; but the development of standardized procedures and databases which will permit comparisons of gel patterns among different laboratories remains an elusive goal. At present, one of the more effective ways of identifying individual proteins and individual sequences of DNA and RNA fragments separated by gel electrophoretic techniques is through the use of blotting techniques. These are discussed next.

4-8.   Southern blot identifies separated fragments of DNA

The term *blotting* as applied to electrophoretically separated biomolecules, refers to the transfer of the separated biomolecules, or of enzymatically cleaved fragments, to specially treated paper. The transfer from gel to paper may be accomplished either by passive diffusion or by electrophoresis. Not surprisingly, electrophoretic transfer is often more rapid and more efficient than diffusion.

Southern blotting (named after Edwin M. Southern, currently at Oxford University), was introduced to solve the problem that, when DNA, or fragments of DNA, were separated according to size by agarose gel electrophoresis (small fragments migrate more rapidly than large), it was not possible to identify the separated components while they were embedded in the gels. Moreover, then available methods such as slicing the gels and eluting the various components from the slices were not satisfactory. In the method of Southern, varying sizes of DNA fragments, produced by the action of restriction endonucleases, are separated by electrophoresis in agarose gels. When desired, the sizes of the fragments could be estimated from calibration curves of log molecular weight (or log # of kilobases) versus mobility of standards. After the electrophoresis is completed, the gels are denatured in buffered ethidium bromide and the resulting single strands are transferred by blotting to strips of cellulose nitrate paper. Each blot, i.e., each gel transfer, calls for setting up a four-layered sandwich consisting of the following parts, from bottom to top: *first* a large piece of thick filter paper soaked in buffer;

second two pieces of clear plastic are placed on the filter paper with a space between them just large enough to hold the agarose gel; *third* the strip of cellulose nitrate paper is placed on top of the gel with its edges resting on the two pieces of plastic; then *fourth* two strips of filter paper moistened with buffer are placed on the plastic with their edges overlapping the cellulose nitrate by about 5 mm; finally, the strips and the cellulose nitrate are covered with dry filter paper. We show key stages of this arrangement in Figure 4-2.

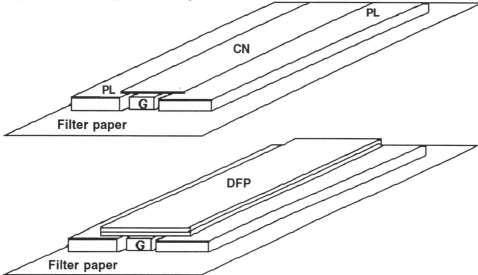

Figure 4-2. Diagram of the *3rd and 4th steps* in setting up a Southern blot:

- *Step 3 (top)*: a strip of cellulose nitrate **CN**, is placed on the gel **G** which is held between two pieces of clear plastic **PL**.
- Then (not shown) moist strips of filter paper are placed on the plastic and allowed to overlap on to the cellulose nitrate paper. Finally,
- *Step 4 (bottom)*: dry filter paper **DFP**, is placed on top of the moist filter paper to cover both the cellulose nitrate and the moist strips.

After the transfer is completed, the cellulose nitrate paper is heated to immobilize the DNA fragments. Later on, the cellulose nitrate paper with DNA fragments is placed in a solution containing radioactive *probes* (single-stranded sequences of deoxyribonucleotides known to be complementary to the immobilized fragments). The labeled probes bind to the immobilized DNA fragments by complementary base-pairing and permit the DNA fragments to be visualized by autoradiography (the detection of radioactivity on x-ray film). Southern's blot technique makes possible the detection of polymorphic forms of a given DNA. The polymorphisms may result from deletions, insertions, or other mutations in that DNA. Restriction cleavage of such modified DNA may produce new size ranges of restriction fragments, and, consequently, changed banding patterns known as *restriction fragment length polymorphisms* (RFLP). An adaptation of Southern's technique permits detection and identification of small fragments not only of DNA but also of RNA. We discuss this adaptation, commonly known as *Northern blotting*, next.

4-9.   Northern blot identifies separated fragments of RNA

Southern's technique does not work with RNA since ribonucleotides do not bind to cellulose nitrate paper. Therefore, in order to apply blotting techniques to the detection and identification of RNA, it is necessary to use specially treated paper to which ribonucleotides bind following transfer from agarose gels. Diazo-benzyloxymethyl (DBM) paper fulfills the requirement for binding ribonucleotides; this binding occurs, apparently, first by ionic linkages and then by the formation of covalent bonds.

It is desirable, before electrophoresis in the agarose gels, to purify the RNA. For the successful transfer of agarose separated bands of RNA from gels to paper, the RNAs should first be denatured. Therefore, after purification, and either before or during electrophoresis, the RNA should be treated with a denaturing agent; methylmercuric hydroxide or glyoxal both have been used successfully as denaturing agents. Gels are prepared for transfer by washing out, or by removing chemically, components which may interfere with subsequent steps on the DBM paper. Transfer is carried out in the manner described by Southern and, after transfer, unreacted diazo groups are removed from the paper. According to the original work of Southern, transfer or blotting is accomplished with a four-layered sandwich. The *first* or bottom layer consists of two sheets of paper soaked with buffer. Then *second,* the gel is placed on top of the sheets of wet filter paper. *Third,* DBM paper saturated with buffer is placed on the gel and *fourth,* three layers of dry filter paper plus dry paper towels and a plexiglass weight are placed on top.

Separated bands of RNA are detected by autoradiography following hybrid-ization of the bands to ^{32}P-labeled probes. This *Northern blotting* to DBM paper is effective in detecting small quantities of RNA in mixtures of unrelated fragments of RNA. The usefulness of the procedure is enhanced by the fact that it also works well with single-stranded (denatured) fragments of DNA less than 100 bases long.

4-10. Western blot detects proteins separated by electrophoresis

Proteins separated by gel electrophoretic methods, like fragments of DNA and RNA, are embedded in the gels. Moreover, the probes that may be used successfully for identifying the separated proteins may not be capable of penetrating the gels to react with the separated proteins. Accordingly, blot techniques also have been applied to the detection and identification of electrophoretically separated proteins. Both one- and two-dimensional techniques may be used for the separation of proteins into the individual polypeptides to be subjected to blotting techniques. Diazophenylthio, or DPT paper, and diazobenzyloxymethyl or DBM paper, in addition to cellulose nitrate paper have all been used for blotting applied to proteins, a procedure which has come to be known as *Western blotting*.

As with the Southern blotting of DNA and Northern blotting of RNA, Western blotting of electrophoretically separated polypeptides onto specially treated paper retains the resolution with which the original polypeptides were separated. With separated polypeptides, also, a sandwich type arrangement is used for the transfer of separated polypeptides from gel to paper. Once the transfer has taken place, it is necessary to identify, and sometimes isolate, an individual protein, or individual proteins, on the paper. Identification is accomplished with probes which react specifically with individual polypeptides. Antibodies, which are discussed in Chapter 6, and which may be labeled either with 125I, or with fluorophores (defined in Chapter 7), have been used as probes for such identification. A useful feature of this approach is that once a polypeptide has been blotted onto a suitable type of paper, the identity of the polypeptide may be checked more than once with different probes or with the same probe; it is necessary only to remove the previous probe before testing with the new one. The Western blotting of proteins onto appropriate paper complements nicely investigations into ligand binding (Chapter 5) and the separation and quantitative assay of individual antibodies from serum (Chapter 6).

4-11.   DNA sequences can be multiplied ten million times by the PCR

Transport processes have played a big role not only in the identification of nucleic acid fragments during sequencing but, also, in the identification of products formed during nucleic acid synthesis. In this section, we discuss one method of DNA synthesis, the *polymerase chain reaction* (PCR) developed at the Cetus Corporation by Kary Mullis and colleagues. The PCR technique can be used to enrich specific sequences of DNA to an extent which makes feasible a large number of analytical, clinical, forensic and research activities which previously could not be attempted. During studies with PCR, agarose gel electrophoresis and other transport tools are used to monitor the results of amplification (synthesis).

The *polymerase chain reaction* (PCR) is a method of enzymatic synthesis used for the *in vitro* multiplication of a nucleic acid sequence of usually less than 3000 bp. The polymerase chain reaction also may be used to modify, or to add new sequences to, a previously multiplied (amplified) sequence. The sequence to be amplified (the *target sequence*) may be a segment of a gene or it may be a cloned nucleic acid sequence carried by a bacteriophage or other vector. Therefore, the target sequence does not have to be present in pure form. It is necessary, however, that the base sequences at opposite ends and on opposite strands of the target sequence be known. Knowing the terminal sequences is necessary to permit synthesis of complementary oligonucleotides of approximately 20 bp which will serve as primers. Because of their complementarity, these primers hybridize to opposite ends, and on opposite strands of, the target sequence. When the PCR was first developed, the enzyme used was the Klenow fragment of *Escherichia coli* DNA polymerase 1 (see section 4-3). Currently, however, the enzyme of choice is a highly thermostable DNA polymerase from the thermophilic bacterium *Thermus aquaticus (Taq)*. With the introduction of *Taq* polymerase, the PCR is now capable of multiplying DNA sequences by a factor of 10^7 with high specificity.

Three sequential steps (Figs. 4-3 a and 4-3 b), repeated many times, are used to carry out the polymerase chain reaction: *denaturation* of double-stranded DNA, *annealing* of the oligonucleotide primers, and *extension* of the annealed primers attached to the target sequence. Each set of three repetitive steps is referred to as a *reaction cycle*.

In the first step, *denaturation by heating*, the double-stranded DNA is separated into two single strands. This separation is performed in a large molar excess of the two primer oligonucleotides and the four deoxynucleoside triphosphates.

In the second step, *annealing of the two oligonucleotide primers to opposite ends of the target sequence,* the mixture from step 1 is cooled and each of the two primers is allowed to hybridize with its complementary sequence. The two primer oligonucleotides hybridize to opposite ends of the separated strands of DNA so that the 3'-ends of the primers point toward each other, permitting synthesis (extension) to take place across the segment between the primers (Figure 4-2 a).

In the final step, *extension of the annealed oligonucleotide primers*, the *Taq* polymerase catalyzes the extension of each primer in such a manner that the extension product of one strand can serve as a template for the synthesis of the other strand. Moreover, in successive cycles, the original strands are separated from the extension products, and each separated strand then serves as a template for a new cycle of extension. Consequently, the quantity of template DNA is doubled in each cycle, and thirty cycles can amplify a small amount of target DNA $2^{30} = 1,073,741,824$ times (i.e., more than a billion times)!

After extension of the annealed primers has been completed, the products of the reaction are heat-denatured and the three-step process is repeated until the desired degree of amplification has been achieved.

The desired amplified product of the polymerase chain reaction corresponding to the target sequence is a double-stranded DNA with four defined 5'-ends; that is, the amplified product is a discrete dsDNA with four 5'-ends corresponding to the 5'-ends of the two oligonucleotides used as primers. Along with the specific double-stranded DNA with four defined 5'-ends corresponding to the target sequence, various intermediate sequences, and some sequence artifacts, are produced in the polymerase chain reaction. Specifically, during the first, and each subsequent reaction cycle, as each oligonucleotide primer is extended on the template, an intermediate, single-stranded DNA molecule of undefined length is formed. That is, during extension, a single-stranded DNA is formed in which the end corresponding to the sequence without annealed primer is undefind. Both the undefined *"long products"* and the defined and amplified target sequence can act as templates for the appropriate oligonucleotide primer during subsequent reaction cycles. Therefore, the long products accumulate linearly in proportion to the number of reaction cycles. Thus, there are 2 long products after the 1st cycle; then 4, 8, 16, 32, 64, 128, 256, 512 and 1024 long products after 2, 3, 4, 5, 6, 7, 8, 9, and 10 cycles ... and so on. The first defined products are found after the 3rd cycle.

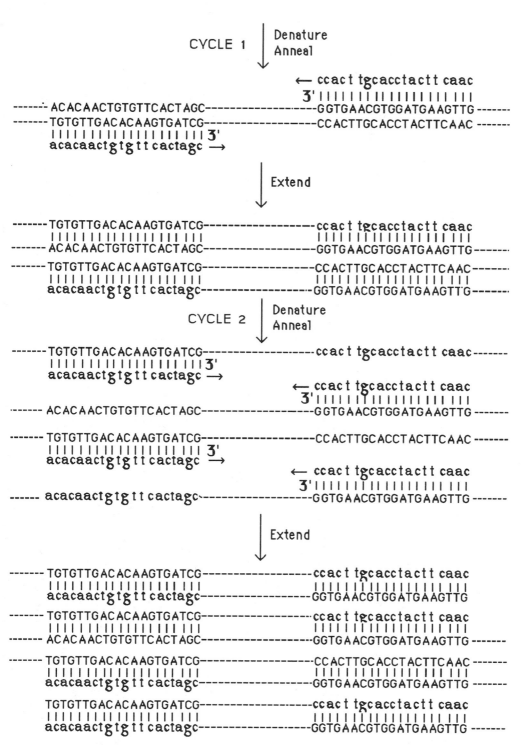

Fig. 4-3 (a). First two cycles of a PCR reaction (after Mullis and Faloona 1987).

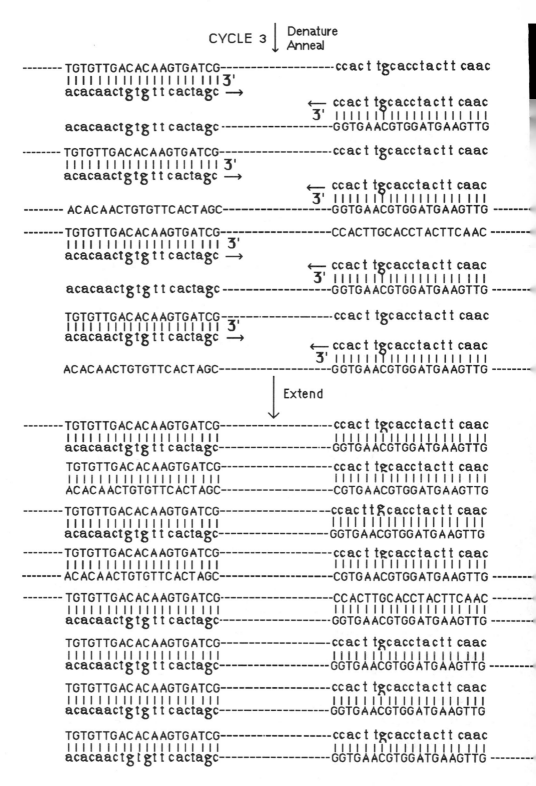

Fig. 4-3 (b). Third cycle of a PCR reaction (after Mullis and Faloona, 1987).

Since "substrate" of the *Taq* polymerase (the original DNA plus long products) increases with each reaction cycle, there comes a point during synthesis at which the polymerase is completely saturated with "substrate" and the enzyme reaction cannot proceed any faster (discussed in section 5-3). When this point is reached, the activity of the polymerase becomes limiting and the reaction rate begins to form a "plateau." Products of the polymerase chain reaction are analyzed by appropriate electrophoretic procedures; that is, as we mentioned at the start of this section, transport methods are used in evaluating PCR experiments.

4-12. Centrifuges are classified on the basis of maximum rotor speeds

In the biomedical sciences, centrifugal force is often expressed as the *relative centrifugal force,* abbreviated RCF. RCF is the acceleration of a centrifugal field relative to the acceleration of gravity. RCF therefore is expressed as acceleration of the centrifugal field times gravity of the earth's field or, in shorthand notation, RCF * g (e.g. 5 000 * g). Mathematically, the relative centrifugal force is defined in terms of the speed of rotation:

$$RCF = \omega 2r/g, \tag{7}$$

where ω = angular velocity in radians sec-1 = 2π * rev.sec-1,
  r = distance of the particle from the axis of rotation (cm or in.),
and g = a constant for the acceleration of gravity, (980.7 cm sec-2).

Equation (7) may be reduced to

$$RCF = 11.18 \, (r) \left[\frac{N}{1000} \right]^2, \tag{8}$$

where N = rev./min
and r = the distance (centimeters) from the axis of rotation.

In equation (8), velocity is expressed in rev./min because it usually is inconvenient to measure velocity in radians sec-1. In reverse, equation (8) may be used to calculate rev./min from RCF given r; that is, rev./min = 1000 * $\sqrt{(RCF/(11.18r))}$ when r is the radius in cm.

During centrifugation (which also is a transport process), a centrifugal force is applied to particles in a biologically appropriate suspending medium. Centrifugal force causes the suspended particles to be transported outward from the axis (center) of rotation. In general, centrifuges used in the biomedical laboratory may be classified into four types: *desk-top centrifuges, high-speed centrifuges, preparative ultracentrifuges,* and *analytical ultracentrifuges.* Depending on the rotors used, modern *desk-top centrifuges* may attain relative centrifugal forces of 5 000 * g; *high-speed centrifuges* may attain forces of 50 000 * g; *preparative ultracentrifuges* may attain forces of 100 000 * g, and *analytical ultracentrifuges* speeds of 500 000 * g. In sections 4-13 to 4-18, we will be concerned first with analytical and then with preparative ultracentrifuges; but because ultracentrifugation frequently calls for subcellular fractions (or preparations of macromolecules) obtained by differential centrifugation, we will discuss, also, precautions which should be followed during differential centrifugation with high-speed centrifuges.

4-13.   Analytical centrifuges measure physical properties very accurately

The analytical ultracentrifuge measures, very accurately, such properties as the shape, size, density and molecular weight of macromolecules. The first analytical ultracentrifuge was built by Svedberg and colleagues in 1925. Two major types of measurement are possible with such instruments: (1) sedimentation velocity, in which the velocity of sedimentation is studied as a function of centrifugal force, and (2) sedimentation equilibrium, in which centrifugal force is just sufficiently large to balance the opposing tendency of the molecules to diffuse from a region of high concentration to one of low concentration. *Sedimentation velocity* is used for determinations of molecular shape and sedimentation coefficients; *sedimentation equilibrium* is used for measuring molecular weight and related properties such as molecular associations, states of aggregation, interactions among macromolecules in solution, and assembly from subunits. Measurements of molecular size, shape, associations, and interactions are important for quantitative work such as constructing mathematical models of molecular processes, predicting the biological-effects of drugs and other molecules, and studying stereospecific, binding interactions.

4-14. Sedimentation velocity yields the coefficient *s* and molecular wt.

Sedimentation velocity may be expressed in terms of the *sedimentation coefficient, s*, which is equal to velocity divided by centrifugal force per unit of mass

i.e.,   s $= \dfrac{dx/dt}{\omega^2 x}$,   (9)

where   x   = the distance from the center of rotation,
        dx/dt   = velocity (rate at which the solute is sedimented),
        ω = velocity of rotor in radians s^{-1},
and   s   = sedimentation coefficient (with the dimensions of time).

Since, by integration

dx/x   = d lnx, equation (9) may be arranged to yield

$$\frac{d\,\ln x}{dt} = \omega^2 s, \tag{10}$$

$$d\,\ln x = \omega^2 s\,dt, \tag{11}$$

and $\ln x = \omega^2 s\,(t). \tag{12}$

Therefore, according to equation (12), s may be obtained from linear least-squares analyses or from plots of $\ln x$ against time. The slope of the curve will be

$$\text{slope} = \omega^2 s, \tag{13}$$

$$s = \text{slope}/\omega^2. \tag{14}$$

By convention, determinations of sedimentation coefficients (an important use of sedimentation velocity) are corrected to the values which they would have in water at 20°C. Such corrected values are designated $s_{20,w}$.

The value of the sedimentation coefficient, s, depends both on molecular weight and on molecular shape. Therefore s alone cannot be used to obtain the molecular weight of a macromolecule (another important use of sedimentation velocity experiments). If however, the diffusion coefficient and the density of the solvent are both known, the molecular weight can be calculated from:

$$M = \frac{RTs}{D(1-\upsilon r)}, \tag{15}$$

where M = molecular weight; R, the gas constant = $8.3144 * 10^7$ erg K^{-1} mol^{-1}; T = temperature °K; s = the sedimentation coefficient; D = the diffusion coefficient; υ = the partial specific volume and ρ = the density of the solvent. *The partial specific volume*, υ, is the change in volume which occurs when 1.0 g of a solute is added to a large excess of solvent; for proteins υ usually falls between 0.70 and 0.75 cm^3 g^{-1}.

4-15.   Sedimentation equilibrium has a thermodynamic basis

In sedimentation equilibrium, the rotor of the analytical centrifuge is run at a lower speed than in sedimentation velocity. At equilibrium, the tendency of a macromolecule to diffuse to the top of the ultracentrifuge cell will be counterbalanced by the centrifugal force; and because such counterbalancing of diffusion and centrifugal force is an equilibrium process, molecular weight can be derived from thermodynamic considerations.

Before the start of centrifugation, the macromolecule to be studied is placed in solution and thus is randomly distributed in the centrifugal cell. As the centrifuge rotor is operated at a speed less than that required to sediment the macromolecule, molecules of the sample are transported toward the bottom of the centrifuge cell. The result is that concentration of the macromolecule increases exponentially toward the bottom of the cell and decreases toward the top. In other words, a concentration gradient is formed from top to bottom but, sedimentation does not occur. Because a concentration gradient is formed from the bottom of the centrifuge cell, a diffusional (i.e., randomizing) force is developed in the opposite direction (toward the top of the cell). That is, there is a tendency for molecules of the sample to move from a region of high chemical potential to one of low chemical potential. Equilibrium is attained when the two forces counterbalance each other exactly. At equilibrium, net transport (or movement) of solutes will not occur, and the solutes will exist in a concentration gradient which *increases* from top to bottom. Since, at equilibrium, concentrations C_1 and C_2 of the macromolecule differ at two points x_1 and x_2 from the center of rotation, there is a difference in free energy ($\Delta G_{centrifugation}$) between x_1 and x_2 caused by the centrifugal field. At equilibrium:

$$\Delta G_{concentration} + \Delta G_{centrifugation} \quad = \quad 0 \qquad (16)$$

When a correction term is applied to the centrifugal force for buoyancy of the macromolecule in solution, the molecular weight may be calculated from equation (17) if solute concentration is measured as a function of distance from the rotor:

$$M \quad = \quad \frac{2RT \ln(c_2/c_1)}{(1-\upsilon\rho)\omega^2 (x_2^2 - x_2^1)}, \qquad (17)$$

where M is the molecular weight of the macromolecule, c_1 is the concentration at a distance x_1 from the center of rotation; c_2 is the concentration at a distance x_2 from the center of rotation, and where R, T, υ, ρ, and ω are defined as they were in equation (15). Molecular weights determined by sedimentation equilibrium are very accurate. But this accuracy should be considered in relation to the lower cost of the other transport methods discussed in this chapter. Not only is a modern, computerized, optically and electronically stable, analytical ultracentrifuge such as the *Beckman Optima XL-A* costly (price ca. $100,000) but, achieving equilibrium may take from a few hours to several days depending on the lengths of the centrifuge cells and the sizes of the molecules. Nevertheless, if one wishes to perform rigorous, quantitative studies of macromolecular interactions at chemical equilibrium, use of the analytical ultracentrifuge may be indispensable.

4-16. Preparative centrifuges can separate and purify macromolecules

Although chromatographic, electrophoretic and other methods of separation are important tools in the biomedical laboratory, they have complemented, but have not replaced, centrifugation in the armamentarium of bioanalytical methods used to separate, purify and characterize biomolecules. The distinction between analytical and preparative ultracentrifuges is based not only on the relative centrifugal forces attainable (section 4-12), but also on the basis of the sizes of samples that may be handled. Thus, sample sizes used with analytical ultracentrifuges generally are less than 1 ml but sample sizes used with preparative ultracentrifuges may range from a total volume of 5 ml to a volume of 2 l. The continuing popularity of centrifugal methods is due to their wide applicability and effectiveness. Separation of cellular components by centrifugation depends on one or both of two properties: on differing rates of sedimentation and on differing buoyant densities of biomolecules in density gradient, suspending media. In one of the most widespread uses of preparative ultracentrifuges, subcellular fractions, previously subjected to differential centrifugation (section 4-18) in high-speed refrigerated centrifuges, may be fractionated further to obtain such components as microsomes (fragments of the endoplasmic reticulum) or, alternatively, soluble cytoplasmic components. In addition, preparative ultracentrifuges may be used for the purification of cell membranes, the Golgi apparatus, peroxisomes, lysosomes, and other subcellular fractions. In other important applications, the preparative ultracentrifuge may be used to characterize macromolecules on the basis of their sedimentation rates, sedimentation coefficients, molecular weights, buoyant densities, and on the basis of interactions with other components during centrifugation. All of these applications may call for the use of density gradient, suspending media.

Four types of rotors are used with preparative ultracentrifuges: swinging-bucket, fixed-angle, vertical and zonal rotors. *Swinging-bucket or swing-out* rotors provide good separations in rate-zonal centrifugation; *fixed-angle rotors* provide good pellets of subcellular fractions and good isopycnic separations; *vertical rotors* provide good isopycnic and rate-zonal separations and zonal rotors provide good rate-zonal separations.

Rate-zonal centrifugation of biomolecules is conducted on a density gradient medium of density less than the densities of the biomolecules. Centrifugation is stopped after the biomolecules are separated but before the biomolecules are pelleted. The separated biomolecules therefore form discrete bands or zones, each zone representing a characteristic rate of sedimentation. The separated zones are obtained by fractionation of the gradient.

Isopycnic separations are those in which components are separated on the basis of density. Such separations also require the use of density gradients but here, the objective is to have the components form bands at their isopycnic positions whereas, in rate-zonal centrifugation, the objective is to maintain the cellular components in narrow, discrete bands as they are being separated.

Preparative ultracentrifuges may be used to estimate the sedimentation coefficients of macromolecules; but results are less accurate than those obtained with the analytical ultracentrifuge. In the preparative ultracentrifuge, the procedures used for estimating sedimentation coefficients vary with the type of rotor being used; in section 4-17, we will discuss one of these procedures, the use of isokinetic gradients in swinging-bucket rotors to estimate sedimentation coefficients.

4-17. Isokinetic gradients may be run in swinging bucket rotors

Isokinetic density gradient centrifugation was first described in 1967 by the American biochemist Hans Noll who derived the theory and provided supporting experimental evidence for the validity of the method. An *isokinetic gradient* is one in which a linear increase in centrifugal force, acting on a biomolecule at increasing distance from the center of rotation, causes the molecule to sediment at a constant rate. Sedimentation occurs at constant velocity because any increase in centrifugal force is balanced by a corresponding increase in density and viscosity of the medium.

There are several advantages to centrifugation in isokinetic gradients, among which are the following: (a) the sedimentation coefficient of a biomolecule in an isokinetic gradient is proportional to the distance moved by the molecule; (b) in isokinetic gradients, sedimentation coefficients also are proportional to the rate of sedimentation; (c) isokinetic gradients yield improved resolution compared with gradients which are not isokinetic; and (d) estimation of sedimentation coefficients can be made comparatively easily by direct calculation or by comparison with standards. Two sets of calculations are involved in working with isokinetic gradients: calculation of the gradient and calculation of the sedimentation coefficient. We will present the calculation of isokinetic sucrose gradients first and, for convenience, will divide the calculation into four steps labeled *density, viscosity, radial distance* and *volume*.

A.   Calculation of the Gradient

1.   Density: The density of the sucrose gradient is calculated as a function of temperature, and of chosen concentrations of sucrose, using a polynomial equation derived by Barber (1966):

$$\rho_{T,M} = (B_1 + B_2T + B_3T^2) + (B_4 + B_5T + B_6T^2)Y + (B_7 + B_8T + B_9T^2)Y$$

where $\rho_{T,M}$ = density of the sucrose solution, g cm^{-3}
  B = empirically determined constants
  T = temperature, °C and
  Y = weight fraction of sucrose
    = wt. %/100 = [g sucrose/(g sucrose + g water)].

2.   Viscosity: The viscosity of the sucrose gradient solution is also calculated as a function of temperature and of chosen concentrations of sucrose, using the appropriate polynomial equation from Barber (1966):

$$\eta_{T,M} = 10^{(A + B/T + C)},$$

where $\eta_{T,M}$ = viscosity of sucrose solution at a given temperature, °C,
  A = a polynomial defined by Barber,
  B = another polynomial defined by Barber,
  C = $G_1 - G_2 [1 + (y/G_3)^2]^{1/2}$;
where G_1, G_2 and G_3 are constants given by Barber, and

$$y = \frac{Y/M_1}{Y/M_1 + (1-Y)/M_2} ,$$

where
- y = mole fraction of sucrose,
- Y = weight fraction of sucrose (defined in step 1, above),
- M_1 = molecular weight of sucrose, and
- M_2 = molecular weight of water.

3. **Radial distance:** The radial distance r, of a given concentration of sucrose from the center of rotation in an isokinetic gradient, is calculated by substituting the density and viscosity values from steps 1 and 2 into the equation:

$$\eta_{(m) r}/\eta_t = r(D_p - D_{m(r)})/r \, _t(D_p - D_t)$$

so that
$$r = \frac{\eta_r \times r_t (D_p - D_t)}{\eta_t \times (D_p - D_r)}$$

where
- $\eta_{(m) r}$ = the viscosity of the sucrose medium at distance r,
- η_t = the viscosity of the sucrose at top of the gradient,
- $D_{(m) r}$ and D_t = are corresponding densities, and,
- D_p = density of the particle, g cm^{-3}.

4. **Volume:** Radial distance is converted to volume of a cylinder, the centrifuge tube:

volume $= \pi r^2 *$ radial distance,

where r $=$ radius of the centrifuge tube.

B.   Calculation of Sedimentation coefficient:

The Svedberg coefficient, S of the sedimented particle, is calculated from the distance the particle has traveled in the isokinetic gradient. The relevant equation is:

$$S_{20,w} = (R/t) (A/K)\omega^2$$

where
- R = distance traveled by the particle, cm
- t = time, sec
- A = $(D_p - D_{20,w})/\eta_{20,w}$
- K = $(D_p - D_t)r_t/\eta_t$ and
- ω = angular velocity, rad s^{-1}.

In addition to the isokinetic technique, sedimentation coefficients may also be determined in linear gradients from precomputed tables (McEwen, 1966), or from use of a mathematical function derived by Young and Krumlauf (1981) and described by Rickwood (1984).

4-18.   Differential centrifugation may yield functional organelles

Differential centrifugation is a procedure in which a homogenate is subjected to successive centrifugations, at increasing speeds, for defined periods of time in order to obtain subcellular fractions for further study. All operations should be carried out in the cold (i.e., at approximately 4°C) to preserve biological activity. Differential centrifugation causes heavier particles to be pelleted or sedimented, leaving lighter particles behind in the supernatant fractions. Schemes for the isolation of cell walls, chloroplasts, nuclei, mitochondria, lysosomes, microsomes, ribosomes, polysomes, and other fractions are given in many textbooks of biochemistry and will not be repeated here. Instead, we will outline the sequence of steps involved in differential centrifugation and we will indicate some precautions which should be taken to ensure proper use and interpretation of the results.

First, a cellular homogenate is prepared, taking into account the fact that biological activity depends on such factors as the proper pH, osmolarity, temperature, and ionic strength. *Then,* the homogenate is centrifuged, successively, at increasing speeds that are likely to sediment the desired fractions. *Next,* the desired fractions are collected, identified by testing *in vitro* for the presence of marker enzymes, and then checked for the presence of contaminating cellular components. *Finally,* the fraction selected as having the right marker enzymes is tested for the ability to carry out reactions or activities associated with a given process.

The six points which follow should be borne in mind in order to obtain reliable results with fractions from differential centrifugation:

• The size, shape and morphological characteristics of organelles and subcellular fractions are not exactly the same in all cells.
• Consequently, a given organelle or subcellular component is not always pelleted at the same centrifugal speed.
• The morphology of organelles or subcellular fractions *in vitro* may be different from that *in vivo*.
• The absence of an enzyme from a fraction *in vitro* does not mean that the same enzyme is absent *in vivo* since the enzyme may have been lost from the subcellular fraction during fractionation.
• The presence of an enzyme activity *in vitro* does not mean necessarily that the activity is associated with that fraction *in vivo*.
• And, finally, fractions obtained by differential centrifugation usually consist of mixtures of subcellular components. These fractions do not contain pure, single subcellular fractions.

4-19.   Electrophoresis in solution is analogous to sedimentation velocity

Electrophoresis is the transport of electrically charged particles through a buffered electrolyte under the influence of an applied electric field. In a general way, moving boundary, or free-zone electrophoresis (section 2-14) is analogous to sedimentation velocity (section 4-14). In both processes, a force is applied to molecules in solution and the force is opposed by the viscosity of the solution. Moreover, the *electrophoretic mobility of transported biomolecules is the counter-part of the sedimentation coefficient* since both are measurements of velocity per unit field under defined conditions. Thus, if a solution containing more than one biomolecule with different sedimentation coefficients were subjected to sedimentation velocity, their rates of sedimentation would depend, in part, on the magnitude of the differences in their sedimentation coefficients. And, if a solution of biomolecules with different electrophoretic mobilities were subjected to moving boundary electrophoresis, they too would migrate at rates which depend partly on differences in their electrophoretic mobilities. Other factors influencing both sedimentation rate and electrophoretic mobility are the sizes and shapes of the biomolecules. In moving boundary electrophoresis the size of the charges carried by the biomolecules, and the amount of current used, also influence mobility.

Until recently, free-zone electrophoresis (section 2-14) had fallen into disuse because moving boundary electrophoresis was the only free-zone method available; but sedimentation velocity in the ultracentrifuge continued to be popular. Among the disadvantages of moving boundary electrophoresis are its slowness, low sensitivity, and poor resolution, caused by Joule heating, i.e., the heating of the conducting medium by the applied electric current. Joule heating leads to convection currents in the buffered electrolyte because the interior of the solution (around the electrodes) becomes hotter than the solution in the vicinity of the walls of the reaction vessel. Hence, convection currents are formed and these currents disrupt the separation process. The slowness, low sensitivity, and peak broadening which hurt moving boundary electrophoresis are solved in capillary zone electrophoresis, (CZE).

4-20. CZE separates biomolecules with great speed and sensitivity

Capillary zone electrophoresis, also known as *capillary electrophoresis* (CE) and *high-performance capillary electrophoresis* (HPCE), is a free-zone method of electrophoresis (cf. section 2-14). CE is carried out in the microchannel of capillary tubes of 10 to 100 cm lengths and inside diameters of 20 to 200 μm. Therefore, the tubes used for CZE have a very large surface-to-volume ratio. This large ratio improves the dissipation of heat to such an extent that good separations may be obtained in uncoated and unpacked capillaries without placing the apparatus in a thermostat and without using refrigeration or other means of cooling. In a manner reminiscent of the differences in effectiveness between HPLC and traditional liquid chromatography, CZE improves resolution over that in moving boundary electrophoresis by lowering the dispersion of separated bands and by lowering the rates of diffusion of solutes from regions of high to regions of low concentration. Moreover, CZE is rapid and sensitive: separations of amino acids, peptides, or nucleotides may take 30 minutes, and samples detected can be as small as 30 pl. But while proteins, and fragments of DNA up to 700 kilobases also can be separated effectively, contamination of the columns from adsorption of these macromolecules is a problem.

An electrokinetic phenomenon of some importance in CZE is electroösmosis also referred to as electroendosmosis. This process was first described in 1809 by F. Reuss and somewhat later by R. Porret in 1816. Columns used in CZE are made of fused silica. Consequently, silanol groups (-SiOH) on the surface of the column ionize to SiO⁻ at the interface between the buffered electrolyte and the wall of the column. Then, when the voltage is applied, the electrolyte, (which has charges opposite to those on the wall), flows toward the cathode. *This flow of solvent is electroösmosis.* The flow is relatively weak in gels but quite strong in CZE - on the order of 1 mm sec^{-1}. The velocity profile of electroösmotic flow along the length of the capillary (an almost flat "plug flow") results in very sharp sample peaks. Therefore, some investigators and manufacturers exploit the occurrence of electroösmosis and use it both as a pump to "inject" samples and to carry migrating solutes toward the cathode. Conversely, other investigators coat the capillary tubes both to reduce electroösmosis and to avoid interactions between biomolecules and the fused silica of the walls in CZE. When the tubes are coated (usually with uncharged polymers), techniques such as electrofocusing, which depend solely on electrophoretic mobility, can be carried out more successfully than in untreated capillaries.

4-21. Basic knowledge may be used to improve CZE separations

In CZE, as in chromatography, understanding the experimental implications of available knowledge will permit an investigator to modify an existing procedure to obtain a desired type of separation. Toward this end, we have mentioned above that minimizing Joule heating contributes to the impressive separations obtained in CZE. Other points that should be considered are

• Electroösmotic mobility varies inversely with viscosity but varies directly with dielectric constant and *zeta potential* (the potential drop between the fixed charges of a biomolecule and the buffered electrolyte in which it is placed).
• Zeta potential may be changed by changing the composition and pH of a buffered electrolyte (cf. section 2-14).
• The separation of charged molecules depends both on differences in the charges and in the frictional resistance of the molecules.
• Positively charged components emerge before negatively charged ones which migrate more quickly or slowly than the electroösmotic flow depending on the sizes of their charges. Neutral molecules migrate as a group with the electroösmotic flow.
• Predictable changes in the ionization and mobilities of the components being separated may be made by changing the pH of the buffered electrolyte.
• Since free-zone electrophoresis takes place in one phase, methods of improving resolution and selectivity are more limited in CZE than in chromatography where there are two phases.
• Micelles (e.g. sodium dodecyl sulfate) may be used to create a pseudophase in which neutral molecules may be separated by partitioning between a polar electrolyte solution and less polar micelles.
• Packings used in gel electrophoresis and gel filtration may enhance resolution and sensitivity in CZE.

4-22. CZE may begin a new era in bioanalytical chemistry

It has been estimated that CZE can yield 2 000 to 2 million theoretical plates (section 3-4). And CZE separations having 1 million theoretical plates have been reported. For comparison, gas-liquid chromatography can yield an impressive, but still relatively low 2 000 to 200 000 theoretical plates. Also impressive is that high-performance liquid chromatography can separate nonvolatile compounds while approaching the efficiency of gas-liquid chromatography. But, HPLC calls frequently for the use of nonaqueous solvents, whereas most biochemical reactions take place in an aqueous environment. Moreover, CZE may be carried out with sample volumes in the picoliter range, and may even be capable of separating and detecting the complex components of single cells. These volumes and quantities are much smaller than can be separated and detected by other methods. Additional advantages of CZE are that it can be automated to perform multiple analyses, that it gives reproducible results, and that it is less demanding of time and manual work than other electrophoretic methods currently in use. In spite of its promise, CZE needs improvement in important areas of the technique. These areas include injection, separation, and detection.

4-23. DNA sequencing yields too much information for manual analysis

The DNA of the bacteriophage φX-174 has a molecular weight of $3.4 * 10^6$ and consists of 5 386 bases. Compared with the sizes of other DNA molecules, the DNA of φX-174 is small. Clearly therefore, determinations of the nucleotide sequence of any DNA molecule can yield an enormous volume of information. And analyzing information from the mapping or sequencing of all but the smallest polynucleotide fragments is virtually impossible using manual methods.

Fortunately, computers are well suited for storing, analyzing and comparing data from the sequencing of nucleic acids and proteins. As a result, the biomedical scientist who has access to a sufficiently powerful computer (and to the increasingly sophisticated, privately and commercially available, software packages on sequence-computing) can perform several otherwise tedious tasks. These include

- entering and editing base sequences;
- searching DNA molecules for homologous sequences;
- searching for promoter sites, polyadenylation, splice sites, palindromes, recognition sites of restriction endonucleases, and other features;
- detecting secondary structures in RNA;
- aligning sequences of DNA and proteins;
- translating sequences of DNA into coding regions; into the amino acid sequence of proteins or, in reverse, translating the amino acid sequence of proteins into that of DNA; and
- designing, analyzing and simulating PCR experiments.

Currently, large databases of nucleotide sequences such as GenBank®, a registered trademark of the U.S. Department of Health and Human Services, and the EMBL Data Library from the European Molecular Biology Laboratory are available. These databases may be searched by investigators for comparisons. In addition, a large database on the amino acid sequences of proteins, the Protein Identification Resource (PIR) is maintained by the National Biomedical Research Foundation (NBRF) and is also available for making computerized searches.

Other databases are available on topics such as immunoglobulins, restriction enzymes, AIDS, carbohydrates, enzymes, kinetics, and metabolic pathways. We cite a few references to these databases at the end of this chapter.

4-24. Worked examples

Example 4-1. Nucleotide fragments of DNA of known sizes were subjected to electrophoresis on agarose gels with the following results:

Mobility, mm	Number of Base Pairs
20	9950
21	8595
22	7425
23	6410
24	5540
25	4785
26	4135
27	3571
28	3084
29	2667
30	2302

What were the sizes of unknown bands of mobilities 23.5 and 25.5 mm run under the same conditions as the nucleotides of known sizes?

Solution:

The calibration curve below shows that nucleotides of mobilities 23.5 and 25.5 mm correspond to sizes of 6 000 and 4 500 base pairs respectively.

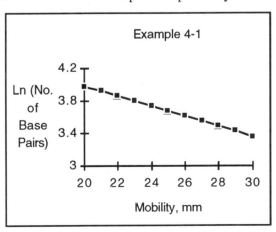

Example 4-2. Four standard proteins were subjected to gel filtration on sephadex G-200 with the following results.

V_e/V_0	0.52	0.74	0.95	1.15
Mol wt	500 000	350 000	250 000	180 000

What is the molecular weight of a protein which had a V_e/V_0 value of 0.85?

Solution:

From the standard curve below, a V_e/V_0 data point of 0.85 corresponds to a value of log 5.46.
Antilog 5.46 = 288 403; ∴ Mol wt of protein = 288 400 g mol^{-1}.

Example 4-3. Sodium dodecyl sulfate polyacrylamide gel electrophoresis of seven standard proteins gave the following results:

Protein	Mobility	Molecular weight
Thyroglobulin	0.02	165 000
Phosphorylase a	0.20	100 000
Bovine serum albumin	0.34	68 000
Ovalbumin	0.50	43 000
Trypsin	0.72	23 300
Myoglobin	0.83	17 200
Cytochrome c	0.97	11 700

Estimate the molecular weight of a polypeptide which had a mobility of 0.47 when it was run under the same conditions as the standard proteins.

Solution:

From the curves below a mobility of 0.47 ≡ to a molecular wt of 45 000 g mol^{-1}.

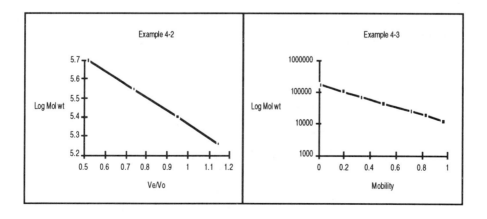

Example 4-4. Convert 48 000 rev./min to radians/sec.

From equation (7):

$$\omega = 2\pi * \text{rev./sec,}$$

$$48\ 000\ \text{rev min}^{-1} = 2\pi * \frac{48\ 000\ \text{rev min}^{-1}}{60\ \text{sec min}^{-1}},$$

$$= 5026\ \text{radians sec}^{-1}.$$

Example 4-5. In an ultracentrifuge run at 48 000 rev min^{-1}, the boundary of a macromolecule moved from 5.65 to 6.40 cm in 190 min. What is the sedimentation coefficient of the macromolecule?

From equation (10):

$$\frac{d\ \ln x}{dt} = \omega^2 s,$$

$$\frac{\Delta\ (\ln x)}{\Delta t} = \omega^2 s,$$

$$s = \frac{\Delta(\ln x)}{\omega^2 \Delta t},$$

$$s = \frac{\ln\ (6.40/5.65)}{\left[2\pi * \dfrac{48\ 000\ \text{rev min}^{-1}}{60\ \text{sec min}^{-1}} \right]^2 * 190\ \text{min} * 60\ \text{sec min}^{-1}},$$

$$s = 4.33 * 10^{-13}\ \text{sec.}$$

Example 4-6. How long would it take the boundary of a protein of sedimentation coefficient 4.33 s to be moved from 5.65 to 6.40 cm in an ultracentrifuge which is run at 48 000 rev min^{-1}?

From equation (10):

$$\frac{d\ \ln x}{dt} = \omega^2 s,$$

$$\Delta t = \frac{\Delta\ \ln x}{\omega^2 s},$$

$$s = \frac{\ln\ (6.40/5.65)}{\left[2\pi * \dfrac{48\ 000\ \text{rev. min}^{-1}}{60\ \text{sec min}^{-1}} \right]^2 * 4.33 * 10^{-13}\ \text{sec}},$$

$$= 11\ 393\ \text{sec,}$$

$$= 190\ \text{min.}$$

Example 4-7. A sephadex G-25 column had the following characteristics: volume of packed gel V_p = 60 ml; the inner volume, V_i = 35 ml and the void volume, V_o = 55 ml. Calculate the partition coefficients of three biomolecules with elution volumes, V_e , of 65, 88 and 129 ml.

From equation (4):

$$K = \frac{(V_e - V_o)}{(V_t - V_o)},$$

$$K = \frac{(65 - 55) \text{ ml}}{(150 - 55) \text{ ml}} = 0.11.$$

$$K = \frac{(88 - 55) \text{ ml}}{(150 - 55) \text{ ml}} = 0.35.$$

$$K = \frac{(129 - 55) \text{ ml}}{(150 - 55) \text{ ml}} = 0.78.$$

Example 4-8. Subjecting a protein to sedimentation-velocity at 20°C in an analytical ultracentrifuge gave the following results: sedimentation coefficient, s = 14.16 * 10^{-13} sec; diffusion coefficient, D = 3.4 * 10^{-7} cm^2 sec^{-1}; partial specific volume v = 0.75 ml g^{-1}; density of solvent, ρ = 1.0 g ml^{-1} and R = 8.3144 * 10^7 erg K^{-1} mol^{-1}. What was the molecular weight of the protein? (Recall for dimensional analysis that, in SI, 1 erg has dimensions of g cm^2 sec^{-2}).

From equation (15):

$$\text{Mol wt} = \frac{RTs}{D(1 - v\rho)},$$

$$= \frac{8.3144 * 10^7 \text{ erg K}^{-1} \text{ mol}^{-1} * 293 \text{ K} * 14.16 * 10^{-13} \text{ sec}}{3.4 * 10^{-7} \text{ cm}^2 \text{ sec}^{-1} (1.0 - 0.75 \text{ cm}^3 \text{ g}^{-1} * 1.0 \text{ g cm}^{-3})},$$

$$= 405\ 829 \text{ g mol}^{-1}.$$

Example 4-9. In an experiment with an analytical ultracentrifuge, sedimentation of an enzyme subunit to equilibrium yielded the following data: temperature = 20°C; x_1 = 5.84 cm; x_2 = 6.63 cm; c_1= 4.11 (arbitrary units); c_2 = 14.89 arbitrary units); v = 0.743 ml g^{-1}; ρ = 1.0 g ml^{-1}; ω = 15 000 rev min^{-1} and R = 8.3144 * 10^7 erg K^{-1} mol^{-1}. What was the molecular weight of the enzyme subunit? (Recall for dimensional analysis that 1 erg has dimensions of g cm^2 sec^{-2}).

From equation (17): $\quad$ Mol wt = $\dfrac{2 RT \ln (c_2/c_1)}{(1 - v\rho)\ \omega^2\ (x_2^2 - x_1^2)},$

$$= \frac{2 * 8.3144 * 10^{7} \text{erg K}^{-1} \text{mol}^{-1} * 293 \text{ K} * \ln (14.89/4.11)}{(1.0 - 0.743 \text{ ml g}^{-1} * 1.0 \text{ g ml}^{-1})(1571 \text{ rad sec}^{-1})^2 (6.63 \text{ cm}^2 - 5.84 \text{ cm}^2)},$$

$$= 10\ 037 \text{ g mol}^{-1}.$$

Example 4-10. You subjected a microsomal protein to sedimentation velocity in the analytical ultracentrifuge at a rotor speed of 56 250 rev min^{-1}. with the following results:

t (sec)	720	1440	2880	3600	4800	5760	6120
x_2/x_1	1.0477	1.0550	1.0697	1.077	1.0892	1.0990	1.1027

What was the sedimentation coefficient of the protein?

Solution:

t (sec)	720	1440	2880	3600	4800	5760	6120
Ln x_2/x_1	0.04659	0.05354	0.06737	0.07417	0.08544	0.0944	0.09776

The graph shows that, when $\Delta x = (5\,000\text{-}3\,000)$ sec, the slope of the curve $= 1 * 10^{-5}$ sec^{-1}.

From equations (12) and (14)

$$\begin{aligned}
\ln \quad x &= \quad \omega^2 s(t), \\
s &= \quad slope/\omega^2, \\
1\ radian &= \quad 2\pi\ rev\ sec^{-1}, \\
\omega &= \quad 5890\ radians\ sec^{-1}, \\
\therefore \quad s &= \quad (1 * 10^{-5}/5890^{\,2})\ sec \quad = \quad 2.83 * 10^{-13}\ sec.
\end{aligned}$$

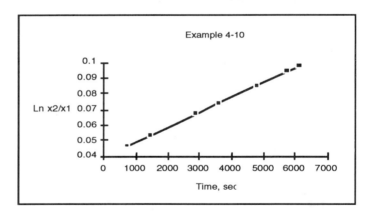

4-25. Spreadsheet solutions

	A	B	C	D	E	F	G	H	I	J
1	EXAMPLE 4-1									
2	INPUT		INPUT							OUTPUT
3	Mobility	x	#BP	y= Log BP	x^2	x*y	y^2			
4	20	20	9950	3.997823	400	79.96	15.98		SLOPE	-0.06355816
5	21	21	8595	3.934246	441	82.62	15.48		INTCP.	5.268910553
6	22	22	7425	3.870696	484	85.16	14.98		CORREL.	-0.9999998
7	23	23	6410	3.806858	529	87.56	14.49			
8	24	24	5540	3.74351	576	89.84	14.01		UNKNOWNS:	
9	25	25	4785	3.679882	625	92	13.54		23.5	5961
10	26	26	4135	3.616476	676	94.03	13.08		25.5	4448
11	27	27	3571	3.55279	729	95.93	12.62			
12	28	28	3084	3.489114	784	97.7	12.17			
13	29	29	2667	3.426023	841	99.35	11.74			
14	30	30	2302	3.362105	900	100.9	11.3			
15										
16	Count	11								
17	Σ	275	58464	40.47952	6985	1005	149.4			
18	(Σx OR Σy)^2	75625		1638.592						
19	Correl	-76.9	34.785	2.210874	76.905					
20										
21	EXAMPLE 4-2									
22	INPUT		INPUT							OUTPUT
23	Ve/Vo	x	Mol Wt	y= Log MW	x^2	x*y	y^2			
24	0.52	0.52	500000	5.69897	0.2704	2.963	32.48		SLOPE	-0.70339677
25	0.74	0.74	350000	5.544068	0.5476	4.103	30.74		INTCP.	6.064915931
26	0.95	0.95	250000	5.39794	0.9025	5.128	29.14		CORREL.	-0.9999897
27	1.15	1.15	180000	5.255273	1.3225	6.044	27.62			
28									UNKNOWNS:	
29	Count	4							0.85	293109
30	Σ	3.36	1E+06	21.89625	3.043	18.24	120			
31	(Σx OR Σy)^2	11.29		479.4458						
32	Correl	-0.62	0.9394	0.660751	0.6207					
33										
34	EXAMPLE 4-3									
35	INPUT		INPUT							OUTPUT
36	Mobility	x	Mol Wt	y= Log MW	x^2	x*y	y^2			
37	0.02	0.02	165000	5.217484	0.0004	0.104	27.22		SLOPE	-1.21382683
38	0.2	0.2	100000	5	0.04	1	25		INTCP.	5.242057645
39	0.34	0.34	68000	4.832509	0.1156	1.643	23.35		CORREL.	-0.9999684
40	0.5	0.5	43000	4.633468	0.25	2.317	21.47			
41	0.72	0.72	23000	4.361728	0.5184	3.14	19.02		UNKNOWNS:	
42	0.83	0.83	17200	4.235528	0.6889	3.515	17.94		0.47	46942
43	0.97	0.97	11700	4.068186	0.9409	3.946	16.55			
44										
45	Count	7								
46	Σ	3.58	427900	32.3489	2.5542	15.67	150.6			
47	(Σx OR Σy)^2	12.82		1046.452						
48	Correl	-6.15	2.2501	2.731332	6.1458					

SPREADSHEET FORMULAS

	A	B	C	D
1	EXAMPLE 4-1			
2	INPUT		INPUT	
3	Mobility	x	#BP	y= Log BP
4	20	20	9950	=LOG10(C4)
5	21	21	8595	=LOG10(C5)
6	22	22	7425	=LOG10(C6)
7	23	23	6410	=LOG10(C7)
8	24	24	5540	=LOG10(C8)
9	25	25	4785	=LOG10(C9)
10	26	26	4135	=LOG10(C10)
11	27	27	3571	=LOG10(C11)
12	28	28	3084	=LOG10(C12)
13	29	29	2667	=LOG10(C13)
14	30	30	2302	=LOG10(C14)
15				
16	Count	=COUNT(B4:B14)		
17	Σ	=SUM(B4:B14)	=SUM(C4:C14)	=SUM(D4:D14)
18	(Σx OR Σy)^2	=B17*B17		=D17*D17
19	Correl.	=B16*F17-B17*D17	=SQRT(B16*E17-B18)	=SQRT(B16*G17-D18)
20				
21	EXAMPLE 4-2			
22	INPUT		INPUT	
23	Ve/Vo	x	Mol Wt	y= Log MW
24	0.52	0.52	500000	=LOG10(C24)
25	0.74	0.74	350000	=LOG10(C25)
26	0.95	0.95	250000	=LOG10(C26)
27	1.15	1.15	180000	=LOG10(C27)
28				
29	Count	=COUNT(B24:B27)		
30	Σ	=SUM(B24:B27)	=SUM(C24:C27)	=SUM(D24:D27)
31	(Σx OR Σy)^2	=B30*B30		=D30*D30
32	Correl	=B29*F30-B30*D30	=SQRT(B29*E30-B31)	=SQRT(B29*G30-D31)
33				
34	EXAMPLE 4-3			
35	INPUT		INPUT	
36	Mobility	x	Mol Wt	y= Log MW
37	0.02	0.02	165000	=LOG10(C37)
38	0.2	0.2	100000	=LOG10(C38)
39	0.34	0.34	68000	=LOG10(C39)
40	0.5	0.5	43000	=LOG10(C40)
41	0.72	0.72	23000	=LOG10(C41)
42	0.83	0.83	17200	=LOG10(C42)
43	0.97	0.97	11700	=LOG10(C43)
44				
45	Count	=COUNT(B37:B43)		
46	Σ	=SUM(B37:B43)	=SUM(C37:C43)	=SUM(D37:D43)
47	(Σx OR Σy)^2	=B46*B46		=D46*D46
48	Correl.	=B45*F46-B46*D46	=SQRT(B45*E46-B47)	=SQRT(B45*G46-D47)

SPREADSHEET FORMULAS

	E	F	G	H	I	J
1						
2						OUTPUT
3	x^2	x*y	y^2			
4	=B4*B4	=B4*D4	=D4*D4		SLOPE	=(B16*F17-B17*D17)/(B16*E17-B18)
5	=B5*B5	=B5*D5	=D5*D5		INTCP.	=(D17*E17-B17*F17)/(B16*E17-B18)
6	=B6*B6	=B6*D6	=D6*D6		CORREL	=B19/E19
7	=B7*B7	=B7*D7	=D7*D7			
8	=B8*B8	=B8*D8	=D8*D8		UNKNOW	
9	=B9*B9	=B9*D9	=D9*D9		23.5	=10^((J4*I9)+J5)
10	=B10*B10	=B10*D10	=D10*D10		25.5	=10^((J4*I10)+J5)
11	=B11*B11	=B11*D11	=D11*D11			
12	=B12*B12	=B12*D12	=D12*D12			
13	=B13*B13	=B13*D13	=D13*D13			
14	=B14*B14	=B14*D14	=D14*D14			
15						
16						
17	=SUM(E4:E1	=SUM(F4:F14	=SUM(G4:G1			
18						
19	=C19*D19					
20						
21						
22						OUTPUT
23	x^2	x*y	y^2			
24	=B24*B24	=B24*D24	=D24*D24		SLOPE =	=(B29*F30-B30*D30)/(B29*E30-B31)
25	=B25*B25	=B25*D25	=D25*D25		INTCP. =	=(D30*E30-B30*F30)/(B29*E30-B31)
26	=B26*B26	=B26*D26	=D26*D26		CORREL	=B32/E32
27	=B27*B27	=B27*D27	=D27*D27			
28					UNKNOW	
29					0.85	=10^((J24*I29)+J25)
30	=SUM(E24:E	=SUM(F24:F2	=SUM(G24:G			
31						
32	=C32*D32					
33						
34						
35						OUTPUT
36	x^2	x*y	y^2			
37	=B37*B37	=B37*D37	=D37*D37		SLOPE =	=(B45*F46-B46*D46)/(B45*E46-B47)
38	=B38*B38	=B38*D38	=D38*D38		INTCP. =	=(D46*E46-B46*F46)/(B45*E46-B47)
39	=B39*B39	=B39*D39	=D39*D39		CORREL	=B48/E48
40	=B40*B40	=B40*D40	=D40*D40			
41	=B41*B41	=B41*D41	=D41*D41		UNKNOW	
42	=B42*B42	=B42*D42	=D42*D42		0.47	=10^((J37*I42)+J38)
43	=B43*B43	=B43*D43	=D43*D43			
44						
45						
46	=SUM(E37:E	=SUM(F37:F4	=SUM(G37:G			
47						
48	=C48*D48					

SPREADSHEET SOLUTIONS

	A	B	C	D	E	F	G
1	EXAMPLE 4-4				EXAMPLE 4-8		
2							
3		INPUT	OUTPUT			INPUT	OUTPUT
4	rev./min	48000			R	8.31E+07	
5	rad/sec		5027		T	293	
6					S	1.416E-12	
7	EXAMPLE 4-5				D	3.4E-07	
8					part spec vol	0.75	
9		INPUT	OUTPUT		density	1	
10	x1	5.65			COMPUTE:		3.45E-02
11	x2	6.4					8.5E-08
12	rev./min	48000			Mol. Weight		405829
13	min	190					
14	S		4.32734E-13				
15							
16	EXAMPLE 4-6						
17							
18		INPUT	OUTPUT				
19	x1	5.65					
20	x2	6.4					
21	rev./min	48000					
22	S	4.33E-13					
23	sec		11393				
24	min		190		EXAMPLE 4-9		
25							
26	EXAMPLE 4-7					INPUT	OUTPUT
27					R	8.31E+07	
28		INPUT	OUTPUT		T	293	
29	Vp	60			c2	14.89	
30	Vi	35			c1	4.11	
31	Vo	55			part spec vol	0.743	
32	Vt		150		density	1	
33	Ve(1)	65			rev./min	15000	
34	Ve(2)	88			x2	6.63	
35	Ve(3)	129			x1	5.84	
36	K(1)		0.11		COMPUTE:		6.27E+10
37	K(2)		0.35				6.25E+06
38	K(3)		0.78		Mol. Weight		10037
39							

SPREADSHEET FORMULAS

	A	B	C
1	EXAMPLE 4-4		
2			
3		INPUT	OUTPUT
4	rev./min	48000	
5	rad/sec		=2*PI()*(B4/60)
6			
7	EXAMPLE 4-5		
8			
9		INPUT	OUTPUT
10	x1	5.65	
11	x2	6.4	
12	rev./min	48000	
13	min	190	
14	S		=LN(B11/B10)/((2*PI()*(48000/60))^2*190*60)
15			
16	EXAMPLE 4-6		
17			
18		INPUT	OUTPUT
19	x1	5.65	
20	x2	6.4	
21	rev./min	48000	
22	S	0.000000000000433	
23	sec		=LN(B20/B19)/(((2*PI()*B21/60)^2)*B22)
24	min		190
25			
26	EXAMPLE 4-7		
27			
28		INPUT	OUTPUT
29	Vp	60	
30	Vi	35	
31	Vo	55	
32	Vt		=B29+B30+B31
33	Ve(1)	65	
34	Ve(2)	88	
35	Ve(3)	129	
36	K(1)		=(B33-B31)/(C32-B31)
37	K(2)		=(B34-B31)/(C32-B31)
38	K(3)		=(B35-B31)/(C32-B31)
39			

SPREADSHEET FORMULAS

	E	F	G
1	EXAMPLE 4-8		
2			
3		INPUT	OUTPUT
4	R	83144000	
5	T	293	
6	S	0.000000000001416	
7	D	0.00000034	
8	part spec vol	0.75	
9	density	1	
10	COMPUTE:		=(F4*F5*F6)
11			=(F7*(1-F8*F9))
12	Mol. Weight		=G10/G11
13			
14			
15			
16			
17			
18			
19			
20			
21			
22			
23			
24	EXAMPLE 4-9		
25			
26		INPUT	OUTPUT
27	R	83144000	
28	T	293	
29	c2	14.89	
30	c1	4.11	
31	part spec vol	0.743	
32	density	1	
33	rev./min	15000	
34	x2	6.63	
35	x1	5.84	
36	COMPUTE:		=(2*(F27*F28)*LN(F29/F30))
37			=((1-F31*F32)*(1571)^2*(F34^2-F35^2))
38	Mol. Weight		=G36/G37
39			

SPREADSHEET SOLUTIONS

	A	B	C	D	E	F	G	H	I	J
1	EXAMPLE 4-10									
2										
3										
4	INPUT		INPUT							OUTPUT
5	t (sec)	x	x2/x1	y= LNx2/x	x^2	x*y	y^2			
6	720	720	1.048	0.0466	518400	33.55	0		SLOPE	9.47E-06
7	1440	1440	1.055	0.05354	2E+06	77.099	0		INTCP.	0.039938
8	2880	2880	1.07	0.06738	8E+06	194.05	0		CORRE	0.999978
9	3600	3600	1.077	0.07418	1E+07	267.05	0.01			
10	4800	4800	1.089	0.08544	2E+07	410.13	0.01		SOLUTION:	
11	5760	5760	1.099	0.0944	3E+07	543.75	0.01		S	2.73E-13
12	6120	6120	1.103	0.09776	4E+07	598.3	0.01			
13										
14	Count	7								
15	Σ	25320	7.54	0.5193	1E+08	2123.9	0.04			
16	(Σx OR Σy)	6E+08		0.26967						
17	Correl	1719	13473	0.12757	1718.8					
18										

SPREADSHEET FORMULAS

	A	B	C	D
1	EXAMPLE 4-1			
2				
3				
4	INPUT		INPUT	
5	t (sec)	x	x2/x1	y= LNx2/x1
6	720	720	1.0477	=LN(C6)
7	1440	1440	1.055	=LN(C7)
8	2880	2880	1.0697	=LN(C8)
9	3600	3600	1.077	=LN(C9)
10	4800	4800	1.0892	=LN(C10)
11	5760	5760	1.099	=LN(C11)
12	6120	6120	1.1027	=LN(C12)
13				
14	Count	=COUNT(B6:B12)		
15	Σ	=SUM(B6:B12)	=SUM(C6:C12)	=SUM(D6:D12)
16	(Σx OR Σy)^2	=B15*B15		=D15*D15
17	Correl	=B14*F15-B15*D15	=SQRT(B14*E15-B16	=SQRT(B14*G15-D16)
18				

SPREADSHEET FORMULAS

	E	F	G	H	I	J
1						
2						
3						
4						**OUTPUT**
5	x^2	x*y	y^2			
6	=B6*B6	=B6*D6	=D6*D6		SLOPE	=(B14*F15-B15*D15)/(B14*E15-
7	=B7*B7	=B7*D7	=D7*D7		INTCP.	=(D15*E15-B15*F15)/(B14*E15-
8	=B8*B8	=B8*D8	=D8*D8		CORREL.	=B17/E17
9	=B9*B9	=B9*D9	=D9*D9			
10	=B10*B10	=B10*D10	=D10*D10		SOLUTION:	
11	=B11*B11	=B11*D11	=D11*D11		S	=J6/5890^2
12	=B12*B12	=B12*D12	=D12*D12			
13						
14						
15	=SUM(E6:E12)	=SUM(F6:F12)	=SUM(G6:G12)			
16						
17	=C17*D17					
18						

4-26.   Review questions

In questions 4-1 to 4-12, each of the transport methods designated A to E is followed by a numbered list of words, phrases or statements and by the letters a, b, c, d, or e.

A. Gel filtration
B. NaDodSO$_4$ Gel electrophoresis
C. Gel filtration and gel electrophoresis
D. Sedimentation velocity
E. Sedimentation equilibrium

Choose the appropriate letter (a, b, c, d, or e) to indicate whether the numbered word, phrase, or statement is associated with method A, B, C, D, or E.

4-1.	Sephadex	a, b, c, d, e
4-2.	Very accurate determination of molecular weight	a, b, c, d, e
4-3.	Diffusion coefficient	a, b, c, d, e
4-4.	Allows determination of sedimentation coefficients	a, b, c, d, e
4-5.	Large molecules are excluded	a, b, c, d, e
4-6.	All polypeptides are negatively charged	a, b, c, d, e
4-7.	Polypeptide chain-amphiphile complexes	a, b, c, d, e
4-8.	Hydrostatic pressure is the driving force	a, b, c, d, e
4-9.	Polyacrylamide is used for separation	a, b, c, d, e
4-10.	Void volume	a, b, c, d, e
4-11.	Exclusion limit	a, b, c, d, e
4-12.	SI unit of electric current	a, b, c, d, e

Please answer questions 4-13 to 4-20 with true or false.

4-13.   Void volume is related to molecular weight in gel filtration.
4-14.   Elution volume is related to molecular weight in gel filtration.
4-15.   Pore size in agarose gels is controlled by the degree of cross-linking.
4-16.   Pore size in sephadex columns is controlled by the degree of cross- linking.
4-17.   In NaDodSO$_4$ polyacrylamide gel electrophoresis, the mobilities of proteins depend both on the charges and on the molecular weights of the proteins.
4-18.   As in kinetics, time is a variable in transport processes.
4-19.   The exclusion limit of a gel filtration column is the maximum molecular size which can pass through its pores.
4-20.   The partial specific volume is the change in volume which occurs when 1.0 mole of a solute is added to a large excess of solvent.

Define the following terms:

4-21. Molecular sieve
4-22. Mobility in $NaDodSO_4$ gels
4-23. Inner volume of sephadex gel bed
4-24. Partition coefficient in gel filtration
4-25. Sedimentation coefficient

4-27. Review problems

4-26. Sephadex G-200 has a fractionation range of 5 000 to 600 000. What would be the order of elution of the following proteins on a column of this gel: catalase (250 000), cytochrome c (13 160), ferritin (452 000), myoglobin (16 900), bovine serum albumin (68 500), and yeast alcohol dehydrogenase (151 000)?

4-27. Given that the elution volumes of the six proteins of question 4-26 are catalase 0.94, cytochrome c 2.77, ferritin 0.58, myoglobin 2.60, bovine serum albumin 1.74, and yeast alcohol dehydrogenase 1.26. What is the molecular weight of an unknown protein of elution volume 0.83?

4-28. If an analytical ultracentrifuge will be operated at 60 000 rev min^{-1} to sediment cytochrome c which has a sedimentation coefficient s of 1.71, what will be the rate of movement of cytochrome c in the centrifugal field?
    If the cytochrome c were located initially 5 cm from the center of rotation, how long wold it take for the protein to be sedimented to 6.5 cm from the center of rotation?

4-29. Given that the density of water is 1.0 g cm^{-3} at 20°C, calculate the molecular weight of a protein which has the following properties in aqueous solution: s = 7.12; diffusion coefficient = 5.76 * 10^{-7} cm^2 sec^{-1} and partial specific volume = 0.72 cm^3 g^{-1}.

4-30. If the boundary of a protein moves from 6.00 to 6.81 cm in 180 min at a rotor speed of 55 000 rev./min, what is its sedimentation coefficient at 20°C?

4-28. Additional problems

4-31. A protein solution was subjected to sedimentation velocity in the analytical ultracentrifuge at 20°C. The following data were obtained:

sedimentation coefficient, s	: 12.14*10^{-13} S
diffusion coefficient, D	: 3. 90*10^{-7} cm^2 sec^{-1}
partial specific volume, ν	: 0.72 ml g^{-1}
density of solvent, ρ	: 1.0 g ml^{-1}
the gas constant, R	: 8.3144*10^{7}ergs deg K^{-1}mol^{-1}

What was the molecular weight of the protein?

4-32. Calculate the molecular weight of an enzyme which gave the following results when subjected to sedimentation equilibrium:

temperature	:	10°C
radial distance, x_1	:	6.817 cm
radial distance, x_2	:	7.095 cm
concentration of solute, c_1	:	3.08 m mol l^{-1}
concentration of solute, c_2	:	13.48 m mol l^{-1}
partial specific volume, n	:	0.725 ml g^{-1}
density of solvent, ρ	:	1.0 g ml^{-1}
rotor speed, ω	:	12 500 rev min^{-1}
the gas constant, R	:	8.3144*10^{7}ergs deg K^{-1}mol^{-1}

4-33. NaDodSO$_4$ polyacrylamide gel electrophoresis of seven standard proteins gave the following results:

Protein	Mobility	Molecular weight
Thyroglobulin	0.03	165 000
Phosphorylase a	0.19	100 000
Bovine serum albumin	0.31	68 000
Ovalbumin	0.48	43 000
Trypsin	0.71	23 300
Myoglobin	0.85	17 200
Cytochrome c	0.98	11 700

What are the molecular weights of two mitochondrial proteins which gave mobilities of 0.28 and 0.91 when run under the same conditions as the standards?

4-34. A calibration curve for gel filtration was prepared from seven standard proteins on Sephadex G-200 with the following results:

Protein	V_e/V_o	Molecular weight
1	0.52	500 000
2	0.74	350 000
3	0.94	250 000
4	1.15	180 000
5	1.26	150 000
6	1.75	68 000
7	2.61	17 000

What are the molecular weights of two polypeptides which had V_e/V_o values of 0.82 and 1.19 when they were eluted from Sephadex G-200 under the same conditions as the standards?

4-35. A macromolecular complex from fungal mitochondria was subjected to sedimentation velocity at the constant velocity of 70 000 rev min^{-1} in an analytical ultracentrifuge and the radial distance of the complex (from the center of rotation) was measured as a function of time with the following results:

Time (min)	0	10	20	30	40	50	60
x (cm)	5.01	5.14	5.26	5.38	5.51	5.64	5.77

Calculate the Svedberg coefficient of the complex.

4-36. Subjecting a protein to sedimentation velocity at 25°C in an analytical ultracentrifuge yielded the following data:

sedimentation coefficient, s	: 14.16*10^{-13} S
diffusion coefficient, D	: 3. 40*10^{-7} cm^2 sec^{-1}
partial specific volume, v	: 0.75 ml g^{-1}
density of solvent, ρ	: 1.0 g ml^{-1}
the gas constant, R	: 8.3144*10^7 ergs deg K^{-1}mol^{-1}

Estimate the molecular weight of the protein.

4-37. In an experiment with an analytical ultracentrifuge, sedimentation of a polypeptide to equilibrium gave the following results:

temperature	: 20°C
radial distance, x_1	: 5.340 cm
radial distance, x_2	: 6.630 cm
concentration of solute, c_1	: 4.11 m mol l^{-1}
concentration of solute, c_2	: 14.89 m mol l^{-1}
partial specific volume, v	: 0.743 ml g^{-1}
density of solvent, r	: 1.0 g ml^{-1}
rotor speed, ω	: 15 000 rev min^{-1}
the gas constant, R	: 8.3144*10^{7}ergs deg K^{-1}mol^{-1}

Calculate the molecular weight of the polypeptide.

4-38. A series of reconstitution experiments was carried out in your laboratory. In one of these experiments, a microsomal protein was subjected to sedimentation velocity in the ultracentrifuge at a constant rotor speed of 56 250 rev min^{-1}. The following results were obtained when the radial distance of the protein (from the center of rotation) was measured as a function of time:

Time (min)	0	10	20	50	60	80	100	12(
x (cm)	6.03	6.04	6.05	6.07	6.08	6.09	6.10	6.1

What was the Svedberg coefficient of the protein?

Suggested Reading

Alberty, R.A. and A. Cornish-Bowden 1993. The pH Dependence of the Apparent Equilibrium Constant, K', of a Biochemical Reaction. *Trends in Biochem. Sci.*, 18: 288-291.

Alwine, J. C., D. J. Kemp and G. R. Stark 1974. Method for Detection of SpecificRNAs in Agarose Gels by Transfer to Diazobenzyloxymethyl-Paper and Hybridization With DNA Probes. *Proc. Natl. Acad. Sci. USA,* 74: 5350-5354.

Alwine, J. C., D. J. Kemp, B. A. Parker, J. Reiser, J. Renart, G. R. Stark, and G. M. Wahl: 1979. *Detection of Specific RNAs or Specific Fragments of DNA by Fractionation in Gels and Transfer to Diazobenzyloxymethyl Paper.* Methods in Enzymology, Academic Press, New York, 68: 220-233.

Appenzeller, T. 1990. Democratizing the DNA Sequence. *Science* 247: 1030-1032.

Arnheim, N. and H. Erlich 1992. *Polymerase Chain Reaction Strategy.* Annu. Rev. Biochem. 61: 131-156.

Aune, K.C. 1978. *Molecular Weight Measurements by Sedimentation Equilibrium: Some Common Pitfalls and How to Avoid Them.* Methods in Enzymology, Academic Press, New York, 48: 163-4185.

Barker, W.C., D.G. George and L.T. Hunt 1990. *Protein Sequence Database.* Methods in Enzymology, Academic Press, New York, 183: 31-49.

Barber, E. J. 1966. *Calculation of Density and Viscosity of Sucrose Solutions as a Function of Concentration and Temperature.* Natl. Cancer Inst. Monograph, 21: 219-239.

Bhagwat, A., 1992. *Restriction Enzymes: Properties and Use.* Methods in Enzymology, Academic Press, New York, 216: 199-224.

Bleasby, A.., P. Griffiths, D. Hines, S. Marshall, L. Staniford, K. Hoover and Kristofferson. 1993. The BIOSCI Newsgroups - Computer Networks Changing Biology. *Trends in Biochem. Sci.* 18: 310-311.

Bly, D. D. 1970. Gel Permeation Chromatography. *Science* 168: 527-533.

Burks, C., J.W. Fickett, W.B. Goad, M. Kanehisa, F.I. Lewitter, W.P. Rindone, C.D. Swindell, C.-S. Tung and H.S. Bilofsky 1985. The GenBank Nucleic Acid Database. *Comput. Appl. Biosci.* 1: 225-233.

Burks, C., M.J. Cinkosky, P. Gilna, J. E.-D. Hayden, Y. Abe, E.J. Atencio, S. Barnhouse, D. Benton, C.A. Buenafe, K.E. Cumella, D.B. Davison, D.B. Emmert, M.J. Faulkner, J.W. Fickett, W.M. Fischer, M. Good, D.A. Horne, F.K. Houghton, P.M. Kelkar, T.A. Kelley, M. Kelly, M.A. King, B.J. Langan, J.T. Lauer, N. Lopez, C. Lynch, J. Lynch, J.B. Marchi, T.G. Marr, F.A. Martinez, M.J. McLeod, P.A. Medvick, S.K. Mishra, J. Moore, C.A. Munk, S.M. Mondragon, K.K. Nasseri, D. Nelson, W. Nelson, T. Nguyen, G. Reiss, J. Rice, J. Ryals, M.D. Salazar, S.R. Stelts, B.J. Trujillo, L.J. Tomlinson, M.G. Weiner, F.J. Welch, S.E. Wiig, K. Yudin, and L.B. Zins 1990. *GenBank : Current Status and Future Directions.* Methods in Enzymology, Academic Press, New York, 183: 3-22.

Cannon, G.C 1987. Sequence Analysis on Microcomputers. *Science* 238: 97-103.

Caskey, C. T. 1987. Disease Diagnosis by Recombinant DNA Methods. *Science* 236: 1223-1229.

Chrambach, A. & D. Rodbard 1971. Polyacrylamide Gel Electrophoresis. *Science* 172: 440-451,

Du, Z., L. Hood and R.K. Wilson 1993. *Automated Fluorescent DNA Sequencing of Polymerase Chain Reaction Products.* Methods in Enzymology, Academic Press, New York, 218: 104-121

Dunbar, B.S., H. Kimura and T.M. Timmons 1990. *Protein Analysis Using High-Resolution Two-Dimensional Polyacrylamide Gel Electrophoresis.* Methods in Enzymology, Academic Press, NY, 182: 441-459.

Duncan, R. and J. W. B. Hershey 1984. Evaluation of Isoelectric Focusing Running Conditions During Two-Dimensional Isoelectric Focusing/Sodium Dodecyl Sulfate-Polyacrylamide Gel Electrophoresis: Variation of Gel Patterns With Changing Conditions and Optimized Isoelectric Focusing Conditions. *Anal. Biochem.* 138:144-155.

Eby, M.J. 1989. The Reality of Capillary Electrophoresis. *Biotechnology* 7: 903-911.

Fleischer, S. and M. Kervina 1974. *Subcellular Fractionation of Rat Liver.* Methods in Enzymology, Academic Press, New York, 31: 6-41.

Fischer, L. 1969. *An Introduction to Gel Chromatography*, Elsevier, New York.

Garfin, D.E. 1990. *One Dimensional Gel Electrophoresis.* Methods in Enzymology, Academic Press, New York, 182: 425-441.

Garfin, D.E. 1990. *Isoelectric Focusing.* Methods in Enzymology, Academic Press, New York, 182: 459-477.

Garrels, J. I. 1980. Computer-Analyzed Two-Dimensional gel Electrophoresis of Proteins. *Trends Biochem. Sci.* 5: 281-283.

Garrels, J. I. 1983. *Quantitative Two-Dimensional Gel Electrophoresis of Proteins.* Methods in Enzymology, Academic Press, New York, 100: 411-423.

Geiger, M.J., M. Bull, D.J. Eckels and J. Gorski 1993. *Amplification of Complementary DNA from mRNA of Unknown 5' Ends by One-Way Polymerase chain Reaction.* Methods in Enzymology, Academic Press, New York, 218: 321-335.

Gordon, M.J., X. Huang, S.L. Pentoney, Jr. and R.N. Zare 1988. Capillary Electrophoresis. *Science* 242: 224-228.

Guyer, R.L. and D.E. Koshland, Jr. 1989. The Molecule of the Year. *Science* 246: 1543-1546.

Gyllensten, U.B. and M. Allen 1993. *Sequencing of in Vitro Amplified DNA.* Methods in Enzymology, Academic Press, New York, 218: 3-16.

Hearst, J.E. and C.W.Schmid 1973. *Density Gradient Sedimentation Equilibrium.* Methods in Enzymology, Academic Press, New York, 27: 111-127.

Kahn, P and G. Cameron 1990. *EMBL Data Library.* Methods in Enzymology, Academic Press, New York, 183: 23-31.

Kibby, M. R. 1986. Four Spreadsheet Templates for the Laboratory. *Comput. Appl. Biosci.* 2: 1-4.

Korn, L.J., C.L. Queen and M.N. Wegman 1977. Computer Analysis of Nucleic Acid Regulatory Sequences. *Proc. Natl. Acad. Sci. USA* 74: 4401-4405.

Landers, J.P. 1993. Capillary Electrophoresis: Pioneering New Approaches for Biomolecular Analysis. *Trends in Biochem. Sci.* 18: 409-414.

Lawrence, J.G., D.L. Hartl and H. Ochman 1993. *Sequencing Products of Polymerase Chain Reaction.* Methods in Enzymology, Academic Press, New York, 218: 26-35.

Lehninger, A. L., D. J. Nelson and M. M. Cox 1992. *Principles of Biochemistry*, 2d. ed., Worth, New York.

Lewis, R. 1993. *DNA Sequencing Software Teases Meaning from Genes.* The Scientist, May 31.

Lewitter, F.I. and W.P. Rindone 1987. *Computer Programs for Analyzing DNA and Protein Sequences.* Methods in Enzymology, Academic Press, N.Y., 155: 582-593.

Luckey, J.A., H. Drossman, T. Kostichka and L.M. Smith 1993. *High-Speed DNA Sequencing by Capillary Gel Electrophoresis.* Methods in Enzymology, Academic Press, New York, 218: 154-172.

Mahler, H. R. and E. H. Cordes 1971. *Biological Chemistry, 2d ed.,* Harper & Row, N.Y.

Mangalan, H. 1993. Computer Corner: Striding the Turf of the Gang of Four. *Trends Biochem. Sci.* 18: 187-188.

Maxam, A. M. and W. Gilbert 1977. A New Method for Sequencing DNA. *Proc. Natl. Acad. Sci. USA,* 74: 560-564.

Maxam, A. M. and W. Gilbert 1980. *Sequencing End-Labeled DNA With Base-Specific Chemical Cleavages.* Methods in Enzymology, Academic Press, N. Y., 65: 499-560.

McCarty, K. S., D. Stafford and O. Brown 1968. Resolution and Fractionation of Macromolecules by Isokinetic Sucrose Density Gradient Sedimentation. *Anal. Biochem.* 24: 314-329.

Morrow, J. F., 1979. *Recombinant DNA Techniques.* Methods in Enzymology, Academic Press, NY, 68: 1-24.

Mullis, K.B. and F.A. Faloona 1987. *Specific Synthesis of DNA in Vitro via a Polymerase-Catalyzed Chain Reaction.* Methods in Enzymology, Academic Press, NY, 155:335-350.

Nathans, D. and H. O. Smith 1975. *Restriction Endonucleases in the Analysis and Restructuring of DNA Molecules.* Annu. Rev Biochem., 44: 273-293.

Neville, D. M. and Glossmann 1974. *Molecular Weight Determination of Membrane Protein and Glycoprotein Subunits by Discontinuous Gel Electrophoresis in Dodecyl Sulfate.* Methods in Enzymology, Academic Press, New York, 32: 92-102.

Nielsen, T. B. and J. A. Reynolds 1978. *Measurements of Molecular Weights by Gel Electrophoresis.* Methods in Enzymology, Academic Press, New York, 48: 3-10.

Noll, H., 1967. Characterization of Macromolecules by Constant Velocity Sedimentation. *Nature* 215: 360-363.

Ochman. H., F.J. Ayala and D.L. Hartl 1993. *Use of Polymerase Chain Reaction to Amplify Segments of Outside Boundaries of Known Sequences.* Methods in Enzymology, Academic Press, NY, 218: 309-321.

O'Farrell, P. H. 1975. High Resolution Two-Dimensional Electrophoresis of Proteins. *J. Biol. Chem., 250:* 4007-4021.

Olsen, D.B., G. Wunderlich, A. Uy and F. Eckstein 1993. *Direct Sequencing of Polymerase Chain Products.* Methods in Enzymology, Academic Press, New York, 218: 79-92.

Olson, M., L. Hood, C. Cantor and D. Botstein 1989. A Common Language for Physical Mapping of the Human Genome. *Science,* 245: 1434-1435.

Reid, E. and R. Williamson 1974. *Centrifugation.* Methods in Enzymology, Academic Press, NY, 31: 713-733.

Reiland, J. 1971. *Gel Filtration.* Methods in Enzymology, Academic Press, NY, 22: 287-321.

Renart, J. and I. V. Sandoval 1984. *Western Blots.* Methods in Enzymology, Academic Press, NY, 104: 455-460.

Reynolds, R. A. 1979. *The Role of Micelles in Protein-Detergent Interactions.* Methods in Enzymology, Academic Press, New York, 61: 58-62.

Richterich, P. and G.M. Church 1993. *DNA Sequencing with Direct Transfer Electrophoresis and Nonradioactive Detection.* Methods in Enzymology, Academic Press, New York, 218: 187-222.

Rickwood, D. 1984. *Centrifugation , 2nd ed.*, IRL Press, Washington, DC.

Rivas, G. and A.P. Minton 1993. New Developments in the Study of Biomolecular Associations via Sedimentation Equilibrium. *Trends in Biochem. Sci.* 18: 284-287.

Saiki, R.K., D.H. Gelfand, S. Stoffel, S.J. Scharf, R. Higuchi, G.T. Horn, K.B. Mullis, H.A. Erlich 1988. Primer-Directed Enzymatic Amplification of DNA with a Thermostable DNA Polymerase. *Science* 239: 487-491.

Salser, W. A.: DNA *Sequencing Techniques.* Annu. Rev. Biochem., 43: 923-965, 1974..

Sanger, F., S. Nicklen and A. R. Coulson 1977. DNA Sequencing With Chain-Terminating Inhibitors. *Proc. Natl. Acad. Sci. USA* 74: 5463-5467.

Schachman, H. K. 1957. *Ultracentrifugation, Diffusion and Viscometry.* Methods in Enzymology, Academic Press, New York, 4: 32-103.

Scholfield, J.P., D.S.C. Jones and V. Vaudin 1993. *Fluorescent and Radioactive Solid-Phase Dideoxy Sequencing of Polymerase Chain Reaction Products in Microtiter Plates.* Methods in Enzymology, Academic Press, New York, 218: 93-103.

Sinsheimer, R. L. 1977. *Recombinant DNA.* Annu. Rev. Biochem. 46: 415-438.

Smith, A. J. H. 1980. *DNA Sequence Analysis by Primed Synthesis.* Methods in Enzymology, Academic Press, NY, 65: 560-580.

Smith, R.J. 1984. *The Analysis of Nucleic Acid Sequences.* In Microcomputers in Biology, ed. C.R. Ireland and S.P. Long. IRL Press, Washington, DC.

Southern, E. M. 1975. Detection of Specific Sequences Among DNA Fragments Separated by Gel Electrophoresis. *J. Mol. Biol.* 98: 503-517.

Southern, E. M. 1979. *Gel Electrophoresis of Restriction Fragments.* Methods in Enzymology, Academic Press, NY, 68: 152-176.

Stellwagen, E. 1990. *Gel Filtration.* Methods in Enzymol, Academic Press, NY,182:317-328.

Teller, D. C. 1973. *Characterization of Proteins by Sedimentation Equilibrium in the Analytical Ultracentrifuge.* Methods in Enzymology, Academic Press, NY, 27: 346-441.

Thornton, J.I.. 1989. DNA Profiling. New tool links evidence to suspects with high certainty. *Chemical & Engineering News,* Nov 20: 18-30.

Tiselius, A. 1957. *Electrophoresis.* Methods in Enzymology, Academic Press, N.Y., 4: 3-20.

Tsang, V.C.W., J.M. Peralta and A.R. Simons 1983. *Enzyme-Linked Immunoelectrotransfer Blot Techniques (EITB) for Studying the Specificities of Antigens and Antibodies Separated by Gel Electrophoresis.* Methods Enzymol., Academic Press, NY. 92: 377-391.

Van Holde, K. E. 1971. *Physical Biochemistry.* Prentice-Hall, New Jersey.

Walton, K.M. and J.E. Dixon 1993. *Protein Tyrosine Phosphatases:* Annu. Rev. Biochem. 62: 101-120.

Weber, K., J.R. Pringle and M. Osborn 1972. *Measurement of Molecular Weights by Electrophoresis on SDS-Acrylamide Gel.* Methods in Enzymology, Academic Press, New York, 26: 3-27.

Weber, K. and M. Osborn, 1969..*The Reliability of Molecular Weight Determinations by Dodecyl Sulfate-Polyacrylamide Gel Electrophoresis.* J. Biol. Chem. 244: 4406-4412.

Wu, R., 1978. *DNA Sequence Analysis.* Annu. Rev. Biochem. 47: 607-634.

Kinetics and Binding Equilibria

5-1.   Enzyme kinetics has many biomedical applications

Enzyme kinetics is concerned with the rates of enzyme-catalyzed reactions. It is regarded sometimes as an area of biochemistry which is of interest only to specialists. But enzyme kinetics, and related topics such as ligand binding and signal transduction across the cell membrane, are widely applicable and essential for understanding the many biological processes which depend on quantitative measurement of enzyme reaction rates. Data on enzyme kinetics help investigators to understand cellular metabolism, to understand the nature of the reactions being catalyzed, and to deduce the mechanisms of these reactions. In addition, experimental modification of the rates of enzyme reactions can help to explain aspects of the regulation of metabolism and can provide insights into processes such as cellular differentation, cell growth and gene expression. Rates of enzyme reactions also are important to clinicians because velocities of the catalyzed reactions are changed in some pathological conditions and are therefore important in the diagnosis of several diseases. Moreover, many anticancer drugs, antibiotics, antiviral drugs, insecticides, and herbicides function by inhibiting enzymes and by modifying their reaction rates. More generally, enzyme kinetics and binding equilibria are important aspects of immunochemistry, pharmacology, microbiology and toxicology. In the industrial world, the pharmaceutical industry uses enzyme reactions for producing pharmaceuticals, as do the makers of beer, cheese, wines, baked foods, laundry detergents, and the new products produced by biotechnology companies.

This chapter deals with the effects of substrate concentration, inhibitor concentration and temperature on the velocities of enzyme catalyzed reactions. We will demonstrate computations for K_m, the *Michaelis-Menten constant*; V_{max}, the *maximal velocity*; and K_i, the *inhibitor constant*, all of which will be defined below. These three parameters are important tools for comparing enzyme activity under differing conditions and for understanding how enzyme activity is regulated. Later in this chapter we will demonstrate computations of another parameter, E_a, the *energy of activation*, defined as the minimum energy which a molecule must have to pass into the activated state. Determination of the energy of activation is useful because it can indicate when phase changes and rate-limiting steps occur in enzyme reactions. We will conclude the chapter with discussions of binding equilibria, enzyme phosphorylation/dephosphorylation, and a brief presentation on chemical kinetics. First, however, let us review briefly, some aspects of enzyme catalysis.

5-2. Enzymes are unique catalysts

Enzymes have enormous catalytic power and exceptional biochemical specificity. They are frequently specific for a single substrate, they often catalyze only one reaction, their products are formed with a high degree of stereospecificity and the catalyzed reactions often proceed with few side reactions. Enzymes thus are the most efficient catalysts known. Until recently, scientists felt that all *known* enzymes are protein. But in the late 1970s, Thomas Cech and Sidney Altman showed that some RNA molecules or *ribozymes* have catalytic activity. Now we know that ribozymes can catalyze peptide bond formation and the making and breaking of aminoacyl tRNAs (they have aminoacyl esterase activity). Therefore we can no longer assume that proteins catalyze all biochemical reactions. Some enzymes perform their catalytic feats alone, others need the help of inorganic ions or of organic biomolecules to perform their functions. The helper inorganic ions are known as cofactors, the organic assistants as *coenzymes*. Sometimes, cofactors and coenzymes are bound tightly to the enzymes and are then referred to as *prosthetic* groups. An enzyme without its prosthetic group is known as an *apoenzyme*, an enzyme with its prosthetic group is known as a *holoenzyme*. Some water-soluble vitamins are precursors of, or are chemically the same as, coenzymes, which explains the need for these vitamins in trace amounts in the diets of animals. When enzymes catalyze reactions, they act in organized sequences with other enzymes instead of acting individually.

Substrates, and prosthetic groups when present, are bound to the active site. *The active site* of an enzyme, E, is a three-dimensional region which reacts with substrate S to form the enzyme-substrate complex, ES. A substrate usually is bound to its enzyme by relatively weak forces; but, nevertheless, binding occurs with considerable specificity. This specificity, coupled with the topology of the active site, results in the substrate being held in the proper spatial orientation for undergoing its reaction. Because the orientation of the substrate on the active site allows the reaction to proceed in the most favorable way, and because the energy barrier which molecules must surmount in order to undergo a chemical reaction is lowered by enzymes, the rates of biochemical reactions can be increased by factors of 10^9 to 10^{12}.

Hundreds of enzymes participate in metabolism at any given time. For metabolic processes to be run smoothly and efficiently, the rates of the many reactions catalyzed by enzymes have to be regulated to meet changing needs. *Regulatory enzymes* increase or decrease the velocities of metabolic pathways by undergoing two types of modification: *allosteric modification (conformational changes*, section

5-11) from the effects of small, modulator molecules; or *covalent modification*, the reversible phosphorylation of proteins with protein kinases and phosphatases (section 5-12).

In working with enzymes, it is natural to be concerned with methods of expressing data on the rates of enzyme catalyzed reactions. Three units, enzyme unit, specific activity, and molecular activity are of interest in this context:

• *Enzyme unit*: The SI definition of an enzyme unit is that amount of enzyme which catalyzes the transformation of 1 mole of substrate per second at 30°C. This unit, also known as the *katal*, has not gained wide acceptance and editors of journals permit the use of other units as long as they are properly defined in each paper, e.g. an enzyme unit may be defined as the quantity of enzyme catalyzing the transformation of 1 micromole of substrate per minute at 25°C.
• *Specific activity*: the number of units per milligram of enzyme protein.
• *Molecular activity*: the number of molecules of substrate transformed per minute per molecule of enzyme.

5-3.   The plot of [S] versus velocity is a rectangular hyperbola

At fixed levels of enzyme concentration, most enzymes show *saturability* when their velocities are studied as functions of substrate concentration. That is, velocity increases initially as substrate concentration is increased. But, as shown in Figure 5-1, the velocity eventually approaches a maximum value, and this value cannot be increased further regardless of the level to which substrate concentration is increased. Early investigators correctly interpreted such saturation curves as that of Figure 5-1 to mean that the enzyme reacted with substrate to form an *enzyme-substrate complex*, ES.

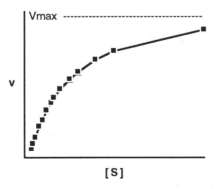

Figure 5-1. Effect of substrate concentration on the velocity of an enzyme reaction

The enzyme-substrate complex ES then broke down to form products:

$$E + S \underset{k_2}{\overset{k_1}{\rightleftharpoons}} ES \xrightarrow{k_3} E + P \tag{1}$$

In equation (1), k_1, k_2 and k_3 are *rate constants* (defined in section 5-14). According to the equation, an enzyme, E, reacts with its substrate S, to form the enzyme-substrate complex, ES. The enzyme can either dissociate to form E and S or it can go on to form product and free enzyme. We assume in this reaction that substrate concentration is much greater than that of enzyme, that the concentrations of enzyme and enzyme-substrate complex are constant during the experiment and that the rates of formation and breakdown of ES are equal. The reaction is at equilibrium under these conditions. In the derivation of the Michaelis-Menten equation which follows, $[E_\tau]$ represents the total enzyme concentration, i.e. the sum of free and combined enzyme so that $[E_\tau] - [ES]$ is the concentration of free or uncombined enzyme. When the rate of formation of ES is equal to the rate of its breakdown in equation (1) then we get the relationship shown in equation (2):

$$k_1([E_\tau] - [ES])[S] = k_2[ES] + k_3[ES], \tag{2}$$

Dividing equation (2) by ES and rearranging we get

$$\frac{[S]([E_\tau] - [ES])}{[ES]} = \frac{k_2 + k_3}{k_1} = K_m, \tag{3}$$

Equation (3) now can be rearranged to yield:

$$[E_\tau][S] = [ES](K_m + [S]), \tag{4}$$

$$[ES] = \frac{[E_\tau][S]}{K_m + [S]}. \tag{5}$$

According to equation (1), the rate constant for decomposition of ES to products is k_3, so that the initial velocity, v_0 is

$$v_0 = k_3[ES]. \tag{6}$$

When [S] is very large, all E in the system is present as ES and the maximal velocity, V_{max}, is attained; i.e.,

$$V_{max} = k_3[E_\tau]. \tag{7}$$

Substituting for [ES] in equation (6) its value from (5)

$$v_0 = \frac{k_3[E_\tau][S]}{K_m + [S]}. \tag{8}$$

And substituting for $k_3[E_t]$ in equation (8) its value from (7), we get

$$\boxed{v_0 = \frac{V_{max} \cdot [S]}{K_m + [S]}} \tag{9}$$

Equation (9), proposed by Leonor Michaelis and Maud Menten in 1913, is known as the Michaelis-Menten equation. When the initial velocity is $1/2 V_{max}$, equation (9) becomes

$$\frac{V_{max}}{2} = \frac{V_{max} \cdot [S]}{K_m + [S]} \tag{10}$$

So that

$$K_m + [S] = 2[S] \tag{11}$$

$$K_m = [S] \tag{12}$$

K_m, therefore, is defined as *the substrate concentration which yields half maximal velocity*; that is, when $v = 1/2 V_{max}$, $K_m = [S]$. The Michaelis-Menten equation yields a rectangular hyperbola because it can be transformed to equation (13) which is equivalent to the curve of the function of a rectangular hyperbola, equation (14):

$$v = V_{max} \frac{1}{K_m / [S] + 1} \tag{13}$$

$$y = a * \frac{1}{X} \tag{14}$$

5-4. K_m and V_{max} may be estimated from plots of kinetic data

Manual determination of kinetic constants from hyperbolic plots such as Figure 5-1 is difficult. So before microcomputers became widely available, scientists developed several linear forms of the Michaelis-Menten equation to permit graphical determination of the kinetic constants. Today, the linear forms are used also to provide initial estimates of K_m and V_{max} for nonlinear regression (cf. sections 5-8 and 10-17). The best-known linear forms, the Lineweaver-Burk, Eadie-Hofstee and Hanes-Woolf equations, (16) to (18), do not give results of equivalent accuracy; therefore, at various times, it may be desirable to choose one over the other two.

The *Lineweaver-Burk* is perhaps the most widely used linear transformation. It is obtained by taking reciprocals of equation (9):

$$\frac{1}{v} = \frac{K_m + [S]}{V_{max}\,[S]} = \frac{K_m}{V_{max}[S]} + \frac{[S]}{V_{max}[S]} \tag{15}$$

$$\frac{1}{v} = \frac{K_m}{V_{max}} * \frac{1}{[S]} + \frac{1}{V_{max}} \tag{16}$$

The *Eadie-Hofstee* equation may be obtained by multiplying equation (16) by V_{max} and rearranging the result to yield:

$$v = -K_m * \frac{v}{[S]} + V_{max} \tag{17}$$

The *Hanes-Woolf* equation is the third linear form of equation (9). It may be obtained from multiplying equation (16) by [S] and rearranging to give:

$$\frac{[S]}{v} = \frac{1}{V_{max}} * [S] + \frac{K_m}{V_{max}} \tag{18}$$

Figures 5-2 to 5-4 show generalized plots according to these three linear equations.

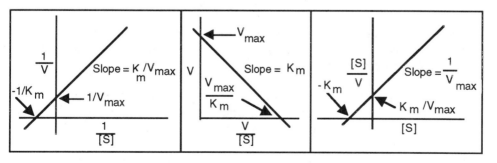

Figure 5-2. The Lineweaver-Burk   Figure 5-3. The Eadie- Hofstee   Figure 5-4. The Hanes-Woolf

5-5.   Characteristic types of enzyme inhibitors can be distinguished

Enzyme inhibitors are substances which react with enzymes or with enzyme-substrate complexes, prevent the enzymes and their substrates from undergoing their normal reactions, and thus decrease the velocities of the catalyzed reactions. Many drugs are enzyme inhibitors. *Biomedical scientists recognize three major types of enzyme inhibition: competitive, noncompetitive, and uncompetitive inhibition.* In each type of inhibition, the inhibitor forms a unique complex with the enzyme or with the enzyme-substrate complex. Such inhibitor complexes decrease velocity by decreasing the rate of formation, or the rate of breakdown, of ES. *Competitive inhibitors* form complexes by combining reversibly with the active site of the enzyme. *Noncompetitive inhibitors* react reversibly both with an essential group other than the active site and with the enzyme-substrate complex. *Uncompetitive inhibitors* combine only with the enzyme-substrate complex. Equations (19) to (26) describe how inhibitors react with enzymes or with ES complexes and in addition, define the inhibitor constants for each type of inhibition:

Competitive   : $E + I \rightleftharpoons EI$ (19)

$$K_i = \frac{[E][I]}{[EI]} \qquad (20)$$

Noncompetitive   : $E + I \rightleftharpoons EI$ (21)

$$ES + I \rightleftharpoons ESI \qquad (22)$$

$$K_i^{E_i} = \frac{[E][I]}{[EI]} \qquad (23)$$

$$K_i^{ESI} = \frac{[ES][I]}{[ESI]} \qquad (24)$$

Uncompetitive   : $ES + I \rightleftharpoons ESI$ (25)

$$K_i = \frac{[S][I]}{[ESI]} \qquad (26)$$

Thus, in equations (19) to (26), the equilibrium constant, K_i, is defined as the dissociation constant of the enzyme-inhibitor complex or the dissociation constant of the enzyme-substrate-inhibitor complex.

An important development in the field of drugs and inhibitors is that scientists are now using computers to design new drugs; this confers important benefits such as increased productivity and safety, coupled with decreased time and cost. The drugs usually are small molecules. At the end of this chapter, we list four reviews related to the rapidly developing field of *computer-aided, rational drug design*: Hansch and Klein, 1991, Livingston, 1991, Martin,1991 and Krieger, 1993.

5-6. K_i can be calculated from Lineweaver-Burk plots

Each of the three major types of inhibition produces characteristic changes in kinetics that can be recognized by graphical analysis on Lineweaver-Burk, double reciprocal plots. Double-reciprocal equations for the three types of inhibition are:

Competitive

$$\frac{1}{v} = \frac{K_m}{V_{max}}\left(1 + \frac{[I]}{K_i}\right)\frac{1}{[S]} + \frac{1}{V_{max}} , \tag{27}$$

Noncompetitive

$$\frac{1}{v} = \frac{K_m}{V_{max}}\left(1 + \frac{[I]}{K_i}\right)\frac{1}{[S]} + \frac{1}{V_{max}}\left(1 + \frac{[I]}{K_i}\right) , \tag{28}$$

Uncompetitive

$$\frac{1}{v} = \frac{K_m}{V_{max}}\left(\frac{1}{[S]}\right) + \frac{1}{V_{max}}\left(1 + \frac{[I]}{K_i}\right) . \tag{29}$$

In competitive inhibition, the slope is increased by the term $(1 + [\,I\,]/K_i)$, K_m is increased for the inhibited reaction but V_{max} values of the controlled and inhibited reactions are equal (cf. section 10-2). In noncompetitive inhibition, both the slope and intercept are increased by the same term, i.e., $(1 + [\,I\,]/K_i)$; therefore K_m values of the controlled and inhibited reactions are equal and V_{max} of the inhibited reaction decreases. In uncompetitive inhibition, the intercept is increased by $(1 + [\,I\,]/K_i)$, the slopes of the controlled and inhibited plots are equal but V_{max} of the inhibited is decreased.

If we represent the slopes and intercepts of the controlled reactions in equations (27) to (29) by m and b respectively, and the slopes and intercepts of the inhibited reactions by m_i and b_i respectively, then K_i may be calculated from the two equations (30) and (31):

K_i for competitive and noncompetitive inhibition:

$$K_i = \frac{(m * [\,I\,])}{(m_i - m)} \tag{30}$$

K_i for noncompetitive and uncompetitive inhibition:

$$K_i = \frac{(b * [\,I\,])}{(b_i - b)} \tag{31}$$

Figures 5-5, 5-6, and 5-7 are generalized, Lineweaver-Burk plots of competitive, noncompetitive and uncompetitive inhibition. These plots correspond to equations (27), (28) and (29) respectively.

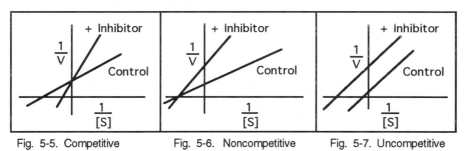

Fig. 5-5. Competitive   Fig. 5-6. Noncompetitive   Fig. 5-7. Uncompetitive

5-7. The energy of activation is used extensively in the sciences

The *energy of activation*, E_a, is an energy barrier which reacting molecules have to surmount in order to undergo a chemical reaction. The rates of enzyme-catalyzed reactions may be increased by factors of 10^9 to 10^{12} (cf. section 5-2) primarily because this energy barrier is lowered by enzymes. As far back as 1889, the Swedish chemist Svante A. Arrhenius derived, from experimental observations, a relationship which permits calculation of energies of activation. This relation, equation (32), has now been applied to an extraordinarily wide range of chemical and biochemical phenomena, some of which include investigations of diffusion coefficients, the rates of chemical reactions, the kinetic theory of gases, the conductivity of semiconductors, the useful lives of electronic devices, statistical mechanics, and quantum mechanical tunneling. Estimations of energies of activation may be made from plots of the *Arrhenius equation,* which relates velocity to the energy of activation as follows:

$$k = Ae - E_a/RT \tag{32}$$

where
k = velocity (rate constant),
E_a = energy of activation,
R = the gas constant = 1.98717 cal mol^{-1} K^{-1},
   = 8.31433 joule mol^{-1} K^{-1} (1 cal = 4.184 joules),
T = absolute temperature,

and
A = a constant which represents the frequency of collisions and is known as the *preexponential factor.*

We should note here that the term $e^{-E_a/RT}$ in equation (32) is similar to the term $e^{-\Delta e_i/kT}$ in the *Boltzmann distribution law*, i.e. equation (4) in Chapter 7. To be used for Arrhenius plots, equation (32) usually is converted to a linear form:

$$\log k = -E_a \cdot \frac{1}{2.303RT} + \log A \tag{33}$$

Equation (33) indicates that a plot of log k against 1/2.303RT should give a straight line if a biochemical reaction obeys the Arrhenius equation under the stated experimental conditions, i.e., if E_a and A remain approximately constant during the conditions of the experiment. These conditions usually involve carrying out measurements of the rates of reaction within an appropriate range of temperatures between 0°C and 45°C. When equation (33) is obeyed, the slope of the straight line which results, a negative slope, is $-E_a$, as shown in Figure 5-8.

The exponential relationship between k and E_a, equation (32), explains the experimentally established fact that a small decrease in E_a, or a small increase in temperature, causes a large increase in velocity, k. For example at 25°C, a decrease in E_a of 1.4 k cal mol^{-1} can increase k (the velocity constant) 10.6 times. And, if we assumed that E_a of a reaction with rate constant k_1 were 12.8 k cal mol^{-1} at

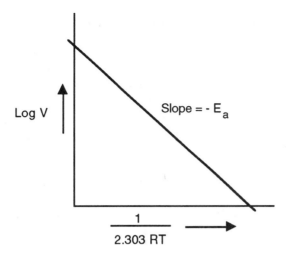

Figure 5-8. Determination of activation energy from a generalized Arrhenius plot

20°C, then at 30°C the ratio of the new rate constant k_2 to the old rate constant i.e. k_2/k_1, would be 2.07. Algebraic transformation of the Arrhenius equation, equation (32), leads to the two equations, (34) and (35), which permit us to calculate increases in velocity constants as a result of decreasing E_a or increasing temperature:

Transformation of Arrhenius equation for computing increase in rate from decreasing E_a:

$$\frac{k_2}{k_1} = \exp\left[\frac{E_{a2} - E_{a1}}{RT}\right]. \tag{34}$$

Transformation of Arrhenius equation for computing increase in rate from increasing T:

$$\frac{k_2}{k_1} = \exp E_a\left[\frac{T_2 - T_1}{R(T_1 T_2)}\right]. \tag{35}$$

According to equation (32), the velocity of a reaction also may be increased by increasing the value of A, the *preexponential factor,* which may be regarded, in physical terms, as the frequency with which molecules, in the proper orientation, collide to produce a chemical reaction. To summarize the foregoing in biomedical terms, an enzyme may speed up the rate of a reaction in one of three ways: (1) by decreasing E_a when the enzyme binds substrate to form an enzyme-substrate complex, (2) by increasing A because the enzyme holds its substrate in a favorable orientation for reaction, or (3) by following both (1) and (2) simultaneously.

In addition to using the Arrhenius equation quantitatively for the calculation of energies of activation, and for computing the velocities of reactions and related quantities, biomedical scientists use the equation qualitatively (graphically) for diagnostic purposes, e.g. for diagnosing when phase changes, rate-limiting steps, denaturation, or other changes, occur during biochemical reactions. Three types of results, first described in words below, and then shown graphically in Figure 5-9, are of interest in this context:

Type 1. *Occurrence of a break in the graph.* Type 1 results suggest that a phase change has occurred, for example, in the lipid environment of a membrane located enzyme.

Type 2. *Change in slope without a break in the curve.* Such changes indicate that a new step has become rate-limiting.

Type 3. *Precipitous decrease in slope of the curve.* Sudden decreases in slope suggest that denaturation has taken place.

Because Arrhenius plots have diagnostic value, we suggest that least-squares estimations of energies of activation generally should be supplemented by graphical analysis, especially if the correlation coefficient and other statistical criteria suggest that a poor least-squares fit of the data may have occurred.

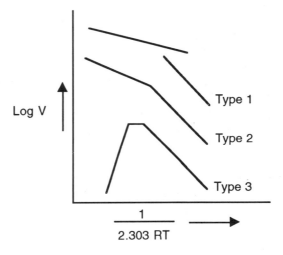

Figure 5-9. Diagnostic uses of Arrhenius plot

5-8. Statistical estimation of K_m and V_{max} has some advantages

Carefully performed calculations of K_m and V_{max} by the three linear transformations of the Michaelis-Menten equation (section 5-4) yield three different estimates. The reason for the differing results is that each transformation has an inherent bias and rearranges the data and experimental error in a different way. To avoid introduction of new types of error, scientists have developed statistical methods which, some claim, yield less distortion and less biased estimates of K_m and V_{max} than the three linear transformations. The usefulness of the statistical methods is increased because they also provide measures of the reliability with which K_m and V_{max} are estimated. Generally, one carries out a statistical estimation of K_m and V_{max} in three steps: (1) provisional ("initial") estimation of K_m and V_{max}; (2) fine adjustment of the provisional estimates; and (3) calculation of the standard errors of estimating K_m and V_{max} (cf. section 10-17).

Provisional (or initial) estimation of K_m and V_{max}, according to Wilkinson (1961), requires that a weighted least-squares analysis (section 10-14) be performed on one of the three linear transformations. Wilkinson proposed that the factor v_i^4 (velocity to the fourth power) be used for statistical weighting of the Lineweaver-Burk equation. Such weighting yields two equations for calculating provisional or initial estimates of K_m and V_{max}:

$$K^\circ_m = \frac{\Sigma v_i^4 \Sigma v_i \frac{v_i^2}{s_i} - \Sigma v_i v_i^2 \Sigma v_i^2 \frac{v_i^2}{s_i}}{\Sigma v_i v_i^2 \Sigma \left(\frac{v_i^2}{s_i}\right)^2 - \Sigma v_i \frac{v_i^2}{s_i} \Sigma v_i^2 \frac{v_i^2}{s_i}}. \tag{36}$$

$$V^\circ_{max} = \frac{\Sigma v_i^4 \Sigma \left(\frac{v_i^2}{s_i}\right)^2 - \left(\Sigma v_i^2 \frac{v_i^2}{s_i}\right)^2}{\Sigma v_i v_i^2 \Sigma \left(\frac{v_i^2}{s_i}\right)^2 - \Sigma v_i \frac{v_i^2}{s_i} \Sigma v_i^2 \frac{v_i^2}{s_i}}. \tag{37}$$

Here, K°_m is the provisional estimate of K_m, and V°_{max} is the provisional estimate of V_{max}. If we let $X = v_i^2$; $Y = v_i^2/s_i$ and $v_i =$ the experimentally determined velocity, then the equations for calculating the provisional estimates are simplified to:

$$K^\circ_m = \frac{\Sigma X^2 \Sigma v_i Y - \Sigma v_i X \Sigma XY}{\Sigma v_i X \Sigma Y^2 - v_i Y \Sigma XY}. \tag{38}$$

$$V^\circ_{max} = \frac{\Sigma X^2 \Sigma Y^2 - (\Sigma XY)^2}{\Sigma v_i X \Sigma Y^2 - \Sigma v_i Y \Sigma XY}. \tag{39}$$

Fine adjustment of the provisional estimates is the second step in the Wilkinson, statistical estimation of K_m and V_{max}. It is carried out by generating and solving an approximately linear form of the Michaelis-Menten equation by the Gauss-Newton method. The Gauss-Newton method, (described in section 10-17), is an *iterative* procedure, i.e. it is a process in which one recalculates estimates of the parameters repetitively to obtain progressively better estimates. Wilkinson obtained the approximately linear equation for estimating K_m and V_{max}, equation (40), by Taylor expansion of the Michaelis-Menten equation and by discarding higher order terms. Fitting the resulting, approximately linear equation is equivalent to fitting a bilinear equation, such as that in section 10-17; but in (40), the parameter a is absent:

$$y = a + b_1 x_1 + b_2 x_2$$

$$v \approx \frac{V \; V^\circ s}{V^\circ K_m + s} + (K^\circ_m - K_m) \frac{V^\circ s}{(K_m + s)^2} \tag{40}$$

If one uses good provisional estimates of K_m and V_{max} for fine adjustment, excellent fits to the Michaelis-Menten equation may be obtained with only one or a few cycles of iteration. Cleland (1977) has proposed that the two Michaelis-Menten constants are well determined if the standard errors of K_m and V_{max} are less than 25% of the values of the two constants. Six equations, (41) to (46), are required for the fine adjustment of the provisional estimates according to Wilkinson (1961):

$$f \quad = \quad V°s/(K°_m + s), \tag{41}$$

$$f' \quad = \quad -V°s/(K°_m + s)^2, \tag{42}$$

$$b_1 \quad = \quad \frac{\Sigma f'^2 \Sigma v_i f - \Sigma f f' \Sigma v_i f'}{\Sigma f^2 \Sigma f'^2 - (\Sigma f f')^2}, \tag{43}$$

$$b_2 \quad = \quad \frac{\Sigma f^2 \Sigma v_i f - \Sigma f f' \Sigma v_i f}{\Sigma f^2 \Sigma f^2 - (\Sigma f f')^2}, \tag{44}$$

$$\boxed{\begin{aligned} K_m &= K°_m + b_2/b_1 \\ V_{max} &= b_1 V° \end{aligned}} \qquad \begin{aligned} &(45) \\ &(46) \end{aligned}$$

Computation of the standard errors of estimating K_m and V_{max}, by using equations (47) to (49), is the third step in Wilkinson's procedure. First, we calculate the variance (defined in section 10-7):

$$s^2 \quad = \quad \frac{\Sigma v_i^2 - b_1 \Sigma v_i f - b_2 \Sigma v_i f'}{n-2}, \tag{47}$$

where s^2 is the variance and s is the standard deviation. It follows from (47) that:

$$\text{S.E. of } K_m \quad = \quad \frac{s}{b_1}\left[\frac{\Sigma f^2}{\Sigma f^2 \Sigma f'^2 - (\Sigma f f')^2}\right]^{\frac{1}{2}}, \tag{48}$$

and
$$\text{S.E. of } V_{max} \quad = \quad V°s\left[\frac{\Sigma f'^2}{\Sigma f^2 \Sigma f'^2 - (\Sigma f f')^2}\right]^{\frac{1}{2}}. \tag{49}$$

Several authors have cautioned (cf. section 10-17) that statistical estimation of K_m, V_{max}, and other parameters cannot improve bad research data. Accordingly, biomedical data subjected to nonlinear regression also should be subjected to graphical analysis to assess whether gross deviations from the fitted equation have occurred.

5-9.   The direct linear plot is another way to estimate K_m and V_{max}

The *direct linear plot* was introduced by Eisenthal and Cornish-Bowden as an alternative to the Lineweaver-Burk, Eadie-Hofstee and Hanes-Woolf linear transformations of the Michaelis-Menten equation. The procedure involved in using the direct linear plot derives from rearranging the Michaelis-Menten equation so as to yield

$$V_{max} \; = \; v \; + \; \frac{v}{[S]} K_m \qquad\qquad (50)$$

This equation may be solved both algebraically and graphically. To follow the algebraic approach used in *Basic Biochemical Laboratory Procedures and Computing* to solve for K_m and V_{max}, let us recall that experimentally, data for estimating K_m and V_{max} are obtained by making n measurements of velocity at n concentrations of substrate. The n pairs of [S] and v which result yield $_nC_r$ combinations of simultaneous equations corresponding to equation (50). For example, the first two simultaneous equations in the $_nC_r$ combinations may be represented as:

$$V_{max} \; = \; v_1 \; + \; \frac{v_1}{[S_1]} K_m \qquad\qquad (51)$$

$$V_{max} \; = \; v_2 \; + \; \frac{v_2}{[S_2]} K_m \qquad\qquad (52)$$

This first pair, and each of the subsequent pairs of the $_nC_r$ simultaneous equations, may be solved algebraically by standard methods to yield $_nC_r$ values of K_m and V_{max}, where n = the number of data points and r = 2.

Graphically, use of equation (50) for the direct linear plot is equivalent to plotting V_{max} against K_m. The way in which a direct linear plot is constructed may be seen by referring to Figure 5-10. To obtain the plot:

*   Place the coordinates of each value of [S] and v on the K_m and V_{max} axes, respectively.
*   Then, draw straight lines through the coordinates and extend them into the first quadrant so that the line representing the first data pair is drawn with intercept -[S_1] on the K_m axis and intercept v_1 on the V_{max} axis.
*   Draw the second line with intercepts - [S_2] on the K_m axis and v_2 on the V_{max} axis.
*   Continue the process for all data pairs.

Theoretically, after one has drawn lines for all pairs of data, the lines will intersect at a common point in the first quadrant, the coordinates of which yield K_m and V_{max} and thus satisfy the Michaelis-Menten equation for all of the experimental observations (Figure 5-10). In practice, however, intersection of the lines usually does not occur at a single point but occurs over a small range of points. In this case, K_m is the median value along the [S]-axis and V_{max} is the median value along the v-axis (Figure 5-11). These lines of the direct plot are not graphs. Rather, they are more correctly regarded as "construction lines." An important feature of the direct linear plot is that it permits estimation of the errors associated with the estimation of K_m and V_{max}. While the direct linear plot is applicable to problems other than the estimation of K_m and V_{max}, it does have some limitations. Thus, it may be used for the estimation of the Scatchard equation but not for the estimation of the Hill equation - discussed next.

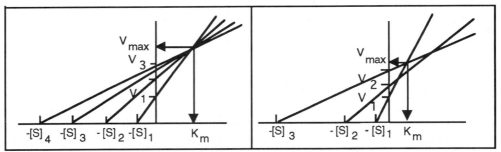

Figure 5-10. A theoretical (perfect) fit   Figure 5-11. A fit that is more likely in the laboratory

5-10.   Ligand-receptor binding starts many biological processes

A *ligand* is an ion or a small molecule which binds to, or interacts with, a macromolecule in a specific manner. A *receptor* is a protein (or other macromolecule) which recognizes, and binds specifically, to ions and small molecules (ligands). Since receptors are high molecular weight compounds, they may have many binding sites and multiple equilibria. In this section, we will consider three constants which quantitate ligand-receptor binding: K_a, the association constant ($\equiv 1/K_d$ the dissociation constant); n, the number of binding sites; and n_H, the Hill coefficient. The specific binding of metal ions, hormones, lipoproteins, drugs, fat-soluble vitamins, antigens and other ligands to receptors, activates second messengers and specific protein kinases. Phosphorylation by these kinases can modify the rates of specific metabolic sequences and switch on key cellular processes (see section 5-12). The study of ligand-receptor binding usually is conducted with radioactive ligands. The source of the receptors may be intact cells, cellular membranes, subcellular fractions, or more recently, synthetic receptors made in the laboratory (for the latter, see references by Borman, and by Spencer, Wandless, Schreiber and Crabtree). To establish that receptors are labeled, three criteria must be met:

   •   *Specificity.*   The ability of a ligand to elicit a specific response should parallel the binding of that ligand to its receptor.

   •   *Sensitivity.*   The concentration of a ligand required for binding should be consistent with the concentration needed to cause the desired biological response.

   •   *Saturability.*   The binding of a ligand to its receptor should exhibit saturability since the number of receptors generally is limited (cf. section 5-3).

Scatchard Plot. First, we consider the simplest case: that *one molecule of radiolabeled ligand reacts with one binding site per receptor* as follows:

$$P + L = PL, \tag{53}$$

where
$$L = \text{free ligand,}$$
$$P = \text{free receptor protein,}$$
$$PL = \text{bound ligand,}$$

and
$$P + PL = \text{total receptor protein} = P_t.$$

The association constant of equation (53) is

$$K_a = \frac{[PL]}{[P][L]}, \tag{54}$$

So that
$$[PL] = K_a[P][L],$$

If the average number of moles of ligand bound to receptor sites on the total protein is $\bar{v}$

Then
$$\bar{v} = \frac{[PL]}{[P_t]}, \tag{55}$$

and
$$\bar{v} = \frac{[PL]}{[P] + [PL]}. \tag{56}$$

Substituting for [PL] in equation (56)

$$\bar{v} = \frac{K_a[P][L]}{[P] + K_a[L][P]}, \tag{57}$$

Dividing equation (57) by [P]

$$\bar{v} = \frac{K_a[L]}{1 + K_a[L]}, \tag{58}$$

If an average of n noninteracting and equivalent sites is occupied on the protein

then
$$\bar{v} = \frac{nK_a[L]}{1 + K_a[L]} \tag{59}$$

Dividing equation (59) by $K_a[L]$ yields the equation of a rectangular hyperbola

$$\bar{v} = \left\{ \frac{n}{1 + \dfrac{1}{K_a[L]}} \right\} \tag{60}$$

This hyperbolic equation can be linearized by taking reciprocals of both sides

$$\frac{1}{\bar{v}} = \frac{1}{n} + \left\{ \frac{1}{n} K_a \cdot \frac{1}{[L]} \right\} \tag{61}$$

The y-intercept of equation (61) yields n from 1/n and the slope yields K_a from $1/nK_a$.

The Scatchard equation, a second transformation of (60), may be derived as follows: From equation (53), $[P_t] = [P] + [PL]$ so that (54) may be rewritten as:

$$K_a = \frac{[PL]}{([P_t] - [PL])[L]} \tag{62}$$

$$[PL] = K_a \{ ([P_t] - [PL])[L] \} \tag{63}$$

Divide equation (63) by $[P_t]$

$$\frac{[PL]}{[P_t]} = K_a \frac{1 - \dfrac{[PL]}{[P_t]}}{}[L] \tag{64}$$

$$\bar{v} = K_a(1 - \bar{v})[L] \tag{65}$$

Divide by [L]:
$$\frac{\bar{v}}{[L]} = K_a(1 - \bar{v}) = K_a - K_a\bar{v} \tag{66}$$

For an average of n noninteracting and equivalent sites occupied on the receptor, we obtain (67) the Scatchard equation. Plotting $\bar{v}/[L]$ against $\bar{v}$ yields y-intercept = nK_a; slope = $-K_a$; and x-intercept = n.

$$\boxed{\frac{\bar{v}}{[L]} = nK_a - K_a\bar{v}} \tag{67}$$

Figures 5-12 to 5-14 show a direct plot from equation (60), a double reciprocal plot from equation (61), and a Scatchard plot based upon equation (67).

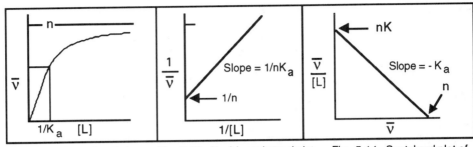

Fig. 5-12. Direct plot of equation (60)   Fig. 5-13. Double reciprocal plot of equation (61)   Fig. 5-14. Scatchard plot of equation (67)

Many authors have cautioned that fitting a least-squares straight line to transformed (e.g., Scatchard) data is based on invalid assumptions; they argue that nonlinear regression on the original nonlinear equation is statistically more correct. We present some pros and cons of this view in section 10-17.

The Hill Plot. Scatchard plots of ligand-receptor data may yield nonlinear curves. The reason may be technical difficulties in performing the experiment. However, the reason also may be that binding one molecule of ligand increases or decreases the binding affinity of a receptor with multiple binding sites for the next molecule of ligand which is bound. The increase of affinity of a receptor for a ligand is referred to as *positive cooperativity,* whereas a decrease in affinity of a receptor for a ligand is referred to as *negative cooperativity.* The *Hill coefficient,* n_H, which measures the cooperativity among binding sites on receptors, *may be obtained from Hill plots of equations for cooperative binding*; e.g. equations (79), (82), and (84) below, which we now derive.

If *a ligand L reacts with multiple sites on a protein* P, and the mean number of molecules of L bound to P is $\bar{\upsilon}$, then

$$\bar{\upsilon} \quad = \quad \frac{\text{number of receptor sites occupied}}{\text{total receptor protein}}. \tag{68}$$

When P contains a *total of n binding sites*, then

$$\frac{\bar{\upsilon}}{n-\bar{\upsilon}} \quad = \quad \frac{\text{number of occupied sites}}{\text{number of unoccupied sites}} \tag{69}$$

If the binding affinity at a single site is independent of the binding affinity at other equivalent sites *(noncooperative binding)* then the resulting equilibrium is

$$\text{Unoccupied sites} + \text{L} \rightleftharpoons \text{occupied sites} \tag{70}$$

The association constant for (70) would then be

$$K_a \quad = \quad \frac{[\text{ occupied sites }]}{[\text{unoccupied sites }][\text{ L }]} \tag{71}$$

Combining equations (69) and (71)

$$K_a = \frac{\bar{v}}{(n-\bar{v})[L]} \tag{72}$$

and

$$\frac{\bar{v}}{n-\bar{v}} = K_a[L] \tag{73}$$

If however, binding at a given receptor site is increased by the binding at other sites *(cooperative binding)*, and if there are *n molecules of L*, and n_H *is a measure of cooperativity*, then

$$\text{unoccupied sites} + nL \rightleftharpoons \text{occupied sites}_{n_H} \tag{74}$$

so that the association (binding) constant will be

$$K_a = \frac{[\text{occupied sites}_{n_H}]}{[\text{unoccupied sites}][L]^{n_H}} \tag{75}$$

$$K_a = \frac{\bar{v}}{(n - \bar{v})[L]^{n_H}} \tag{76}$$

$$\frac{\bar{v}}{n - \bar{v}} = K_a[L]^{n_H} \tag{77}$$

and

$$n_H \log \frac{\bar{v}}{n - \bar{v}} = n_H \log[L] + \log K_a \tag{78}$$

Since $\quad K_a = \dfrac{1}{K_{\text{dissociation constant}}} = \dfrac{1}{K_d}$, then

$$\log \frac{\bar{v}}{n - \bar{v}} = n_H \log[L] - \log K_d \tag{79}$$

and

$$\frac{\bar{v}}{n - \bar{v}} = K_a[L]^{n_H} \quad \text{OR} \quad \frac{\bar{v}}{n - \bar{v}} = \frac{[L]^{n_H}}{K_d} \tag{80}$$

A direct plot of equation (80) yields a sigmoid curve whereas a direct plot of equation (60) yields a rectangular hyperbola. Sigmoid curves also are obtained when the velocities of allosteric enzymes are plotted against substrate concentration, and when the concentration of oxygen is plotted against the degree of saturation of the oxygen-binding sites of hemoglobin as in Figures 5-15 and 5-16. Indeed, equation (81), the original version of (80) was introduced in 1913 by the English scientist *Archibald Hill* (hence *Hill plot*) to explain nonlinear oxygen-hemoglobin saturation curves:

$$\frac{Y}{1-Y} = \frac{pO_2^{\,n_H}}{K_d} \tag{81}$$

This may be transformed to

$$\log \frac{Y}{1-Y} = n_H \log pO_2 - \log K_d \tag{82}$$

In equations (81) and (82), and in Figure 5-15, the concentration of oxygen is expressed in terms of the partial pressure, pO_2, for convenience i.e. because oxygen is a gas. Biochemists use *equation (83), an enzymatic equivalent of (81) the Hill equation, and graphical analysis, as in Figure 5-17, to study the kinetics of allosteric enzymes.*

$$\frac{\overline{\upsilon}}{V_{max} - \overline{\upsilon}} = K_a[S]^{n_H} \tag{83}$$

so, $$\log \frac{\overline{\upsilon}}{V_{max} - \overline{\upsilon}} = n_H \log [S] - \log K_d, \tag{84}$$

where $\overline{\upsilon}$ = velocity of the reaction,
V_{max} = maximum velocity,
$[S]$ = substrate concentration,
and K_d = dissociation constant.

Equation (84) says that a *Hill plot* of $\log \overline{\upsilon}/(V_{max} - \overline{\upsilon})$ versus $\log [S]$ should yield a straight line with slope = n_H and y-intercept = $\log K_d$ so that when $\overline{\upsilon} = 0.5\ V_{max}$ then $n_H \log [S_{0.5}] = -\log K_d$ and $K_d = [S_{0.5}]^{n_H}$ (cf. Figure 5-17). We refer to the x-intercept as $[S_{0.5}]$ or $K_{0.5}$ instead of K_m because the kinetic properties of allosteric enzymes are different from those of Michaelis-Menten enzymes. It is of interest that K_d is not equal to K_m except when $n_H = 1$. When $n_H > 1$, $K_d = [S_{0.5}]^{n_H}$. In a Hill plot of $\log (\overline{\upsilon}/n - \overline{\upsilon})$ versus $\log [L]$ according to equation (79), the slope of the curve is the Hill coefficient n_H, and the y-intercept is equal to $\log K$. The resulting values of the Hill coefficient tell whether or not cooperativity occurs: if there is *no cooperativity* between sites, then $n_H = 1$; if there is *positive cooperativity*, then $n_H > 1$; if there is *negative cooperativity, or heterogeneity among the sites*, then $n_H < 1$. The value of n_H can never be greater than the number of binding sites; and for cases in which $n_H < 1$, distinguishing between negative cooperativity and heterogeneous binding sites requires additional experiments. Figures 5-15 to 5-17 show a direct plot for an oxygen-hemoglobin curve, as well as a direct plot and a Hill plot for an allosteric enzyme.

Let's revisit the view that enzyme kinetics and ligand binding are related topics (section 5-1). Receptors start biological processes by binding specific small molecules to form molecular complexes; so do enzymes, antibodies and other macromolecules. Hence, we may classify enzymes and antibodies as *receptors* and regard substrates and antigens as *ligands*. This receptor-ligand classification does more than organize information. It indicates that since multiple equilibria are involved, one may ignore receptor sites, and then use a strictly thermodynamic approach to derive, in terms of equilibrium constants, a general equation for the mean number of moles of ligand bound to a macromolecule. When desired, one may transform this general equation to equivalent forms for broad or specific use (cf. Klotz, 1986, Chapter 10).

We recall from section 5-2 that *allosteric enzymes* may increase or decrease the rates of metabolic pathways. The rates of such pathways and of individual enzyme reactions also may be controlled by two other processes: *covalent modification* of enzymes and the *induction and repression* of enzymes. We will discuss these three regulatory processes next, in sections 5-11, 5-12, and 5-13.

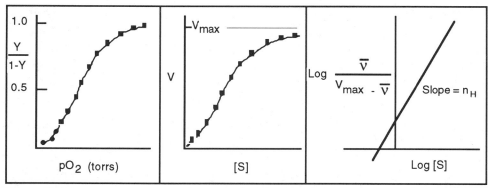

Fig. 5-15. Direct plot for the binding of O_2 to hemoglobin, equation (87)

Fig. 5-16. Direct plot of [S] versus v for an allosteric enzyme, equation (83)

Fig. 5-17. Hill plot for an allosteric enzyme. X-intercept = Log $[S]_{0.5}$; y-intercept = Log K_d, equation (84)

5-11. Allosteric enzymes control rates by conformational change

Allosteric enzymes have two or more subunits and more than one binding site for modulators which modify reaction rates by causing conformational changes in the enzyme. The binding sites may be both distinct and distant from the active site. Although plots of the cooperative binding of oxygen to hemoglobin, and plots of velocity versus substrate concentration for allosteric enzymes yield similar curves, allosteric modulators need not exhibit cooperativity in modifying the kinetics of allosteric enzymes. Two models of allosteric regulation currently enjoy the greatest support; they are the *Concerted Model* of Monod, Wyman and Changeux and the *Sequential Model* proposed by Koshland, Némethy, and Filmer.

Key features of the concerted model

   •   Allosteric enzymes are oligomeric proteins consisting of two or more identical subunits or protomers.
   •   All subunits of the oligomeric protein must exist in one of two conformations, designated T (tensed) or R (relaxed). For example, in an oligomer consisting of four identical protomers, the only allowable conformations are TTTT (T_4) or RRRR (R_4).   Mixed conformations such T_3R, T_2R_2, TR_3 etc., are not allowed.
   •   Progressive binding of a modulator to protomers in the T conformation shifts the equilibrium constant to favor existence of the R conformation.
   •   In both the T and R conformations, each binding site has a characteristic and equivalent binding constant for a specific ligand.

How the Concerted Model Works

The concerted model, Figure 5-18, proposes that as the first molecule of modulator, S, is bound to the enzyme in the T conformation, the equilibrium constant for the transition between the two conformations begins to shift toward the R conformation. The R has a greater affinity for the modulator than the T conformation and, after the fourth molecule of ligand is bound, the equilibrium then shifts predominantly toward the R conformation.   The concerted model is so named because, during the transition between the two conformations, all subunits of the enzyme exist either in the T or the R form; that is, all sububits undergo simultaneous, all-or-none, concerted changes in conformation. Some scientists like this model because it makes several predictions which may be tested experimentally.

Key features of the sequential model

- Allosteric enzymes are oligomers consisting of two or more subunits.
- Subunits of the oligomeric protein may exist in one of two conformational states, designated T (tensed) or R (relaxed) but, unlike the concerted model, mixed conformations such as T_3R, T_2R_2, TR_3 are allowed.
- Progressive binding of a modulator to protomers in the T conformation shifts the equilibrium constant to favor existence of the R conformation.
- There may be intermediate conformations between the T and R conformations, each having a characteristic catalytic activity.

How the Sequential Model Works

The *sequential model*, Figure 5-19, proposes that during transitions between the T and R conformations, the binding of a modulator S to a vacant site on a given subunit causes that subunit to change its conformation. This change in conformation leads to a change in the binding consants of the binding sites on other subunits of the enzyme. The sequential model thus depends on the geometry of interactions between subunits. To some scientists, occurrence of hybrid subunits in the sequential model means that more precise regulation of the activities of allosteric enzymes may be obtained than in the concerted model. But, both the concerted and sequential models have their strong points, and some biomedical scientists believe it to be unlikely that any one model can account effectively for the kinetic behavior of all allosteric enzymes.

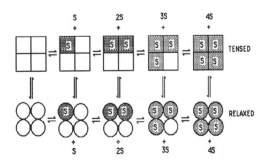

Figure 5-18. The Concerted Model

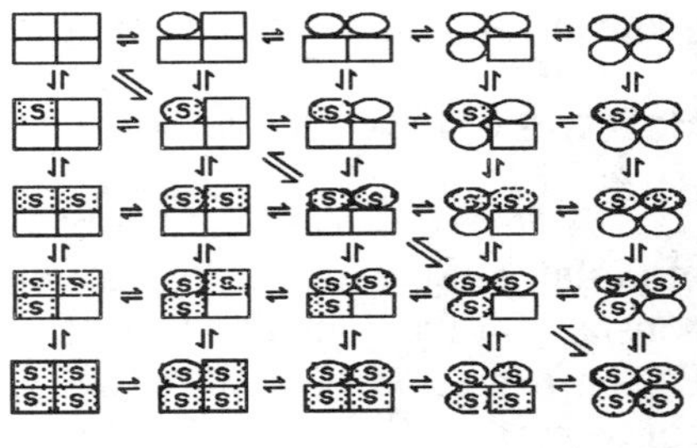

Figure 5-19. The Sequential Model

5-12. Phosphorylated enzymes can change the rates of reactions

Transferases are enzymes which catalyze the transfer of functional groups. *Kinases* (one group of transferases) catalyze the transfer of a phosphate group from ATP to specific substrates. *Protein kinases* therefore catalyze the transfer of phosphate from ATP to proteins; but *protein phosphatases* (enzymes with the reciprocal activity) catalyze the removal of phosphate groups from proteins. These phosphate-transferring enzymes with reciprocal activities *activate or inactivate* many other proteins; that is, these two transferases control the rates of cellular processes by *covalent modification* of proteins. Protein kinases usually have two catalytic (C) subunits and two regulatory (R) subunits. Often, the protein to be phosphorylated through the action of a protein kinase is an enzyme. Often also, phosphorylations occur at the hydroxyl groups of specific serine or tyrosine residues; furthermore, the phosphorylations/dephosphorylations are designed to take place sequentially and repetitively so that first, one enzyme is activated, then the first enzyme activates a second, the second activates a third and so forth. Such linked activation reactions convert proteins into active forms sequentially and lead ultimately to amplification of the initial signal by factors of the order of $>10^6$. Amplification of a single signal thereby can regulate the rate of an entire cellular process. This multistep amplification is often referred to as a *cascade*. Usually, the development of a cascade by protein phosphorylation and dephosphorylation involves the following general reactions:

$$\text{Protein} + 2\text{ATP} \xrightarrow{\text{Protein kinase}} \text{Protein}\text{---}\text{P} + 2\text{ADP}$$

$$\text{Protein}\text{---}\text{P} + 2\text{H}_2\text{O} \xrightarrow{\text{Protein phosphatase}} \text{Protein} + 2\text{P}_i$$

The best known example of this type of bipartisan reaction is covalent modification of *glycogen phosphorylase* by *phosphorylase kinase* and its enzymatic spouse *phosphorylase phosphatase*:

$$\underset{\text{(less active)}}{\text{Phosphorylase b}} \xrightarrow[\text{ATP}]{\text{Phosphorylase kinase ---P}} \underset{\text{(more active)}}{\text{Phosphorylase a} + \text{ADP}}$$

$$\underset{\text{(more active)}}{\text{Phosphorylase a}} \xrightarrow[\text{H}_2\text{O}]{\text{Protein phosphatase}} \underset{\text{(less active)}}{\text{Phosphorylase b} + \text{P}_i}$$

Here, phosphorylase a, the phosphorylated enzyme is active, the dephosphorylated is relatively inactive. Thus, the rate of glycogen breakdown to phosphate esters of glucose by glycogen phosphorylase depends on the ratio of active to inactive forms of the phosphorylase. The rate of glycogen breakdown by phosphorylytic cleavage is further controlled by the hormones *epinephrine and glucagon*.

Hormones, nutrients, neurotransmitters, growth factors (section 6-7), and various extracellular agents initiate intracellular cascades in a three-step process. First they bind to, and activate, specific receptors on the cell surface. Then, the stimulated receptors (section 5-10) activate ion channels or enzymes to produce intracellular *second messengers*. Finally, the second messengers activate cytoplasmic signaling proteins such as the *G-proteins* (GTP-binding and hydrolyzing proteins), protein kinases, and protein phosphatases. For example, epinephrine binds to its receptor on the cell surface and activates the membrane-bound enzyme *adenyl cyclase*; the cyclase then produces the intracellular second messenger *cyclic AMP*, abbreviated cAMP:

$$\text{ATP} \xrightarrow[\text{Adenyl cyclase}]{\text{Mg}^{2+}, \text{ epinephrine or glucagon}} \text{cAMP} + \text{PP}_i.$$

Cyclic AMP activates protein kinase by binding to the inactive protein kinase complex $C_2R_2(MgATP)_2$ consisting of two pairs of subunits and MgATP. When cAMP is bound to $C_2R_2(MgATP)_2$, the complex dissociates into free (active) catalytic subunits and an inactive complex $R_2(cAMP)_4$ as follows:

$$\underset{\text{(inactive kinase)}}{C_2R_2(MgATP)_2 + 4\text{ cAMP}} \longrightarrow \underset{\text{(active kinase)}}{2C(MgATP) + R_2(cAMP)_4.}$$

Protein kinases regulate the rate of *glycogen synthesis* as well as the rate of glycogen breakdown. During glycogen synthesis, the kinase transfers phosphate groups to *glycogen synthase*; but, control of glycogen synthesis is the reciprocal of glycogen breakdown. That is, in glycogen synthesis, the phosphorylated enzyme is inactive, whereas the dephosphorylated is active; moreover epinephrine and glucagon inhibit glycogen synthesis.

$$\underset{\text{(inactive)}}{\text{Glycogen synthase}-\text{P}} \xrightarrow[\text{H}_2\text{O}]{\text{Phosphatase}} \underset{\text{(active form)}}{\text{Glycogen synthase} + \text{P}_i}$$

$$\underset{\text{(active form)}}{\text{Glycogen synthase}} \xrightarrow[\text{ATP}]{\text{Synthase kinase}-\text{P}} \underset{\text{(inactive form)}}{\text{Glycogen synthase}-\text{P} + \text{ADP}}$$

As we have seen, some protein kinases are activated by *cyclic AMP*; other protein kinases, however, require second messengers such as *cyclic GMP, Ca²⁺, double-stranded RNA, diacylglycerol, inositol-1,4,5-triphosphate or even NO•, nitric oxide, a gas*. So important are the protein kinases that approximately 1 % of the total eukaryotic genome may be used to encode them. Moreover, many significant functions of cells depend on the participation of protein kinases. NO•, for example, mediates several activities such as neurotransmission, control of blood pressure, and penile erections. Other second messengers mediate the functions of hormones; the synthesis of proteins; muscle contraction; activation of T cells (section 6-5); transmembrane signaling; cell growth; differentiation; gene expression; and the development of cancers caused by viruses. We provide information on some of these functions in Table 5-1.

Table 5-1. Some cellular functions modulated by kinases

SIGNAL	SECOND MESSENGER	PROTEIN KINASE	CELLULAR RESPONSE
Epinephrine	cAMP	PK-A	Glycogen metabolism
PtdIns(4,5)P2	Ca²⁺ , calmodulin	Myosin light chain -K	Muscle contraction
Growth factors	cGMP, cAMP	PK-C	Gene expression
Antigens	Ins(1,4,5)P3, 1,2-DAG	Protein tyrosine-K	T-cell activation
Phospholipase C	1,2-DAG	PK-C	Differentiation

ABBREVIATIONS: PtdIns(4,5)P2, phosphatidylinositol 4,5-bisphosphate; Ins(1,4,5)P3, inositol 1,4,5-trisphosphate; 1,2-DAG, sn-1,2-diacylglycerol; PK, protein kinase; K, kinase.

Since it is evident that phosphates are central to the regulation of some of the most important processes in cells, one is entitled to ask *what is special or biochemically unique about phosphate?* We provide four answers:

- Phosphoric acid, H_3PO_4, can release three protons in its reactions; cells thus can form two phosphate bonds (e.g., $3' \rightarrow 5'$ nucleotide links) and still have one negative charge.
- The negative charges on phosphorylated biomolecules ensure that they do not diffuse through the lipid-based, biological membranes of cells.
- The negative charges on phosphates also are physically close to the ester bonds; consequently, these charges can repel approaching negative charges which, otherwise, may attack the ester link. This repulsion reduces the rates of hydrolysis (decreases the reactivity) of phosphorylated compounds because it makes them resistant to attack by nucleophiles – in the absence of enzymes.
- Enzymes, however, can increase reaction rates by factors of 10^9 to 10^{12} (section 5-2). Consequently, phosphate esters, which are chemically stable in the absence of an enzyme, are metabolized quite rapidly when their reactions are catalyzed by enzymes. The reactivity of phosphates (with or without enzymes) therefore is well suited to biochemical environments. To summarize:

- protein phosphorylation/dephosphorylation is a major regulatory process;
- the charges on phosphates ensure that they are retained within cells;
- phosphoric acid can form two ester links and still have a negative charge;
- the reactivity of phosphates is ideally suited to their cellular functions.

5-13. Enzyme induction and repression also control reaction rates

We have seen in sections 5-11 and 5-12 that the rates of enzyme-catalyzed reactions may be regulated by allosteric regulation and by covalent modification. In this section, we discuss a third method which cells use to control the rates of enzyme reactions. This method is enzyme induction and enzyme repression.

Repression is a process in which a protein (the *repressor*) binds to a specific *regulatory sequence* of DNA (the *operator*) and prevents that sequence of DNA from being transcribed. *Induction* is the opposite process; here, a specific biomolecule (an *inducer*) binds to and causes the repressor to be removed from the operator site so that a specific gene may be transcribed to mRNA (messenger RNA). The mRNA then diffuses to the ribosomes where it is used as a template for synthesis of the protein coded for by the transcribed gene.

Constitutive enzymes are enzymes of the central metabolic pathways; they are present, and needed, at all times. *Inducible enzymes* are produced only when needed; need is signaled by the appearance of a substrate or related inducer. Enzyme induction may take place when an organism suddenly is presented with a change of diet. Thus, if the organism ordinarily were fed a high carbohydrate, low-protein diet, enzymes for the degradation of amino acids would be present in low amounts. But if, suddenly, the ratio of protein and carbohydrate in the diet were reversed, the organism would respond by increasing the synthesis of enzymes for metabolizing amino acids. The rate of amino acid catabolism thereby is increased.

The foregoing suggests that nature usually is economical and efficient. This is true. Consequently, it would be wasteful for cells to have large amounts of enzymes or other biomolecules present at all times, even when they are not needed. When, however, the demand for certain enzymes or other biomolecules changes, cells should respond; and they do. In these circumstances, they use the process of induction for increasing, and its counterpart, the process of repression, for decreasing the amounts of enzymes in cells.

Although we have noted that biochemical systems use biomolecules such as phosphates in unique ways to perform their cellular functions, we do not wish to imply that the principles of biochemistry are fundamentally different from the principles of chemistry. Any such implication is false. Indeed, the principles behind the chemical reactions carried out by organisms are essentially similar to the principles behind the chemical work of academic, industrial, pharmaceutical, clinical or other laboratories. Further to explore this assertion of the similarity between the chemical fundamentals used by cells and scientific laboratories, we will examine, in an elementary way, some principles of chemical kinetics in the following section.

5-14. Some aspects of elementary chemical kinetics

The reactions of biochemical systems follow the same kinetic and thermodynamic rules as those of chemical systems. When biomedical scientists investigate the turnover of labeled metabolites, follow the velocities of Michaelis-Menten enzymes at low substrate concentrations, or measure the rate of decay of radioisotopes, they are describing first-order reactions. First-order reactions in biology are no different, fundamentally, from the first-order decomposition of nitrogen pentoxide in carbon tetrachloride studied by physical chemists. Similarly, when biomedical scientists measure the rate of renaturation of two strands of DNA to reform a double helix, or the rate of reassembly of head and tail fragments of a bacteriophage to form a functional bacteriophage, they, like the organic chemist who investigates the kinetics of the reaction of methyl iodide with hydroxide ion, are describing second-order reactions. Also falling within the province of chemical kinetics are explanations of why some spontaneous reactions occur without difficulty, whereas other spontaneous reactions (such as the oxidation of paper in a textbook) normally occur only at infinitessimal rates. The study of chemical kinetics is of interest because it can assist investigators in finding the best experimental conditions for obtaining products of interest and because it helps provide an understanding of the nature of the reactions being studied.

Chemical kinetics is that branch of chemistry which deals with the rates and mechanisms of chemical reactions. One of its goals is:
*   to describe the velocities of chemical reactions and the dependence of reaction rates on concentration, temperature, catalysts and other factors. A second goal is:
*   to formulate plausible explanations for relative reaction rates, product distribution, stereochemical configuration and other chemical properties in terms of reaction mechanism. The term *reaction mechanism* embraces several topics. It

includes deducing whether the reaction occurs in one or several steps, or what intermediates may have been present; it also refers to where atoms involved in a reaction are located during the reaction, and to how the postulated intermediates may have led to the experimental results. In other words, a reaction mechanism is a detailed (but inferred) account of the sequence of events by which a reaction occurs.

In this section, we will be concerned mainly with the first goal of chemical kinetics i.e.with the description of rate data. To achieve this objective we will outline how reaction rates are measured; we will define some terms used in chemical kinetics; we will describe equations which permit the estimation of rate constants by linear least squares procedures (or graphically); and we will summarize how tests may be made to distinguish between various levels of reaction order.

Reaction rates are measured In two general ways

The velocities of chemical reactions usually are measured as the change in concentration of a given component per unit time. In kinetic studies, *the progress of reactions may be followed either by measuring the use of reactants or by measuring the rate of appearance of products*. Accepted units of measurement are mol l^{-1} s^{-1} or mol l^{-1} min^{-1}. In biochemistry, where substances of unknown molecular weight and unknown purity frequently have to be used, other units such as μg ml^{-1} min^{-1} or absorbance of NADH sec^{-1} are common. Experimental conditions such as the pH of the assay medium, initial concentrations of reactants and the temperature of the assay system should be specified. When desired, the experimental conditions may be varied individually to shed light on the underlying nature of given reactions. After the reaction rate has been measured, the next step is to formulate a rate law for the reaction. A rate law is an expression which shows how reaction rate varies with the concentration of reactants (or products).

Some terminology of chemical kinetics

We will discuss three terms: kinetic order, rate constants, and molecularity. Not only are rates of reactions measured in terms of the change of concentration but, these rates usually are expressed as functions of concentration. For example,

$$\text{rate} = \text{a proportionality constant} * [\text{reactant or product}]^n. \tag{85}$$

The exponent n in equation (85) is the *order of the reaction* also known as its *kinetic order*. The *kinetic order* is experimentally determined, *varies with experimental conditions*, and it reflects the overall change occurring in a reaction. In Figure 5-20, for example, which shows the reaction of E + S for a Michaelis-Menten enzyme, the reaction is first order at low concentrations of S, is zero order i.e. is independent of [S] at high concentrations when all E is tied up as ES, and is of mixed order at intermediate concentrations. In equation (85), the "proportionality constant" is variously referred to as the *reaction rate constant,* the *rate constant,* or the *specific reaction rate,* and it is represented by the letter k. Rate constants are not pure numbers; they have dimensions which vary depending on the order of each reaction. The magnitude of a rate constant indicates whether the reaction has an in-

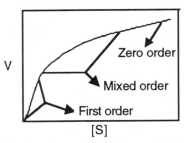

Figure 5-20. Relation of reaction order to substrate concentration

trinsic tendency to proceed rapidly or slowly. That is, large values of k indicate that a reaction will proceed rapidly, whereas smaller values of k mean that the reaction will proceed more slowly. In contrast to kinetic order, which refers to the overall order of a multistep reaction, *molecularity* refers to single, intermediate steps in multistep reactions. A step which depends on the transformation of a single molecule is *unimolecular*. Steps which involve the collision of two molecules are *bimolecular* and those involving three molecules are *termolecular*. An important relationship is that the kinetic order of a single step is the same as the molecularity of that step. But, although a bimolecular reaction is usually second-order and a termolecular reaction is usually third-order, the opposite of these generalizations is not necessarily true.

First-Order Reactions

A *first-order reaction* is one in which the rate of reaction is proportional to the concentration of one reactant. Considering (85), this relation may be written as

$$\text{rate} \quad = \quad k[A]. \tag{86}$$

In differential form, the reaction rate at any time t may be expressed as

$$-\frac{d[A]}{dt} \quad = \quad k[A]. \tag{87}$$

Integration of both sides of (87) yields

$$-\ln\frac{[A]}{[A_0]} \quad = \quad \ln\frac{[A_0]}{[A]} \quad = kt, \text{ so that} \tag{88}$$

$$\boxed{[A] \quad = \quad [A_0]\,e^{-kt}} \tag{89}$$

and

$$\log\frac{[A_0]}{[A]} \quad = \quad \frac{kt}{2.303}, \tag{90}$$

where $[A_0]$ is the concentration of reactant at time zero and $[A]$ is the concentration at time t. Rearrangement of equation (90) yields

$$\boxed{\log[A] \quad = \quad \frac{-kt}{2.303} + \log[A_0]} \tag{91}$$

Equation (91) means that if a reaction obeys first-order kinetics, a plot of the log of concentration of reactant versus time will give a straight line with slope -k/2.303

and intercept log $[A_o]$ as shown in Figure 5-21. Sometimes, it is convenient to develop a first-order rate equation in terms of the amount of product formed rather than concentration of reactant used in given time periods. In these circumstances, if we let $[A_o] = a$ and $x =$ the amount reacting in time t, then

$$-\frac{d(a-x)}{dt} = \frac{kt}{2.303} \quad \text{and} \tag{92}$$

$$\boxed{\log \frac{a}{a-x} = -\frac{kt}{2.303}} \tag{93}$$

Equation (93) is linear in t and it indicates that values of k for first-order reactions may be obtained from plotting log $a/(a-x)$ against time. This yields a straight line, the slope of which is $-k/2.303$ as shown in Figure 5-22.

In addition to the rate constant, k, another important quantity in chemical kinetics is the half-life of a reaction, written as $t_{1/2}$. The *half-life* is the time required for the concentration of a reactant to decrease to one-half its original value. For a first-order reaction, $t_{1/2}$ refers to the time at which $[A] = [A_o]/2$; $t_{1/2}$ is a value of some importance in first-order reactions because of the relation:

$$\ln\frac{[A_o]/2}{[A_o]} = -kt_{\frac{1}{2}}, \tag{94}$$

$$t_{\frac{1}{2}} = \ln 0.5/-k, \tag{95}$$

$$t_{\frac{1}{2}} = \ln 2/k, \tag{96}$$

$$t_{\frac{1}{2}} = 0.693/k. \tag{97}$$

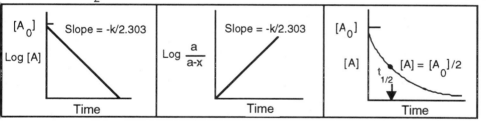

Fig 5-21. Reactant used, (91)   Fig. 5-22. Product formed, (93)   Fig 5-23. Exponential plot, (89)

Like the rate constant k, $t_{1/2}$ measures the inherent tendency of a reaction to proceed rapidly. But, whereas k values are large for rapid reactions, $t_{1/2}$ values are small since reactions which proceed rapidly require short times for one-half of the initial reactant to be used. Rate constants and $t_{1/2}$ values are independent of concentration but are affected by temperature. The value of $t_{1/2}$ may be obtained from the relation $t_{1/2} = \ln 2/k$, equation (96).

Whereas equation (93) is linear, equation (89) is not. Figure 5-23 shows, as equation (89) predicts, that a plot of [A] versus time decreases exponentially to zero starting from $[A_o]$. The negative sign in (89) and similar equations is a convention meant to indicate that the concentration of A decreases with time. Equation (89) is a succinct expression of the fact that organisms use logarithmic functions to carry out certain biological processes. As mentioned at the start of this section, the velocities of enzyme reactions at low substrate concentrations, the turnover of metabolites and the decay of radioisotopes are first-order processes which follow an exponential time course. In addition, the excretion of cellular products, drugs and other substances also follow exponential kinetics indicating that biological systems make frequent use of logarithmic (exponential) processes.

Second-order reactions

A *second-order reaction* is one in which velocity is proportional to the square of the concentration of one reactant or proportional to the concentrations of two reactants. Thus, we may have two general kinds of second-order reactions

$$A + A \longrightarrow Products \tag{98}$$
$$A + B \longrightarrow Products \tag{99}$$

A reaction which is second-order with respect to one reactant, for example equation (98), will yield a rate law of the following type when the reaction is followed in terms of the amount of reactant disappearing at chosen time intervals:

$$\frac{- d[A]}{dt} = k[A]^2 \tag{100}$$

When integrated, equation (100) will yield

$$\frac{1}{[A]} - \frac{1}{[A_0]} = kt \tag{101}$$

$$\boxed{\frac{1}{[A]} = kt + \frac{1}{[A_0]}} \tag{102}$$

where $[A_0]$ is the concentration of reactant at time zero and $[A]$ is the concentration at time t. As shown in Figure 5-24, a plot of $1/[A]$ versus time yields a straight line. The slope of this line is the second-order rate constant, k and the intercept is $1/[A_0]$, the reciprocal of the initial concentration. The dimensions of k are liter $mol^{-1}s^{-1}$.

If a reaction which is second-order with respect to two reactants, for example (99), is followed in terms of the concentration of product formed, it will yield a rate law of the following type

$$\frac{- d[A]}{dt} = k[A][B] \tag{103}$$

If we let a and b represent the initial concentrations of A and B, respectively, and x the concentration of the rate determining reactant used in time t, then

$$\frac{- d[A]}{dt} = k(a - x)(b - x) \tag{104}$$

which, when integrated and rearranged will yield

$$\boxed{\frac{1}{a - b} \ln \frac{b(a - x)}{a(b - x)} = kt} \tag{105}$$

Since (105) is linear, a plot of $\log b(a - x)/a(b - x)$ as ordinate versus time will yield a straight line with slope $= k (a - b)/2.303$ as shown in Figure 5-25. From the slope of this straight line, the second-order rate constant then may be calculated from the relation $k = slope * 2.303/(a - b)$. As was the case for equation (101), the dimensions of k are liter $mol^{-1}s^{-1}$. Equation (104) is based on the assumption that the initial concentrations of the reacting substances are not equal i.e. $a > b$.

If, however, the initial concentrations of the reacting substances are equal, then $a = b$, and the rate equation becomes

$$\frac{d[A]}{dt} = k(a - x)^2 \tag{106}$$

and $$\boxed{\frac{x}{\{a(a-x)\}} = kt} \tag{107}$$

so that experimental data may be treated graphically by plotting the left hand side of (107) against time as in Figure 5-26.

If a second-order reaction such as equation (99) is followed in terms of the concentration of reactant being used, then rate equation (103) reduces to

$$\boxed{\frac{1}{[B_0] - [A_0]} \ln \frac{[A_0][B]}{[B_0][A]} = kt} \tag{108}$$

and a plot of $\ln \frac{[A_0][B]}{[B_0][A]}$ against time will yield a straight line which passes through the origin as in Figure 5-27. The slope of this line is the rate constant, k, with dimensions of liter $mol^{-1}s^{-1}$.

Figures 5-21 to 5-27 thus indicate that tests of the overall order of reactions may be made by linear least squares procedures or by graphical methods. As indicated in section 10-17, nonlinear curve-fitting procedures also may be used. We note here as we do in section 10-17, however, that Duggleby who is a pioneer in the use of nonlinear regression, recently made the suggestion in *Trends Biochem. Sci.* 16: 51-52, 1991 that "most of the supposed advantages of nonlinear regression are illusory and the results obtained from a hand-drawn line after linear transformation may be almost as useful." Above all, we repeat the cautionary note that the use of increasingly sophisticated or complicated methods of analysis cannot improve poor data.

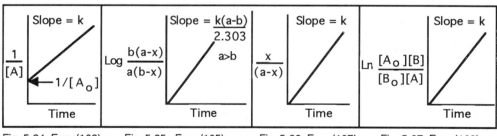

Fig. 5-24. Eqn. (102)   Fig. 5-25. Eqn. (105)   Fig. 5-26. Eqn. (107)   Fig. 5-27. Eqn. (108)

To conclude this chapter, we return to the view (expressed at the beginning of this section) that chemical and biochemical reactions obey the same kinetic and thermodynamic laws; this time, however, we invite the reader to review the reference *Do the laws of chemistry apply to living cells?* by Peter J. Halling. Then, we suggest, it would be interesting to examine whether (a) Halling's views are "novel" and (b) his "argument is correct."

5-15.   Worked examples

Example 5-1.   If 100 pg of an enzyme transform 0.5 n mol of substrate S sec^{-1}, calculate the specific activity of the enzyme.

The solution to this problem is in two parts:
(a) the calculation of enzyme unit and (b) the calculation of specific activity.

(a) From the definitions given in section 5-2, we first set up the following ratio for calculating an enzyme unit.

$$\frac{0.0005\ \mu\ mol\ S\ min^{-1}}{10^{-4}\ mg\ enzyme} = \frac{1.0\ \mu\ mol\ S\ min^{-1}}{x\ mg\ enzyme},$$

$$x = 2 * 10^{-4}\ mg\ enzyme,$$

$$\therefore\ 1\ enzyme\ unit \equiv 2 * 10^{-4}\ mg\ enzyme.$$

(b)   Specific activity   = number of enzyme units per mg enzyme,

= $1/2 * 10^{-4}$ units per mg enzyme,

= 5 000 units per mg enzyme.

Example 5-2.   Assuming that a lysosomal hydrolase obeys Michaelis-Menten kinetics, what fraction of its V_{max} should be obtained at [S] values corresponding to 0.1 K_m, 0.5 K_m and 2.0 K_m?

From equation (9)   v   = $\dfrac{V_{max} \cdot [S]}{K_m + [S]}$,

v/V_{max} = $0.1K_m/(K_m + 0.1K_m)$,

= $0.1K_m/1.1\ K_m$,

= 0.09.

v/V_{max} = $0.5K_m/(K_m + 0.5K_m)$,

= $0.5K_m/1.5\ K_m$,

= 0.33.

v/V_{max} = $2.0K_m/(K_m + 2.0K_m)$,

= $2.0K_m/3.0K_m$,

= 0.67.

Example 5-3.   If a hepatic enzyme which had a V_{max} of 180 m mol min^{-1} gave a velocity of 60 m mol min^{-1} at a substrate concentration of 5 x 10^{-2} M, what was K_m of the enzyme assuming that the Michaelis-Menten equation is obeyed?

60 x (K_m + [S])   = $V_{max} \cdot [S]$,

60 m mol min^{-1} * K_m = $(180 * 50) - (60 * 50)$ m mol min^{-1} mM,

60 K_m = 6000 mM,

K_m = 100 mM,

K_m = 0.1 M.

Example 5-4. An enzyme which obeys Michaelis-Menten kinetics gave the following values: V_{max} = 3275 μ mol min^{-1} and K_m = 2.5 x 10^{-4} M at a substrate concentration of 5 x 10^{-4} M. What was the observed velocity?

From equation (9)

$$v = \frac{V_{max} \cdot [S]}{K_m + [S]},$$

$$v = \frac{3275 * 5 * 10^{-4} \; \mu \text{ mol min}^{-1} \text{ M}}{(2.5 * 10^{-4} + 5 * 10^{-4}) \text{ M}},$$

$$= \frac{1.6375 \; \mu \text{ mol min}^{-1} \text{ M}}{7.5 * 10^{-4} \text{ M}},$$

$$= 2183 \; \mu \text{ mol min}^{-1}.$$

Example 5-5. Assume that 25 mg of an enzyme of molecular weight 56 000 had a V_{max} of 7.5 m mol substrate min^{-1}. Calculate the turnover number of the enzyme.

The turnover number of an enzyme is the number of molecules of substrate transformed per minute per molecule of enzyme. Thus,

If 0.025 g of enzyme transform 7.5 m mol min $^{-1}$,
Then 56 000 g transform (7.5 *56 000) m mol min^{-1}. g/ 0.025 g,

$$= 420 \; 000 \text{ m mol min}^{-1} . \text{ g/ } 0.025 \text{ g,}$$

$$= 1.68 * 10^7 \text{ m mol min}^{-1}.$$

Example 5-6. The data which follow were obtained for the three linear transformations of the Michaelis-Menten equation in three independent experiments. The substrate concentration was in m mol liter^{-1} and the rate was in μ mol min^{-1}. What were K_m and V_{max} for each set of data?

Lineweaver-Burk: slope = 0.05033; y-intercept = 1.83;
x-intercept = -3.636.
Eadie-Hofstee: slope = -4.90 * 10^{-1}; y-intercept = 6.26 * 10^{-1};
x-intercept = 1.23.
Hanes-Woolf: slope = 1.46; y-intercept = 8.50 * 10^{-1};
x-intercept = -0.582.

Lineweaver-Burk:

x-intercept $= \dfrac{-1}{K_m}$ = -3.636;

$K_m = \dfrac{-1}{-3.636}$ = 2.75 * 10^{-1} m mol liter^{-1};

OR $K_m/V_{max} = \dfrac{\text{slope}}{\text{y-intercept}}$ = $\dfrac{1}{V_{max}} = \dfrac{1}{1.83}$;

$K_m = 0.05033 * \dfrac{1}{1.83}$ = 2.75 * 10^{-1} m mol liter $^{-1}$;

V_{max} = 1/1.83 = 0.546 μ mol liter $^{-1}$.

Eadie-Hofstee:
y-intercept = V_{max} = $0.626\ \mu$ mol liter^{-1},
slope = $-K_m$ = 0.490 m mol liter^{-1}.
Hanes-Woolf:
x-intercept = $-K_m$ = -0.582,
K_m = 0.582 m mol liter^{-1},
slope = $\dfrac{1}{V_{max}}$ = 1.46,
V_{max} = $0.685\ \mu$ mol liter^{-1}.

Example 5-7. Given the following data (1) estimate K_m and V_{max} graphically by the three transformations of section 5-4 then (2) explain why results obtained graphically may differ from those obtained later with a spreadsheet.

[S], m mol liter^{-1}	:	0.13	0.5	0.8	2.0
v, μ mol min^{-1}	:	1.0	2.5	3.0	4.0
1/[S]	:	7.7	2.0	1.25	0.5
1/v	:	1.0	0.4	0.33	0.25
v/[S]	:	7.7	5.0	3.75	2.0
v	:	1.0	2.5	3.0	4.0
[S]	:	0.13	0.5	0.8	2.0
[S]/v	:	0.13	0.2	0.27	0.5

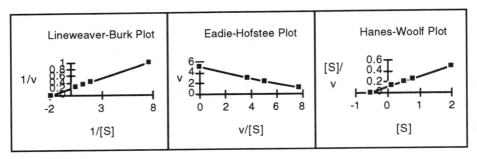

Answers for the three parts of Example 5-7:

Lineweaver-Burk
x-intercept: $-1/K_m$ = -2.0 K_m = 0.5 mM,
y-intercept: $1/V_{max}$ = 0.2 V_{max} = 5.0 μ mol min^{-1}.
Eadie Hofstee
slope : $-K_m$ = -0.5 K_m = 0.5 mM,
y-intercept: V_{max} = 5.0 V_{max} = 5.0 μ mol min^{-1}.
Hanes-Woolf
x-intercept: $-K_m$ = -0.5 K_m = 0.5 mM,
slope: $1/V_{max}$ = 0.2 V_{max} = 5.0 μ mol min^{-1}.

Example 5-8. The permeability of phosphatidylcholine liposomes to 0.3 molar glycerol was measured at temperatures betweeen 33°C and 43°C with the following results:

Rate, cm min^{-1}	1.155	1.646	2.000	2.751
Temperature, °C	33	37	40	43

Estimate the energy of activation for permeation, in k cal from an Arrhenius plot.

Solution.

ln v	0.144	0.498	0.693	1.011
1/RT (k cal)	1.644	1.622	1.610	1.592

The plot of ln v against 1/RT in kilocalories follows. From this plot, the energy of activation is 16.6 k cal mol^{-1}.

Example 5-9. An allosteric enzyme transformed its substrate with a V_{max} of 225 µ mol min^{-1} and gave the following results when velocity was measured as a function of substrate concentration:

[S] x 10^{-5} M	1.58	3.17	3.98	5.02	6.31	7.96	12.86	20.70	66.21
v, µ mol min^{-1}	13	45	64	87	113	139	194	205	223

Estimate $[S_{0.5}]$ and the Hill coefficient for the reaction.

Solution.

We transform the substrate concentration to µmol $liter^{-1}$ for convenience and take the logarithms of these values to obtain:

log [S]	log $v/(V_{max} - v)$
1.2	-1.21
1.5	-0.60
1.6	-0.40
1.7	-0.20
1.8	0.004
1.9	0.20
2.2	0.80
2.3	1.01
2.8	2.05

Then we plot the transformed data as a Hill plot from which we learn that its slope = 2.04, its x-intercept = 1.80 and the y-intercept = -3.67. Consequently, $[S_{0.5}]$ for the enzyme was 63.1 µM and the Hill coefficient, n was 2.

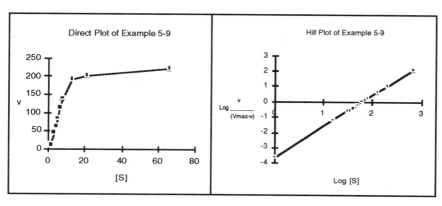

Example 5-10. Assume that the binding of a ligand, L, to a macromolecule gave the following results:

$\bar{v}$	[L]	$\bar{v}$/[L]
0.59	0.20	2.95
0.78	0.30	2.60
0.94	0.41	2.29
1.07	0.52	2.06
1.28	0.77	1.66
1.37	0.92	1.49
1.50	1.21	1.24
1.80	2.65	0.68

Calculate the number of binding sites and the constants of binding and dissociation.

Solution.

From the plot of $\bar{v}$ versus $\bar{v}$/[L] below, n = 2 ; K_a = 1.86; and K_d = 0.54.

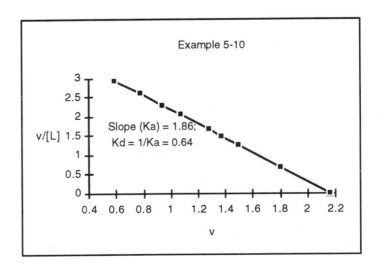

5-16. Spreadsheet solutions

	A	B	C	D	E	F	G
1	EXAMPLE 5-1				EXAMPLE 5-5		
2		INPUT	OUTPUT			INPUT	OUTPUT
3	Amt. subst., μmol	0.0005			Amt. enzyme	0.025	
4	Amt. enzyme, mg	0.0000001			Mol. weight	56000	
5	Enzyme unit		0.0002		Vmax	7.5	
6	Specific activity		5000		Turnover #		1.68E+07
7							
8	EXAMPLE 5-2				EXAMPLE 5-6		
9		INPUT	OUTPUT			INPUT	OUTPUT
10	[S] as Fraction Km	0.1					
11	v/Vmax		0.09090909		LINEWEAVER		
12					Slope	0.05033	
13	[S] as Fraction Km	0.5			y-Intercept	1.83	
14	v/Vmax		0.33333333		x-Intercept	-3.636	
15					Km		0.5464481
16	[S] as Fraction Km	2			Vmax		19.868865
17	v/Vmax		0.66666667				
18					EADIE		
19	EXAMPLE 5-3				Slope	-0.491	
20		INPUT	OUTPUT		y-Intercept	0.626	
21	Vmax	180			x-Intercept	1.23	
22	v	60			Km		-0.491
23	[S], mM	50			Vmax		0.626
24	Km		100				
25							
26	EXAMPLE 5-4				HANES		
27		INPUT	OUTPUT		Slope	1.46	
28	Vmax	3275			y-Intercept	0.85	
29	[S]	0.0005			x-Intercept	-0.582	
30	Km	0.00025			Km		-0.582
31	v		2183.33333		Vmax		0.6849315
32							
33							

Spreadsheet formulas

	A	B	C	D	E	F	G
1	EXAMPLE 5-1				EX. 5-5		
2		INPUT	OUTPUT			INPUT	OUTPUT
3	Amt. subst., μmol	0.0005			Amt. enzyme	0.025	
4	Amt. enzyme, mg	0.0000001			Mol. weight	56000	
5	Enzyme unit		=(1/(B3))*0.0000001		Vmax	7.5	
6	Specific activity		=1/C5		Turnover #		=(F5*F4)/
7							
8	EXAMPLE 5-2				EX. 5-6		
9		INPUT	OUTPUT			INPUT	OUTPUT
10	[S] as Fraction Km	0.1					
11	v/Vmax		=B10/(1+B10)		LINEWEAVER		
12					Slope	0.05033	
13	[S] as Fraction Km	0.5			y-Intercept	1.83	
14	v/Vmax		=B13/(1+B13)		x-Intercept	-3.636	
15					Km		=1/F13
16	[S] as Fraction Km	2			Vmax		=1/F12
17	v/Vmax		=B16/(1+B16)				
18					EADIE		
19	EXAMPLE 5-3				Slope	-0.491	
20		INPUT	OUTPUT		y-Intercept	0.626	
21	Vmax	180			x-Intercept	1.23	
22	v	60			Km		=F19
23	[S], mM	50			Vmax		=F20
24	Km		=(((B21*B23)-(B22*B23))/60)				
25							
26	EXAMPLE 5-4				HANES		
27		INPUT	OUTPUT		Slope	1.46	
28	Vmax	3275			y-Intercept	0.85	
29	[S]	0.0005			x-Intercept	-0.582	
30	Km	0.00025			Km		=F29
31	v		=((B28*B29)/(B30+B29))		Vmax		=1/F27
32							
33							

Spreadsheet solutions

	A	B	C	D	E	F	G	H	I	J
1	EXAMPLE 5-7									
2										
3	LINEWEAVER									
4	INPUT		INPUT							OUTPUT
5	Enter [S]	1/[S]	Enter v	1/v	x^2	xy	y^2		SLOPE	0.104193
6	0.13	7.6923	1	1	59.172	7.692	1		INTCP.	0.19778
7	0.5	2	2.5	0.4	4	0.8	0.16		Km	0.526814
8	0.8	1.25	3	0.33333	1.5625	0.417	0.111		Vmax	5.056119
9	2	0.5	4	0.25	0.25	0.125	0.063		CORRE	0.999905
10										
11	Count	4								
12	Σ	11.442	10.5	1.98333	64.984	9.034	1.334	Σ		
13	(Σx OR Σy)^2	130.93		3.93361						
14	Correl. Calcns.	13.442	11.358	1.18357	13.443					
15										
16	EADIE-HOFSTEE									
17	INPUT		INPUT							OUTPUT
18	Enter [S]	v/[S]	Enter v	v	x^2	xy	y^2		SLOPE	-0.521305
19	0.13	7.6923	1	1	59.172	7.692	1		INTCP.	5.028517
20	0.5	5	2.5	2.5	25	12.5	6.25		Km	-0.521305
21	0.8	3.75	3	3	14.063	11.25	9		Vmax	5.028517
22	2	2	4	4	4	8	16		CORRE	-0.998714
23										
24	Count	4								
25	Σ	18.442	10.5	10.5	102.23	39.44	32.25			
26	(Σx OR Σy)^2	340.12		110.25						
27	Correl. Calcns.	-35.875	8.2956	4.33013	35.921					
28										
29	HANES-WOOLF									
30	INPUT		INPUT							OUTPUT
31	Enter [S]	[S]	Enter v	[S]/v	x^2	xy	y^2		SLOPE	0.198325
32	0.13	0.13	1	0.13	0.0169	0.017	0.017		INTCP.	0.104103
33	0.5	0.5	2.5	0.2	0.25	0.1	0.04		Km	0.524914
34	0.8	0.8	3	0.26667	0.64	0.213	0.071		Vmax	5.042239
35	2	2	4	0.5	4	1	0.25		CORRE	0.999829
36										
37	Count	4								
38	Σ	3.43	10.5	1.09667	4.9069	1.33	0.378			
39	(Σx OR Σy)^2	11.765		1.20268						
40	Correl. Calcns.	1.5594	2.8041	0.55621	1.5596					
41										
42	Lineweaver-Burk Plot				Eadie-Hofstee Plot				Hanes-Woolf Plot	
43										
44										
45										
46										
47										
48										
49										
50										

Lineweaver-Burk Plot

Eadie-Hofstee Plot

Hanes-Woolf Plot

Spreadsheet formulas

	A	B	C	D
1	**EXAMPLE 5-7**			
2				
3	LINEWEAVER-BUR			
4				
5	Enter [S]	I/[S]	Enter v	1/v
6	0.13	=1/A6	1	=1/C6
7	0.5	=1/A7	2.5	=1/C7
8	0.8	=1/A8	3	=1/C8
9	2	=1/A9	4	=1/C9
10				
11	Count	=COUNT(B6:B9)		
12	Σ	=SUM(B6:B9)	=SUM(C6:C9)	=SUM(D6:D9)
13	(Σx OR y)^2	=B12*B12		=D12*D12
14	Correl	=B11*F12-B12*D12	=SQRT(B11*E12-B13)	=SQRT(B11*G12-D13)
15				
16				
17	EADIE-HOFSTEE			
18				
19	Enter [S]	v/[S]	Enter v	v
20	0.13	=C20/A20	1	=C20
21	0.5	=C21/A21	2.5	=C21
22	0.8	=C22/A22	3	=C22
23	2	=C23/A23	4	=C23
24				
25	Count	=COUNT(B20:B23)		
26	Σ	=SUM(B20:B23)	=SUM(C20:C23)	=SUM(D20:D23)
27	(Σx OR y)^2	=B26*B26		=D26*D26
28	Correl	=B25*F26-B26*D26	=SQRT(B25*E26-B27)	=SQRT(B25*G26-D27)
29				
30				
31	HANES-WOOLF			
32				
33	Enter [S]	[S]	Enter v	[S]/v
34	0.13	=A34	1	=B34/C34
35	0.5	=A35	2.5	=B35/C35
36	0.8	=A36	3	=B36/C36
37	2	=A37	4	=B37/C37
38				
39	Count	=COUNT(B34:B37)		
40	Σ	=SUM(B34:B37)	=SUM(C34:C37)	=SUM(D34:D37)
41	(Σx OR y)^2	=B40*B40		=D40*D40
42	Correl	=B39*F40-B40*D40	=SQRT(B39*E40-B41)	=SQRT(B39*G40-D41)
43				

Spreadsheet formulas

	E	F	G	H	I	J
1						
2						
3						
4						**OUTPUT**
5	x^2	xy	y^2		SLOPE	=(B11*F12-B12*D12)/(B11*E12-B13)
6	=B6*B6	=B6*D6	=D6*D6		INTCP.	=(D12*E12-B12*F12)/(B11*E12-B13)
7	=B7*B7	=B7*D7	=D7*D7		Km	=J5*(1/J6)
8	=B8*B8	=B8*D8	=D8*D8		Vmax	=1/J6
9	=B9*B9	=B9*D9	=D9*D9		CORREL	=B14/E14
10						
11						
12	=SUM(E6:E9)	=SUM(F6:F9)	=SUM(G6:G9)			
13						
14	=C14*D14					
15						
16						
17						
18						**OUTPUT**
19	x^2	xy	y^2		SLOPE	=(B25*F26-B26*D26)/(B25*E26-B27)
20	=B20*B20	=B20*D20	=D20*D20		INTCP.	=(D26*E26-B26*F26)/(B25*E26-B27)
21	=B21*B21	=B21*D21	=D21*D21		Km	=J19
22	=B22*B22	=B22*D22	=D22*D22		Vmax	=J20
23	=B23*B23	=B23*D23	=D23*D23		CORREL	=B28/E28
24						
25						
26	=SUM(E20:E23	=SUM(F20:F23	=SUM(G20:G23)			
27						
28	=C28*D28					
29						
30						
31						
32						**OUTPUT**
33	x^2	xy	y^2		SLOPE	=(B39*F40-B40*D40)/(B39*E40-B41)
34	=B34*B34	=B34*D34	=D34*D34		INTCP.	=(D40*E40-B40*F40)/(B39*E40-B41)
35	=B35*B35	=B35*D35	=D35*D35		Km	=J34/J33
36	=B36*B36	=B36*D36	=D36*D36		Vmax	=1/J33
37	=B37*B37	=B37*D37	=D37*D37		CORREL	=B42/E42
38						
39						
40	=SUM(E34:E37	=SUM(F34:F37	=SUM(G34:G37)			
41						
42	=C42*D42					
43						

Spreadsheet solutions

	A	B	C	D	E	F	G	H	I	J
1	EXAMPLE 5-8									
2	INPUT		INPUT							OUTP
3	T,°C	x=1/RT	v	y= Ln v	x^2	x*y	y^2			
4	33	0.0016437	1.155	0.1441	3E-06	2E-04	0.021		SLOPE	-1628
5	37	0.0016225	1.646	0.498348	3E-06	8E-04	0.248		INTCP.	26.913
6	40	0.001607	2	0.693147	3E-06	0.001	0.48		CORREL.	-0.99
7	43	0.0015917	2.751	1.011964	3E-06	0.002	1.024			
8									SOLUTION:	
9	Count	4							Ea (kcal/mol)	-1
10	Σ	0.006465	7.552	2.34756	1E-05	0.004	1.774			
11	(Σx OR Σy)^2	4.18E-05		5.511038						
12	Correl. Calcn.	-9.65E-05	8E-05	1.258382	1E-04					
13										
14	EXAMPLE 5-9									
15	INPUT		INPUT							OUTP
16	[S], E-5	x=log [S]	v	Logv/(V-v)	x^2	x*y	y^2			
17	1.58	1.1986571	13	-1.21239	1.437	-1.453	1.47		SLOPE	2.0240
18	3.17	1.5010593	45	-0.60206	2.253	-0.904	0.362		INTCP.	-3.627
19	3.98	1.5998831	64	-0.40065	2.56	-0.641	0.161		CORREL	0.9981
20	5.02	1.7007037	87	-0.20036	2.892	-0.341	0.04			
21	6.31	1.8000294	113	0.00386	3.24	0.007	1E-05		SOLUTION:	
22	7.96	1.9009131	139	0.208516	3.613	0.396	0.043		x-intcp	
23	12.86	2.109241	194	0.79644	4.449	1.68	0.634		S(0.5)	
24	20.7	2.3159703	205	1.010724	5.364	2.341	1.022		n	2
25	66.21	2.8209236	223	2.047275	7.958	5.775	4.191			
26										
27	Count	9								
28	Σ	16.94738	1083	1.651357	33.77	6.861	7.924			
29	(Σx OR Σy)^2	287.2137		2.726981						
30	Correl. Calcn.	33.758381	4.084	8.281708	33.82					
31										
32	EXAMPLE 5-10									
33	INPUT		INPUT							OUTPU
34	v	x = v	[L]	y=v/[L]	x^2	x*y	y^2			
35	0.59	0.59	0.2	2.95	0.348	1.741	8.703		SLOPE	-1.8790
36	0.78	0.78	0.3	2.6	0.608	2.028	6.76		INTCP.	4.06281
37	0.94	0.94	0.41	2.292683	0.884	2.155	5.256		CORREL.	-0.9999
38	1.07	1.07	0.52	2.057692	1.145	2.202	4.234			
39	1.28	1.28	0.77	1.662338	1.638	2.128	2.763		SOLUTION:	
40	1.37	1.37	0.92	1.48913	1.877	2.04	2.218		Ka	-1
41	1.5	1.5	1.21	1.239669	2.25	1.86	1.537		Kd	-0
42	1.8	1.8	2.65	0.679245	3.24	1.223	0.461		n	2
43										
44	Count	8								
45	Σ	9.33	6.98	14.97076	11.99	15.38	31.93			
46	(Σx OR Σy)^2	87.0489		224.1236						
47	Correl. Calcn.	-16.67398	2.979	5.597552	16.67					
48										

Spreadsheet formulas

	A	B	C	D
1	EXAMPLE 5-8			
2	INPUT		INPUT	
3	T,°C	x=1/RT	v	y= Ln v
4	33	=1/(1.98717*(273.15+A4))	1.155	=LN(C4)
5	37	=1/(1.98717*(273.15+A5))	1.646	=LN(C5)
6	40	=1/(1.98717*(273.15+A6))	2	=LN(C6)
7	43	=1/(1.98717*(273.15+A7))	2.751	=LN(C7)
8				
9	Count	=COUNT(A4:A7)		
10	Σ	=SUM(B4:B7)	=SUM(C4:C7)	=SUM(D4:D7)
11	(Σx OR Σy)^2	=B10*B10		=D10*D10
12	Correl. Calcn.	=B9*F11-B10*D10	=SQRT(B9*E11-B11)	=SQRT(B9*G11-D11)
13				
14	EXAMPLE 5-9			
15	INPUT		INPUT	
16	[S], E-5	x=log [S]	v	Logv/(V-v)
17	1.58	=LOG10(A17*10)	13	=LOG10(C17/(225-C17))
18	3.17	=LOG10(A18*10)	45	=LOG10(C18/(225-C18))
19	3.98	=LOG10(A19*10)	64	=LOG10(C19/(225-C19))
20	5.02	=LOG10(A20*10)	87	=LOG10(C20/(225-C20))
21	6.31	=LOG10(A21*10)	113	=LOG10(C21/(225-C21))
22	7.96	=LOG10(A22*10)	139	=LOG10(C22/(225-C22))
23	12.86	=LOG10(A23*10)	194	=LOG10(C23/(225-C23))
24	20.7	=LOG10(A24*10)	205	=LOG10(C24/(225-C24))
25	66.21	=LOG10(A25*10)	223	=LOG10(C25/(225-C25))
26				
27	Count	=COUNT(A17:A25)		
28	Σ	=SUM(B17:B25)	=SUM(C17:C25)	=SUM(D17:D25)
29	(Σx OR Σy)^2	=B28*B28		=D28*D28
30	Correl. Calcn.	=B27*F32-B28*D28	=SQRT(B27*E32-B29	=SQRT(B27*G32-D29)
31				
32	EXAMPLE 5-1(			
33	INPUT		INPUT	
34	v	x = v	[L]	y=v/[L]
35	0.59	=A35	=0.2	=A35/C35
36	0.78	=A36	0.3	=A36/C36
37	0.94	=A37	0.41	=A37/C37
38	1.07	=A38	0.52	=A38/C38
39	1.28	=A39	0.77	=A39/C39
40	1.37	=A40	0.92	=A40/C40
41	1.5	=A41	1.21	=A41/C41
42	1.8	=A42	2.65	=A42/C42
43				
44	Count	=COUNT(A35:A42)		
45	Σ	=SUM(B35:B42)	=SUM(C35:C42)	=SUM(D35:D42)
46	(Σx OR Σy)^2	=B45*B45		=D45*D45
47	Correl. Calcn.	=B44*F51-B45*D45	=SQRT(B44*E51-B46	=SQRT(B44*G51-D46)
48				

Spreadsheet formulas

	E	F	G	H	I	J
1						
2						
3	x^2	x*y	y^2			OUTPUT
4	=B5*B5	=B5*D4	=D4*D4			
5	=B6*B6	=B6*D5	=D5*D5		SLOPE	=(B10*F10-B11*D10)/(B10*E10-B1
6	=B7*B7	=B7*D6	=D6*D6		INTCP.	=(D10*E10-B11*F10)/(B10*E10-B1
7	=B8*B8	=B8*D7	=D7*D7		CORREL.	=B13/E12
8						
9					SOLUTION	
10	=SUM(E4:E7)	=SUM(F4:F7)	=SUM(G4:G7)		Ea (kcal/mo	=J5/1000
11						
12	=C13*D12					
13						
14						
15						
16	x^2	x*y	y^2			
17	=B20*B20	=B20*D17	=D17*D17			
18	=B21*B21	=B21*D18	=D18*D18			OUTPUT
19	=B22*B22	=B22*D19	=D19*D19			
20	=B23*B23	=B23*D20	=D20*D20		SLOPE	=(B30*F27-B31*D27)/(B30*E27-B3
21	=B24*B24	=B24*D21	=D21*D21		INTCP.	=(D27*E27-B31*F27)/(B30*E27-B3
22	=B25*B25	=B25*D22	=D22*D22		CORREL.	=B33/E29
23	=B26*B26	=B26*D23	=D23*D23			
24	=B27*B27	=B27*D24	=D24*D24		SOLUTION	
25	=B28*B28	=B28*D25	=D25*D25		x-intcp	=-(J21)/J20
26					S(0.5)	=10^J25
27	=SUM(E17:E25)	=SUM(F17:F25)	=SUM(G17:G25)		n	=J20
28						
29	=C33*D29					
30						
31						
32						
33						
34	x^2	x*y	y^2			
35	=B40*B40	=B40*D35	=D35*D35			
36	=B41*B41	=B41*D36	=D36*D36			
37	=B42*B42	=B42*D37	=D37*D37			
38	=B43*B43	=B43*D38	=D38*D38			OUTPUT
39	=B44*B44	=B44*D39	=D39*D39			
40	=B45*B45	=B45*D40	=D40*D40		SLOPE	=(B49*F45-B50*D45)/(B49*E45-B51
41	=B46*B46	=B46*D41	=D41*D41		INTCP.	=(D45*E45-B50*F45)/(B49*E45-B51
42	=B47*B47	=B47*D42	=D42*D42		CORREL.	=B52/E47
43						
44					SOLUTION	
45	=SUM(E35:E42)	=SUM(F35:F42)	=SUM(G35:G42)		Ka	=J40
46					Kd	=1/J40
47	=C52*D47				n	=-J41/J40
48						

5-17. Review questions

5-1.   What effect does substrate concentration have on the velocities of enzyme reactions?

5-2.   What are the dimensions of the Michaelis-Menten constant?

5-3.   What is the energy of activation of a reaction?

5-4.   What does the constant A represent in the Arrhenius equation?

5-5.   What information might graphical analysis of the Arrhenius equation yield that may not otherwise be apparent from least squares analysis?

5-6.   If two enzymes from different sources catalyze the same reaction, which is more efficient, the one with the larger, or the one with the smaller K_m value?

5-7.   What is meant by a linear transformation of the Michaelis-Menten equation?

5-8.   What is an enzyme inhibitor?

5-9.   Compare and contrast three types of inhibition of enzyme reactions.

5-10.   What are the dimensions of K_i?

5-11.   What does K_m/V_{max} represent in the Lineweaver-Burk and Hanes-Woolf plots?

5-12.   Draw curves which illustrate three ways in which Arrhenius plots may be of diagnostic value.

5-13.   Label double-reciprocal plots to show the factors by which the slopes and intercepts are changed by competitive, noncompetitive and uncompetitive inhibition.

5-14.   What is the x-intercept in plots representing each of the three linear transformations of the Michaelis-Menten equation?

5-15.   On what variable or variables does v depend in the Michaelis-Menten equation?

5-16.   On what variable or variables does 1/v depend in the double-reciprocal equation for noncompetitive equation?

5-17.   To what does $1/V_{max}$ correspond in the Lineweaver-Burk and Hanes-Woolf plots?

5-18.   Why is it dimensionally correct to say that, in Lineweaver-Burk plots, the x-intercept is $-1/K_m$, but incorrect to say that in the Hanes-Woolf plot, the x-intercept is $-1/K_m$?

5-19.   To what term is concentration / time equivalent in the Michaelis-Menten equation?

5-20.   How many terms have units of concentration in the double-reciprocal equation for uncompetitive inhibition? Name them.

5-21.   Define the term *allosteric enzyme.*

5-22.   Convert the Hill equation to a form which is suitable for graphical analysis of allosteric enzyme kinetics. Show a plot corresponding to the equation.

5-23.   Give an account of the main features of the concerted and sequential models of how allosteric enzymes regulate reaction rate. Name the scientists who proposed the two models.

5-24.   What constant for allosteric enzymes is equivalent to K_m for Michaelis-Menten enzymes?.

5-25. What do the following values of the Hill coefficient suggest about cooperativity in allosteric enzymes: n = 1; n > 1 and n < 1?

5-26. Explain how you would plot enzyme kinetic data to estimate K_m and V_{max} using the direct plot of Eisenthal and Cornish-Bowden?

5-27. What criteria would you use to demonstrate that receptors are labeled. Explain what each criterion means.

5-28. Show why the exponential relation between k and E_a explains the fact that a small decrease in E_a or a small increase in temperature causes a large increase in k.

5-29. Explain from the Arrhenius equation how an enzyme may increase reaction rate.

5-30. List two goals of chemical kinetics.

5-31. Define the terms *rate constant, order of reaction* and *molecularity*.

5-32. Describe the term *half-life*.

5-33. Describe two general kinetic approaches for following the progress of biochemical reactions.

5-34. What precautions should one follow for successful use of the Wilkinson, statistical estimation of K_m and V_{max}?

5-35. List the major steps that are necessary for statistical estimation of K_m and V_{max} by the method of Wilkinson.

5-36. When would you use statistical estimation of K_m and V_{max}?

5-37. List some biochemical processes which follow first order kinetics.

5-38. List some biochemical processes which follow second order kinetics.

5-39. Define the terms *specific activity* and *molecular activity* of enzymes.

5-40. How do you interpret the rectangular hyperbolic curve for substrate concentration versus the velocity of an enzyme?

5-41. How do you interpret the fact that plots of ligand-receptor data according to the Scatchard equation sometimes give nonlinear curves?

5-42. Define the terms *ligand* and *receptor*.

5-43. Name two ways by which regulatory enzymes may increase or decrease the rates of metabolic pathways.

5-44. Describe the nature and function of the *active site of enzymes*.

5-45. Define the term *ribozyme* and explain why the discovery of ribozymes has led to new thinking about enzymes.

5-46. What are the abscissa and ordinate for each of the following: *the Scatchard plot, a Hill plot for allosteric enzymes* and *the a Hill plot for the binding of oxygen to hemoglobin*?

5-47. How would you calculate k for the second order reaction A + A → Products?

5-48. How would you calculate k for the second order reaction A + B → Products?

5-49. What are *prosthetic groups*? How do they help enzymes to function?

5-50. Define the terms *positive* and *negative cooperativity*.

QUESTIONS 5-51 TO 5-60 CONSIST OF TWO STATEMENTS LINKED BY
THE WORD *BECAUSE*; THE STATEMENTS ARE FOLLOWED BY (a, b, c, d, e).

Circle a if *both* statements are *true and* the *second* is the correct
    explanation of the first;

Circle b if *both* statements are *true* but the second is NOT the
    correct explanation of the *first*;

Circle c if the *first* statement is *true* but the *second* is *false*;

Circle d if the *first* is *false* but the *second* is *true*;

Circle e if *both first* and *second* statements are *false*.

5-51. It is easy for an investigator to recognize competitive inhibition in the laboratory *because* the degree to which an enzyme is inhibited is decreased by increasing the concentration of substrate (a, b, c, d, e).

5-52. Biochemists find determination of the energy of activation useful *because* it can be used to relate phase changes to the preexponential factor (a, b, c, d, e).

5-53. Plots of substrate concentration againt the velocities of enzyme reactions are used frequently *because* they provide effective methods for estimating K_m and V_{max} (a, b, c, d, e).

5-54. Least-squares computations of the energy of activation should be supported with graphical analysis of the same data *because* it is generally a good idea to use different methods to check the validity of a conclusion or result (a, b, c, d, e).

5-55. In double-reciprocal plots of competitive inhibition, the intercept on the $1/v$ axis is greater for the inhibited than for the uninhibited enzyme *because* V_{max} is decreased by the inhibitor and cannot be restored regardless of how high the substrate concentration may be raised (a, b, c, d, e).

5-56. If Eadie-Hofstee plots are used to analyze data for competitive inhibitors, the curve for the inhibited reactions will have the same y-intercept as that for the control *because* V_{max} is unchanged in competitive inhibition (a, b, c, d, e).

5-57. If Eadie-Hofstee plots are used to analyze data for competitive inhibitors the curves for the inhibited reactions will have greater slopes than the slope of the control *because* K_m is increased in competitive inhibition (a, b, c, d, e).

5-58. K_m is decreased in noncompetitive inhibition *because* the inhibitor causes the enzyme to use smaller quantities of substrate (a, b, c, d, e).

5-59. It is inconsequential which of the three linear plots (Lineweaver-Burk, Eadie-Hofstee, or Hanes-Woolf) an investigator chooses for estimating K_m and V_{max} *because* each of the three has advantages and disadvantages (a, b, c, d, e).

5-60. An enzyme may appear not to undergo saturation with its substrate *because* the rate of decomposition of its enzyme-substrate complex may be extremely rapid and therefore difficult to make rate-limiting (a, b, c, d, e).

5-18. Review problems

5-61. Whenever possible, it is desirable that experimental determinations of K_m embrace a range of substrate concentrations from 0.2 to 5.0 K_m. To what fractions of the maximal velocities do the velocities attained at 0.2 K_m; 1.0 K_m; 2.0 K_m and 5.0 K_m correspond?

5-62. Assume that two competitive inhibitors X and Y were used in an experiment. Assume further that K_m for the control was 0.3 mM, that K_m for X was 1.5 mM and that K_m for Y was 2.5 mM. Calculate the velocities of the control and inhibited reactions as fractions of V_{max} for a substrate concentration of 2.2 mM.

5-63. The K_m values of an enzyme which can catalyze the transformation of two stereoisomers, A and B, to the same intermediate were $2.7 * 10^{-3}$ mM and $1.9 * 10^{-2}$ mM respectively. If V_{max} was identical for the two substrates, calculate the velocities of the enzyme reaction at substrate concentrations of $1.0 * 10^{-3}$ M; $8 * 10^{-3}$ mM; $1.0 * 10^{-1}$ mM and 1.5 mM for each of the two isomers, as fractions of V_{max}. Are these results consistent with the Michaelis-Menten equation or not? Explain.

5-64. Calculate the substrate concentration of an enzyme reaction in terms of K_m if the velocity of the reaction is 3/4 of V_{max}.

5-65. Show that if $K_m = 5.82 * 10^{-3}$ M and $V_{max} = 6.2$ mol min^{-1}, then $v = 0.75 * (6.2$ mol min$^{-1})$ and $[S] = 3 * (5.82 * 10^{-3}$ M). Explain how this arithmetic result is related to the algebraic prediction of question 5-64.

5-19. Additional problems

A . Enzyme kinetics and binding equilibria

5-66.   The following data were obtained for an enzyme-catalyzed reaction:

[S], mM:	2.00	2.67	3.33	6.70	13.33	20.00	26.70	33.33	67.00
v, mmols min^{-1}:	2.02	2.56	3.02	4.83	6.82	7.94	8.65	9.12	10.30

Determine K_m, V_{max} and the reliability of the least squares fit of the data for the three linear transformations of the Michaelis-Menten equation.

5-67. In the same experiment from which the data of question 5-66 were taken, three inhibitors labeled X, Y and Z at concentrations of 0.5 mM, 7 µM, and 25 µM respectively, were used to inhibit the enzyme reaction with the following results:

[S], mM	2.00	2.67	3.33	6.70	13.33	20.00	26.70	33.33	67.00
v, (+X)	1.43	1.83	2.22	3.81	5.56	7.15	7.90	8.33	10.10
v, (+Y)	1.35	1.71	2.02	3.17	4.44	5.30	5.82	6.15	7.20
v, (+Z)	1.63	2.04	2.32	3.21	4.03	4.34	4.54	4.76	5.26

As in question 5-66, the units of velocity were mmols min^{-1}. Determine the type of inhibition produced by each inhibitor and calculate K_i for each.

5-68. Subject the data of question 5-66 to the statistical procedure of Wilkinson (1961) to estimate K_m and V_{max}. Then explain your results in the context of the least squares parameters obtained from question 5-66 and the standard errors obtained with the Wilkinson procedure.

5-69. The permeability of phosphatidylcholine liposomes to 0.3 M glycerol was measured at temperatures between 25°C and 43°C with the following results:

Rate, cm min^{-1}	0.537	0.881	1.220	1.700	2.000	2.770
Temperature °C	25	30	33	37	40	43

Estimate the energy of activation for permeation.

5-70. The rate of oxidative deamination of glutamate by mitochondrial glutamate dehydrogenase (EC 1.4.1.4 L-glutamate: NADPH oxidoreductase, deaminating) was studied as a function of temperature. Calculate the energy of activation for the range of temperatures over which the reaction was studied:

Temp., °C	10	20	30	40
Rate†	4.09	5.58	7.19	9.33

† Rate measured in µmol of NADP reduced min^{-1}/µg protein.

Express your answer both in k Joules and k Calories.

5-71. The following data from a simulated experiment represent the binding of a ligand L, to a macromolecule:

v	1.8	2.4	3.0	3.3	3.6	3.9	4.2	4.5	4.8
[L], mM	0.52	0.88	1.40	1.84	2.44	3.36	5.00	8.84	24.60

Calculate the number of binding sites and the constants of binding and dissociation.

5-72. An experiment on the binding of a ligand L to a protein gave the following data:

v	0.59	0.78	0.94	1.07	1.28	1.37
[L], µM	0.20	0.30	0.41	0.52	0.77	0.92

Calculate the number of binding sites and the constants of binding and dissociation.

5-73. The following are selected data from a simulated experiment on the binding of oxygen to hemoglobin:

pO$_2$ (torr)	% Saturation
39.8	9.9
50.2	17.5
81.3	45.6
101.2	60.2
170.0	87.2
436.0	99.0

Use these data to determine the minimum number of subunits in the hemoglobin molecule.

5-74. An allosteric enzyme transformed its substrate with a V_{max} of 225 μmol min^{-1} and gave the following data:

[S] * 10^{-5} M:	1.58	3.17	3.98	5.02	6.31	7.96	12.86	20.70	66.21
v, μmol min^{-1}:	13	45	64	87	113	139	180	207	223

Estimate [S]$_{0.5}$ and the Hill coefficient for the reaction.

B. First-order reactions and rate constants

5-75.   A biomolecule was transformed enzymatically to products and the course of the reaction was followed spectrophotometrically at 303 K. Before the reaction was started, the initial concentration of the biomolecule in buffered, aqueous solution was 15 μg ml^{-1}. During the reaction, the concentration of product formed, μg ml^{-1}, was measured yielding the following results:

Time, min	0	5	10	20	25
Product formed, μg ml^{-1}	0	6.5	11.2	14.1	14.6

Calculate the first order rate constant and the half-life of the reaction.

5-76. The following data were obtained at 298 K for the turnover of a lipid labeled with ^{14}C :

Time, min	0	2	4	12	18	24	42	48	61	95
^{14}C, DPM *10^2	1000	920	860	660	540	440	230	200	120	40

What were the first order rate constant and half-life of the reaction?

5-77.   The course of an NAD-linked dehydrogenase reaction was followed by keeping one reactant at a constant concentration while measuring, (at 340 nm), the decrease in absorbance of the other reactant, NADH. The following is an excerpt of the results obtained at 37°C:

Time, sec	0	10	60	120	360	720	900
Absorbance, NADH	1.66	1.63	1.50	1.35	0.895	0.482	0.354

Determine the pseudo first-order rate constant and $t_{1/2}$ of the reaction.

5-78. The catabolism of an oligonucleotide (the reactant) to its constituent mono-nucleotides (the products) was followed spectroscopically at 298 K with the following results:

Time, min	0	2	5	10	15
[Reactant] , μM	2.5	2.0	1.5	0.72	0.40

Calculate the 1st order rate constant and $t_{1/2}$ for the catabolic reaction.

5-79. Assume that 100 n mol of a biomolecule was converted to a product at 310 K in an irreversible, enzyme-catalyzed reaction. Determine the 1st order rate constant and half-life for the conversion from the following data:

Time, min	0	5	10	20	30
[Product] , mM	20	35	50	65	80

C. Second-order reactions and rate constants

5-80. A reaction is classified as second order if (a) its rate is proportional to the product of the concentrations of two reacting species or (b) the reaction rate is proportional to the square of the concentration of one of the reactants. The following data were obtained for the *in vitro* transformation of a biomolecule in a reaction of the second type, i.e. A + A $\longrightarrow$ Products:

Time, min	0	10	20	30	60	120
[Reactant] , $\mu mol\ L^{-1}$	14.0	9.5	7.1	5.8	3.6	2.1

Calculate the 2nd order rate constant and $t_{1/2}$ for the synthesis of this biomolecule.

5-81. A second order biomolecular reaction of the type A + B $\longrightarrow$ Products, was carried out at 298 K between the two reacting substances which were present in equivalent concentrations (a=b) of 5 $\mu mol\ liter^{-1}$. The following data were obtained:

Time, sec	10	20	30	40	50	60	120
μmol reacting, x	1.32	1.89	2.30	2.67	2.87	3.08	3.78

Calculate the 2nd order rate constant of the reaction.

5-82. From the accompanying data on the concentrations of the rate determining reactant used in specified time periods, calculate the second order rate constant for the reaction between two biomolecules A and B at 30°C :

Time, sec	0	10	15	30	60
[Reaxtant used] , $mmol\ L^{-1}$	0.000	0.008	0.012	0.020	0.030

At the start of the experiment, the concentration of A was 0.1 mmol L^{-1} and that of B was 0.05 mmol L^{-1}.

Suggested Reading

Abraham, R.T., L.M. Kamitz, J.P. Secrist and P.J. Leibson 1992. Signal Transduction Through the T cell Receptor. *Trends Biochem. Sci.*, 17: 434-437.

Alberty, R.A. and A. Cornish-Bowden 1993. The pH Dependence of the Apparent Equilibrium Constant, K', of a Biochemical Reaction. *Trends in Biochem. Sci.* 18: 288-291.

Anderson, S.R. 1991. Time-Resolved Fluorescence Spectroscopy. *J. Biol. Chem.* 266:11405-11408.

Asaoka, Y, S. Nakamura, K. Yoshida and Y. Nishizuka 1992. Protein Kinase C, Calcium and Phospholipid Degradation. *Trends Biochem. Sci.*, 17: 414-417.

Ault, A 1974. An Introduction to Enzyme Kinetics. *J. Chem. Educ.*, 51: 381-386.

Barrow, G. M. 1974. *Physical Chemistry for the Life Sciences, 2d. ed.*, McGraw-Hill, NY

Bell, R.M. and D.J. Burns 1991. Lipid Activation of Protein Kinase C. *J. Biol. Chem.* 266: 4661-4664

Borman, S., November 15, 1993. Synthetic Receptors Make it Possible to Turn Genes on and off at Will. *Chem. & Engineering News* Pages 55 - 57.

Burk, D., 1984. Enzyme Kinetic Constants: the Double Reciprocal Plot. *Trends Biochem. Sci.* 9: 202-204.

Cech, T. R. 1987. The Chemistry of Self-Splicing RNA and RNA Enzymes. *Science* 236: 1532-1539.

Chang, R. 1981. *Physical Chemistry With Applications to Biological Systems. 2d. ed.*, Macmillan, New York.

Cleland, W. W., 1967. *The Statistical Analysis of Enzyme Kinetic Data.* Advances in Enzymology, ed. F. Nord, Wiley, New York, 29: 1-32.

Cleland, W.W. 1979. *Statistical Analysis of Enzyme Kinetic Data.* Methods in Enzymology, Academic Press, New York 63: 103-138.

Cohen, P. 1992. Signal Integration at the Level of Protein Kinases, Protein Phosphatases and their Substrates. *Trends Biochem. Sci.* 17: 408-413.

Cohen, P. and P.T.W. Cohen 1989. Protein Phosphatases Come of Age. *J. Biol. Chem.* 264: 21435-21438.

Cornish-Bowden, A. 1981. *Basic Mathematics for Biochemists*, Chapman and Hall, London.

Culotta, Elizabeth and D.E. Koshland 1992. NO News is Good News. *Science* 258: 1862-1865.

Dahlquist, F. W. 1978. *The Meaning of Scatchard and Hill Plots.* Methods in Enzymology, Academic Press, New York 48: 270-299.

Dickerson, R. E., H. B. Gray, M. Y. Darensbourg and D. J. Darensbourg 1984. *Chemical Principles, 4th ed.*, Benjamin/Cummings, Reading, Massaachusetts.

Dixon, M. and E. C. Webb 1979. *Enzymes. 3d. ed.*, Academic Press, New York.

Dowd, J. E. and D. S. Riggs 1965. A Comparison of Estimates of Michaelis-Menten Kinetic Constants From Various Linear Transformations. *J. Biol. Chem.* 240: 863-869.

Eisenberg, D. and D. M. Crothers 1979. *Physical Chemistry With Applications to the Life Sciences.* Benjamin/ Cummings, Massachusetts.

Erikson, R.L. 1991. Structure, expression, and regulation of protein kinases involved in the phosphorylation of ribosomal protein S6. *J. Biol. Chem.* 266: 6007-6110.

Exton, J.H. 1990. Signaling Through Phosphatidylcholine Breakdown. *J. Biol. Chem.* 26 1-4.

Fantl, W.J., D.E. Johnson and L.T. Williams 1993. *Signalling by Receptor Tyrosine Kinases.* Annu. Rev. Biochem. 62: 453-48.

Gibson, Q.H. 1989. Hemoproteins, Ligands, and Quanta. *J. Biol. Chem.* 264: 20155-20158.

Glick, N. A., D. Landman and B. D. Roufogalis 1979. Correcting Lineweaver-Burk Calculations of V and K_m. *Trends Biochem. Sci.* 4: N82-N83.

Halling, Peter J. 1989. Do the Laws of Chemistry Apply to Living Cells? *Trends Biochem. Sci.* 14: 317-318.

Hansch, C. and T.E. Klein 1991. *Quantitative Structure-Activity Relationships and Molecular Graphics in Evaluation of Enzyme-Ligand Interactions.* Methods in Enzymology, Academic Press, N.Y. 202: 512-543.

Hepler, J.R. 1992. G Proteins. *Trends Biochem. Sci.*, 17: 383-387.

Hershey, J.W.B. 1989. Protein Phosphorylation Controls Translation Rates. *J. Biol. Chem.* 264: 20823-20826.

Jencks, W.P. 1989. How Does a Calcium Pump Pump Calcium? *J. Biol. Chem.* 264: 18855-18858.

Johnson, L.N. and D. Barford 1990. Glycogen Phosphorylase. The Structural Basis of Allosteric Response and Comparison with Other Allosteric Proteins. *J. Biol. Chem.* 265: 2409-2412.

Katz, M. A. 1976. *Calculus for the Life Sciences.* Marcel Dekker, New York.

Kennelly, P.J. and E.G. Krebs 1991. Consensus Sequences as Substrate Specificity Determinants for Protein Kinases and Protein Phosphatases. *J. Biol. Chem.* 266: 15555-15558.

Khorana, H.G. 1992. Rhodopsin, photoreceptor of the rod cell. An Emerging Pattern for Structure and Function. *J. Biol. Chem.* 267: 1-4.

Klotz, I. 1986. *Introduction to Biomolecular Energetics.* Academic Press, Orlando, FL.

Knowles, R.G. and S. Moncada 1992. NO as a Signal in Blood Vessels. *Trends Biochem Sci.* 258: 399–402.

Kolb, A., S. Busby, H. Buc, S. Garges and S. Adhya 1993. *Transcriptional Regulation by cAMP and its Receptor Protein.* Annu. Rev. Biochem. 62:749-796.

Koshland, D. E., G. E. Némethy and D. Filmer 1966. Comparison of Experimental Binding Data and Theoretical Models in Proteins Containing Subunits. *Biochemistry 5*: 365-385.

Koshland, D. E. 1992. The Molecule of the Year. *Science* 258: 1861.

Krieger, J.H. April 26, 1993. Computer-Aided Molecular Design Has Continuing Impact. *Chem. & Engineering News.* Pages 25-29.

Laszlo, J.A. 1987. Determination of stoichiometric association constants by a non-iterative computational method. *Computer Applications in the Biosciences* 3: 351-357.

Lehninger, A. 1975. L. *Biochemistr.* 2d ed., Worth, New York.

Lehninger, A. L., D. J. Nelson and M. M. Cox 1992. *Principles of Biochemistry*, 2d. ed., Worth, New York.

Livingstone, D.J. 1991. *Pattern Recognition Methods in Rational Drug Design.* Methods in Enzymology, Academic Press, N.Y., 203: 613-638..

Lyons, J. M., J. K. Raison and J. Kumamoto 1974. *Polarographic Determination of Phase Changes in Mitochondrial Membranes in Response to Temperature.* Methods in Enzymology, Academic Press, NY, 32: 258-262. **Note**. *This paper contains a section on Arrhenius Plots.*

Majerus, P.W. 1992. *Inositol Phosphate Biochemisty.* Annu. Rev. Biochem. 61: 225-250.

Martin, Y.C. 1991. *Computer-Assisted Rational Drug Design.* Methods in Enzymology, Academic Press, N.Y., 203: 587-693.

Michel, R.H. 1992. Inositol Lipids In Cellular Signalling Mechanisms. *Trends Biochem. Sci.* 17: 274-276.

Montgomery, R., and C. A. Swenson 1976. *Quantitative Problems in the Biochemical Sciences,* 2d. ed., Freeman, San Francisco.

Morris, J. G. 1974. *A Biologist's Physical Chemistry, 2d ed.,* Arnold, Baltimore, Maryland.

Munson, P. J. 1983. *LIGAND : A Computerized Analysis of Ligand Binding Data.* Methods in Enzymology, Academic Press, N.Y., 92: 543-576.

Newton, A.C. and D.S. Williams 1993. Does Protein Kinase C Play a Role in Rhodopsin Desenitization? *Trends in Biochem. Sci.* 18: 275-277.

Nimmo, I. A. and G. L. Atkins 1979. The Statistical Analysis of Non-Normal (Real?) Data. *Trends Biochem. Sci.* 4: 236-238.

Nishizuka, Y. 1992. Signal Transduction: Crosstalk. *Trends Biochem. Sci.,* 17: 367.

Nishizuka, Y. 1992. Intracellular Signaling by Hydrolysis of Phospholipids and Activation of Protein Kinase C. *Science* 258: 607-614.

Noller, H.F., V. Hoffarth and L. Zimniak 1992. Unusual Resistance of Peptidyl Transferase to Protein Extraction Procedures. *Science* 256: 1416-1419.

Orsi, B. A. and K. F. Tipton 1979. *.Kinetic Analysis of Progress Curves.* Methods in Enzymology, Academic Press, N.Y., 63: 159-183.

Peitsch, M.C., C. Borner and J. Tschopp 1993. Sequence Similarity of Phospholipase A_2 Activating Protein and the G-protein β-subunits: a New Concept of Effector Protein Activation in Signal Transduction? *Trends in Biochem. Sci.* 18: 292-294.

Piccirilli, J.A., T. S. McConnell, A. J. Zaug, H. F. Noller and T. R. Cech 1992. Aminoacyl Esterase Activity of the Tetrahymena Ribozyme. *Science* 261: 1420-1424.

Pyle, A.M. 1993. Ribozymes: A Distinct Class of Metalloenzymes. *Science* 261: 709-714.

Rhee, S.G. and K.D. Choi 1992. Regulation of Inositol phospholipid-specific phospholipase C isozymes. *J. Biol. Chem.* 267: 12393-12396.

Roach, P.J. 1991. Multisite and Hierarchal Protein Phosphorylation. *J. Biol. Chem.* 266: 14139-14142.

Rodbard, D. and H. A. Feldman 1975. *Theory of Protein-Ligand Interaction.* Methods in Enzymology, Academic Press, New York, 36: 3-16.

Rodbard, D. and G. R. Frazier 1975. *Statistical Analysis of Radioligand Assay Data.* Methods in Enzymology, Academic Press, New York, 37: 3-22.

Rudolph, F. B. and H. J. Fromm 1979. *Plotting Methods for Analyzing Enzyme Rate Data .* Methods in Enzymology, Academic Press, New York, 63: 138-159.

Scatchard, G. 1949. The Attractions of Proteins for Small Molecules & Ions. *Ann. NY Acad. Sci.*, 51: 660-672.

Schlessinger, J., 1993. How Receptor Tyrosine Kinases Activate Ras. *Trends in Biochem. Sci.* 18: 273-275.

Segel, I. H. 1976. *Biochemical Calculations. 2d. ed.,* Wiley, New York.

Shabb, J.B. and J.D. Corbin 1992. Cyclic nucleotide-binding domains in proteins having diverse functions. *J. Biol. Chem.* 267: 5723-5726.

Smith, D. B. 1979. The Menten of the Michaelis-Menten Equation. *Trends Biochem. Sci.* 4: N150.

Smith, E. L., R. L. Hill, I. R. Lehman, R. J. Lefkowitz, P. Handler and A. White 1983. *Principles of Biochemistry. 7th ed.,* McGraw-Hill, New York.

Soderling, T.R. 1990. Protein Kinases. Regulation by Autoinhibitory Domains. *J. Biol. Chem.* 265: 1823-1826.

Spencer, D.M., T.J. Wandless, S.L. Schreiber and G.R. Crabtree 1993. Controlling Signal Transduction with Synthetic Ligands. *Science* 262: 1019-1024.

Stiles, G.L. 1992. Adenosine Receptors. *J. Biol. Chem.* 267: 6451-6454.

Symons, R.H. 1992. *Small Catalytic RNAs.* Annu. Rev. Biochem. 61: 641-71.

Trowbridge, I..S. 1991. CD45. A Prototype for Transmsmbrane Tyrosine Phosphatases. *J. Biol. Chem.* 266: 23517-23520.

Walton, K.M. and J.E. Dixon 1993. *Protein Tyrosine Phosphatases*: Annu. Rev. Biochem. 62: 101-120.

Westheimer, F.H. 1987. Why Nature Chose Phosphates. *Science* 235: 1173-1178.

Wilkinson, G. N. 1961. Statistical Estimations in Enzyme Kinetics. *Biochem. J.* 80: 324-332.

Zierler, K. 1989. Misuse of Nonlinear Scatchard Plots. *Trends Biochem. Sci.* 14: 314 - 317.

Chapter 6
Molecular Immunology and Immunochemical Assays

6-1.   Immunochemical assays are powerful biomedical tools

The immune system consists mainly of specialized white blood cells and their products. This astonishing system *protects mammalian cells from infections in three ways*:

*   *Humoral immunity*: secreting antibodies into body fluids or "humors" to inactivate potentially infective agents circulating in the body fluids;
*   *Cellular immunity*: attacking ("rejecting") localized masses of foreign cells and tissues; and
*   *Phagocytic removal*: using infection-fighting cells (macrophages and neutrophils) directly to engulf and digest foreign material .

Among the advantages of immunochemical assays are their great specificity and sensitivity. In addition, certain determinations required both in clinical and research laboratories can be carried out with greatest precision by immunochemical methods whereas others cannot be carried out conveniently by any other method. For example, immunochemical methods have been used to concentrate and purify proteins from dilute solution; to isolate proteins by affinity chromatography; to localize desired proteins in sectioned tissues; to study the appearance of proteins during development; and to study the active sites, multiple molecular forms, and conformation of enzymes. In another important development, classical immunization techniques were, until 1975, the only source of antibodies for research and clinical applications; however, the introduction of hybridoma techniques by Cesar Milstein and Georges Köhler has given life scientists a powerful new tool (monoclonal antibodies) which could provide solutions to some difficult clinical and research problems.

As a result of the enormous growth and interest in immunology, many detailed immunology textbooks, many technical review articles, and numerous published proceedings of immunochemical meetings are currently available. Since immunochemical methods are usually based on the reaction of antibodies with macromolecules, we will be concerned mostly with antibodies in this chapter on immunochemical assays; but, to put the immune system in broad perspective, we will also discuss, briefly, cellular immunity and the phagocytic removal of particulate material. Our main purpose in this chapter will be to outline some of the principles behind antigen-antibody reactions, to examine the basis of antibody diversity, and to select, on the basis of their importance and usefulness for quantitative determinations of biomolecules, a few of the more frequently used immunochemical assays. The methods we have selected for discussion are *radioimmunoassay* (RIA), *enzyme-linked immunosorbent assay* (ELISA), and *electroimmunoassay* (EIA).

6-2. The antigen-antibody reaction is the basis of immunoassays

An *antigen* is a species-foreign protein or other macromolecule which, when introduced into a vertebrate by intravenous or intramuscular injection, or by oral, respiratory, enteric or other routes, elicits the synthesis of antibodies. Many antibodies also are formed in response to infection by microorganisms, and a given vertebrate may be made to produce antibodies to several antigens simultaneously.

An *antibody* is a serum protein belonging to one of five classes of immunoglobulins, produced in the blood and in certain tissues of vertebrates, in response to injection of an antigen. The five classes of immunoglobulins are classified as IgG, IgM, IgA, IgD, and IgE, of which IgG is most abundant. Immunologists estimate that all vertebrates are genetically capable of forming an extremely large number ranging between 10^5 and 10^7 different antibodies or immunoglobulins.

An antibody usually reacts only with the antigen which evoked its formation although, sometimes, the antibody may react to a lesser degree with compounds which are chemically similar to the antigen. The interaction between antibody and antigen occurs with high affinity and results in the precipitation of an insoluble antigen-antibody complex. Even though an animal is injected with just one antigen, the resulting "antibody" is not one biomolecule but a heterogeneous mixture of closely related biomolecules. The reason for this heterogeneity is that single antigens usually have multiple binding sites resulting from projecting sequences of amino acids in proteins, from certain sequences of nucleotides in nucleic acids, or from repeating sugar units in linear polysaccharides. In such closely related mixtures of antibodies, the specific sequence on the antigen recognized by a given member of the antibody family is known as an *epitope* or an *antigenic determinant*. Not only do antigens have multiple binding sites but antibodies, in addition, have two binding sites (as we shall see in section 6-3 below). The reaction of antigen and antibody therefore may occur not just in one, but in several different combining ratios depending on the relative amounts of antigen and antibody present and on assay conditions such as temperature, ionic strength, and pH. The interaction between antigen and antibody occurs through noncovalent interactions i.e. hydrophobic interactions, hydrogen bonding, and dipole-dipole interactions. These interactions require that both the antigen and antibody are, initially, in the correct conformation so as to ensure accessibility to the reacting sites. It has been inferred that antibodies are capable of "recognizing" any biomolecule or organic compound which falls within the size range 0.6 to 3.4 nanometer and in any conformation that molecule may assume. Antibodies may, however, be formed against low molecular weight compounds if they are linked to larger, antigenic macromolecules. Moreover, some of the antibodies formed in this manner are directed against the small molecule which is referred to as a *hapten*. The foregoing points on the antigen-antibody reaction suggest that, according to the law of mass action, antigen reacts reversibly with antibody to form an antigen-antibody complex as shown in equation (1):

$$Ag + Ab \; \rightleftharpoons \; AgAb, \tag{1}$$

so that
$$K_a \;=\; \frac{[AgAb]}{[Ag]\,[Ab]}, \tag{2}$$

where
K_a = the equilibrium (association) constant,
$[AgAb]$ = concentration of antigen-antibody complex,
$[Ag]$ = concentration of antigen,
and
$[Ab]$ = concentration of antibody.

The association constants, K_a, for the antigen-antibody reaction range from 10^4 to 10^{10} M^{-1}, or greater, depending on the type of antigen involved. Thermodynamic constants (specifically $\Delta G^{o'}$, $\Delta S^{o'}$ and $\Delta H^{o'}$) undergo relatively small changes during the reaction. Immunologists suggest that these small changes reflect both conformational changes of Ag and Ab when Ag-Ab is formed, and the removal of water from Ab binding sites leading to the formation of hydrophobic interactions.

When increasing quantities of antigen are added to a fixed quantity of antibody in solution, the quantity of antigen-antibody precipitate formed increases to a maximum and then decreases, as shown in Figure 1:

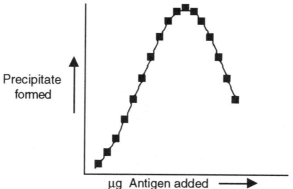

Precipitate formed

μg Antigen added ⟶

Figure 6-1. A generalized antigen-antibody precipitation curve

Curves such as that of Figure 6-1 have been interpreted to mean that, since antigens have multiple binding sites and antibodies have two, then several antigen and antibody molecules may link together to form a single antigen-antibody complex. Maximum precipitation occurs with the highest ratios of antibody to antigen per complex, whereas soluble antigen-antibody complexes are more likely (there is little precipitation) at low ratios of antibody to antigen. These relationships may be expressed as:

$$Ag + Ab \quad \rightleftharpoons \quad Ag \text{ - } Ab,$$
$$Ag \text{ - } Ab + Ab \quad \rightleftharpoons \quad Ag \text{ - } (Ab)_2,$$
$$Ag \text{ - } Ab + Ag \text{ - } Ab \ldots \quad \rightleftharpoons \quad Ag_x \text{ - } Ab_y \ldots$$

6-3. Antibodies consist of four chains in a Y-shaped structure

Investigations of the primary structure of antibodies by enzymatic hydrolysis and ion-exchange chromatography, of their conformation by x-ray diffraction and of their ultrastructure by electron microscopy, have together, revealed that they are constructed of four polypeptide chains, two identical *light* (L) *chains* and two identical *heavy* (H) *chains* (Figure 6-3).

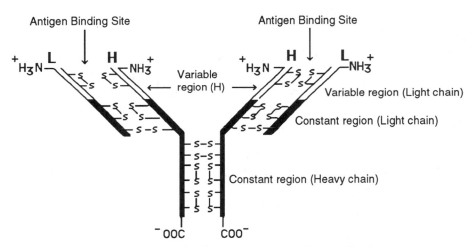

Figure 6-2. Diagram of primary structure of an antibody

Disulfide bonds play a major role in the structure of antibodies. Not only are several intrachain disulfide linkages present in each chain, but the four chains are linked by interchain disulfide bonds at several locations to form a Y-shaped structure. Each chain has a *variable region* in which the sequence of amino acids is variable, and a *constant region* in which the sequence of amino acids is constant. Each light chain has approximately 214 amino acids, and each heavy chain has approximately 430 amino acids. Starting from the N-terminal ends (section 2-15), the variable regions of the light chains consist of the first 107 amino acid residues, whereas the variable regions of the heavy chains consist of the first 121 amino acid residues. Each of the constant regions consists of three clearly defined domains. The two antigen binding sites are located on each arm of the Y-shaped structure, in crevices between the N-terminal ends of the L and H chains; thus, the amino acid sequence in this region is largely responsible for the specificity of a given antibody. The variable sequences of amino acids at these N-terminal ends of antibodies also endow vertebrates with the ability to form large numbers of antibodies, each with differing L and H chains. As a result, no two antibodies would have the same primary structure in the variable region. These points raise a question that has tantalized biomedical scientists for some time. How can one account for the extraordinary ability of vertebrates to form large numbers of diverse and specific antibodies for virtually any group of foreign substance imaginable? We will address this question in the next section.

6-4. Immunological diversity has molecular and cellular explanations

At the molecular level, various hypotheses have been proposed to account for the amazing diversity of antibodies and among these proposals were the *germline* and *somatic mutation* hypotheses. The former proposed that antibody-forming cells inherit from germ cells (i.e. from sperm and egg) all of the genes required for synthesis of the large number of antibodies. According to the somatic mutation hypothesis, sperm and egg contain a limited number of antibody coding genes, but these genes undergo high rates of mutation, leading to the synthesis of large numbers of antibodies. Both hypotheses, especially the first, require that significant proportions of the genome of vertebrates be reserved for the coding of antibodies, and since there is not enough DNA in cells to fulfill this requirement, neither hypothesis is tenable. As an alternative, William Dreyer and J. Claude Bennett proposed in 1965 that two separate genes are combined to code for synthesis of the *variable* (V) or *constant* (C) region of each antibody chain. This proposal became known as the two gene-one polypeptide chain hypothesis, and in 1976, Susumi Tonegawa and colleagues began a series of experiments which have lent strong support to the hypothesis. Initially, Tonegawa prepared restriction maps (section 4-3) of antibody DNA and then identified the fragments from hybridization with antibody mRNA. The antibody DNA was obtained from two sources: from mouse embryo and from tumor cells of mouse B lymphocytes. (Lymphocytes are a class of white blood cells found in lymph vessels and lymph nodes. Lymphocytes have receptors which recognize and bind antigens.) The restriction maps showed that genes coding for the variable and constant regions were separated more widely in embryonic DNA than in DNA of tumor cells. This rearrangemet supported the Dreyer-Bennett hypothesis so Tonegawa next collaborated with Walter Gilbert and Allan Maxam (see section 4-3) to determine the sequence of nucleotides in DNA from the light chain of the mouse tumor cells. The first results revealed that the coding segments for the V and C regions of the light chain were not contiguous but were separated by 1250 base pairs of DNA. To gain more information on this surprising phenomenon, Tonegawa and colleagues determined the nucleotide sequence of DNA for the variable region of the light chain DNA from mouse embryos. They discovered that this V gene did not code for the entire 110 amino acids of the variable region. Instead, the DNA corresponded only to 98 amino acids, and the DNA which coded for the remaining 12 amino acids was missing. Subsequently, Tonegawa and collaborators discovered that the missing DNA for the V region was located 1250 base pairs away, just before the start of the C region of the gene for the light chain. This region of the light chain was referred to as the J region, J representing "joining" region. In later studies, Tonegawa established that coding segments for the V region of the heavy chains also were not contiguous. In this case, however, a third segment, referred to as D (for "diversity") was found between the V and J segments. Thus the evidence suggests that, even though there are relatively small numbers of V and C genes, different V segments combine with different J segments to form coding segments for many thousands of light chains. Similarly, many more thousands of heavy chains are assembled by combining many variants of V, D and J chains. Since any of the thousands of light chains can combine with any of the thousands of heavy chains to form an antibody, the total number of antibodies that may be assembled totals several millions.

At the cellular level, antibody diversity and specificity are accounted for by the fact that embryonic B lymphocytes are the precursors of mature *plasma cells* which have the ability to secrete antibodies into the bloodstream. Both the embryonic B lymphocytes, and the mature plasma cells which actually produce and secrete antibodies, occur in several different cell lines or clones. Each clone has the complete genome of the organism but nevertheless, when stimulated to do so, a clone produces just one type of antibody in a two-stage process. In the first stage, which occurs before a clone comes into contact with its antigen (the antigen-independent stage), the embryonic B lymphocyte produces an antibody bound to its cell surface where the antibody serves as a receptor for antigen when present. In the second stage (the antigen-dependent stage), antigen is introduced into the organism and "selects" one specific clone of embryonic B lymphocytes by binding stereospecifically to the receptor (antibody) on the cell surface. This binding of antigen to membrane-bound antibody stimulates the B lymphocytes to proliferate and to be differentiated into mature plasma cells. The plasma cells, then multiply and produce (by the molecular mechanisms discovered by Tonegawa) antibody molecules that are specific for a given antigen. Antibodies having the same specificities as the surface antibodies on embryonic B lymphocytes are then secreted into the bloodstream by the plasma cells.

6-5. Scientists have simulated antibody diversity in vitro

To obtain antibodies (including monoclonal antibodies, section 6-9), it is necessary, currently, to immunize animals. Scientists have, however, reproduced antibody diversity *in vitro* without using animals or animal cells. One experimental system consisted of filamentous bacteriophages which translate phage genes into surface peptides; i.e., given synthetic nucleic acids (genes) corresponding to antibodies, this system will assemble light and heavy chains into antibodies on the phage surface. The scientists who did this work (Lerner et al., 1992) took four steps to simulate *diversity*: (1) they synthesized antibody genes, (2) they randomized the antigen-binding regions of the genes, (3) they inserted the randomized genes into bacteriophages to be replicated, and (4) they subjected the replicated copies to mutation. Lerner et al. increased the *specificity* in antigen-binding capacity of the synthesized "antibody" by repetitively binding treated phages to and eluting them from, immobilized antigen (cf. affinity chromatography, section 3-9-2). Eventually, workers in this field hope to develop *in vitro* systems which produce as many specific antibodies as animals, without using animal cells or immunizing animals.

6-6.   T lymphocytes mediate cellular immunity

Lymphocytes consist of three types of cells: the *B cells* produced in the bone marrow, *T cells* formed in the bone-marrow but which develop in the thymus gland, and *natural killer* (NK) cells. As we have seen, B cells are mediators of humoral immunity; T cells, on the other hand, are concerned with cellular immunity, the type of immunity which determines the compatibility or rejection of tissues and organs during transplant operations. The functions of NK cells are less well known than those of B and T cells, both of which carry out their functions by recognizing antigens. Whereas antibodies are the antigen-binding proteins of B lymphocytes, the *T cell receptor* (TCR) is the antigen-binding protein of T lymphocytes. Current evidence suggests that TCR (which has been cloned) consists of two polypeptides; these polypeptides, known as α and β, are linked to each other by disulfide linkages and are, in addition, associated with a set of invariant proteins known as CD3. Unlike antibodies, the membrane-bound proteins of the TCR are water insoluble and are produced in small quantities. But, like antibodies, the polypeptides of the T cell receptor contain regions of variable amino acid composition (V-, D-, J- and N- regions). These variable regions combine to yield a large number of binding regions (a number estimated to range between 10^5 to 10^7 different regions) each with different binding specificities. But, although approximately the same number of binding regions is possible both in B and T cells, the phenomenon of somatic mutation in V genes, common in B-lymphocytes, seems to be absent in T cells.

The receptors of T cells recognize foreign peptides as being foreign when these peptides are bound to cell-surface proteins coded for by a set of genes known as the *major histocompatibility complex* or MHC. In general terms it is usually said that T cells help to distinguish "self from nonself" the basis of this statement being that the TCR has evolved to distinguish between "self" molecules and "nonself" molecules at the cell surface in a process that has been called *tolerance induction*. Among current hypotheses for explaining this induction is that, during their maturation, T lymphocytes are subjected to a process of stringent selection. In this process, those T lymphocytes which have receptors that recognize "self" macromolecules are eliminated before they mature. Conversely, during self-induction, those T lymphocytes which do not recognize "self" molecules are retained and become mature T cells. Subsequently, these mature T cells recognize "nonself" antigens associated with cell-surface proteins coded for by the major histocompatibility complex. The sequence of events by which antigens cause TCR to stimulate the activation, growth and differentiation of T cells are complex and incompletely understood. It does appear, however, that the binding of antigen to TCR stimulates *protein tyrosine kinase* (see section 5-12) to catalyze the phosphorylation of various proteins. The phosphorylations lead to the production of the two *second messengers, inositol-1,4,5-triphosphate,* and *1,2-diacylglycerol.* These two *second messengers* then elicit increased Ca^{2+} concentration and the formation of protein kinase C.

6-7.   Phagocytic attack is less specific than antibody protection

Macrophages function in a less specific manner than antibodies and TCR. Macrophages, nevertheless do exhibit specificity in engulfing and digesting AgAb precipitates and other particulate material such as unwanted cells, foreign material, and invading organisms. For example, when a bacterium enters the bloodstream of a mammal, it may be intercepted and bound by IgG. Under these circumstances, receptors on the surfaces of the macrophages recognize the constant regions of the bound IgG and then engulf and destroy the IgG bacterium precipitate. Like TCR, macrophages can distinguish self from nonself. Thus, if one type of macrophage were presented simultaneously with red blood cells from different species, the rate of ingestion of the red blood cells would be in direct proportion to their phylogenetic differences - the greater the phylogenetic difference between the macrophage and the red blood cell, the more rapid would be the rate of ingestion. It also has been reported that macrophages may ingest oxidized particles of LDL cholesterol; such ingestion may, however, ultimately be unfavorable since it may lead to the deposition of atherosclerotic plaques. Neutrophils too may promote unfavorable effects on the heart. After a heart attack, they are attracted, and attach themselves to, damaged heart tissue. Then, these white blood cells release hydrolytic enzymes or free radical producing compounds as if they were attacking a foreign infective agent. More heart tissue thereby is killed, as are some neutophils. This accumulation of dead cells leads to clogging of blood vessels and restriction of blood flow. The result is additional damage to the heart (cf. section 9-5). To conclude this brief survey of specialized cells in the immune system, we observe that several unique substances, produced by these specialized cells, are components of, and play important protective and regulatory roles in, the immune system. Among the most remarkable of these substances are the cytokines, defined by Oppenhein as *"nonimmunoglobulin polypeptide factors secreted by activated lymphocytes."* These factors exert effects in response to immunological reactions induced by tissue injury, by inflammation or by microbial infection.

6-8.   Cytokines regulate certain functions of lymphocytes and other cells

Cytokines (defined in section 6-7 above) were previously called *lymphokines* because it was thought that these polypeptide factors were produced only by lymphocytes. But, since many of these substances are now known to be produced by a wide variety of nonlymphoid cells, the term cytokine has been proposed as a more accurate characterization of the origins of these hormone-like polypeptides. In the immune system, T cells and macrophages produce and secrete such important cytokines as *interleukins 1 and 2 (IL-1 and IL-2), B cell growth factor* (BCGF), *colony stimulating factor* (CSF), and *γ-interferon* (IFN-γ). Thus, IL-1, produced by stimulated macrophages, induces growth, differentiation and numerous metabolic, neurological, and physiological effects; IL- 2, also produced by macrophages, enhances the functions and replication of T lymphocytes; BCGF, produced by T cells, is required for the growth of B cells; CSF, produced by lymphocytes, macrophages, and other cells, stimulates the growth and differentiation of many cells; and

IFN-γ and other interferons suppress viral replication and the proliferation of cells. T cells are activated when they bind antigen and, when stimulated, can attack and kill antigen-bearing cells. Antigen-stimulated T cells can also stimulate production of antibodies by B lymphocytes and stimulate macrophages to engulf foreign cells. Although knowledge of many cytokines has only been discovered in the1980s, this knowledge is already being applied to treat such ailments as leg sores, ulcers of the mouth, heart disease, and arthritis.

As we noted in section 6-2, since antigens have multiple binding sites, antibodies formed by immunization of laboratory animals usually are not single molecular species. When a single species of antibody is needed, it can (at present) be produced most successfully by the technique discussed next.

6-9.   Monoclonal antibodies are made by hybridoma technology

In *hybridoma technology*, a myeloma line of cells is fused with a second type of cell to yield a hybrid cell. The hybrid cell has the capacity to produce proteins of both cells. *Monoclonal antibodies* are produced by fusing B lymphocytes from a suitably immunized animal with cells from a myeloma cell line which can grow in culture. Individual clones which secrete desired antibodies are selected from the hybrid cells by assaying the antibodies secreted into the culture medium. In contrast to antibodies produced *in vivo*, antibodies produced by individually selected clones *in vitro* consist of a single species of antibody induced by a single antigenic determinant. A key factor in production of antibodies by hybridoma techniques is the use of an enzyme-deficient mutant of the myeloma cell line. Such mutant cells will grow slowly in culture before hybridization. When, however, the hybrid cells are fused with cells which contain normal quantities of the enzyme, the hybrids multiply rapidly in culture and, once established, can be cultured for long periods of time. Subsequently, if individual clones which can secrete desired antibodies are selected and cultured, *monospecific or monoclonal antibodies* are found in the culture medium after a period of weeks. These monoclonal antibodies can be recovered by immunoadsorbent techniques and can be duplicated exactly in different laboratories. To date, the hybridoma technique has been used to prepare antibodies against several antigens including viruses, cell surface antigens, glyco-proteins, peptide hormones, enzymes and polysaccharides. In addition, biomedical and industrial scientists now use monoclonal antibodies for the purification of a wide range of biomolecules. A brief list of substances purified in this manner includes enzymes, antigens, interferons, interleukin-2, and blood. Monoclonal antibodies, moreover, are being used for the delivery of drugs to tissues (e.g., for the delivery of the anticancer drug methotrexate to cancerous tissues), for the diagnosis and treatment of plant and animal diseases, for preventing rejection of transplanted kidneys, and to test for the occurrence of pregnancy. Whether produced by classical immunization methods or by hybridoma technology, antibodies provide sensitive and specific techniques for the study of biomolecules both in biomedical and industrial laboratories. We will discuss three of these techniques in sections 6-10 to 6-12.

6-10.   Radioimmunoassay is based on competitive binding

Radioimmunoassay (RIA) depends on competitive binding between labeled and unlabeled samples of a ligand for a limited number of sites on a protein in solution.   In RIA, unlabeled antigen competes with a constant amount of radioactively labeled antigen for a limited number of combining sites on an antibody. Experimentally, competition for the combining sites is achieved by including in the assay solution an appropriate amount of unlabeled antigen, a fixed amount of labeled antigen and a fixed amount of antibody. The soluble reactants are allowed to react and achieve equilibrium and, after equilibration, free and bound labeled antigen are separated and measured.   One obtains data for standard curves by adding known amounts of unlabeled antigen to a constant amount of radioactive antigen and a constant amount of antibody.   The degree of competition between labeled and unlabeled antigen may be determined by interpolation from the calibration curves since the ratio of labeled to unlabeled analyte depends on the amount of unlabeled analyte present. Specifically, the *radioactivity of antibody-bound, labeled antigen* (as counts per min, cpm) is *inversely proportional to the concentration of unlabeled antigen* (Figs. 6-3 and 6-4).

Analyses of RIA data may be carried out by many different curve-fitting methods of varying degrees of convenience and statistical reliability.   The independent variable usually is the concentration of antigen or the logarithm of antigen concentration. Among the more common response or dependent variables used in RIA are bound counts, B; free counts, F; the fraction of total counts bound (B/T); the ratio of bound counts to free counts (B/F); or the ratio of counts bound in the presence of unlabeled antigen to counts bound in the absence of unlabeled antigen (B/B_0). Two reasons for the use of several different response variables and several methods of calculation are that (a) many RIA calibration curves are nonlinear and   (b) different segments of the curves have different levels of experimental error.   Since there is no single method of RIA curve fitting that is generally accepted as "the best," we will demonstrate the preparation of four types of calibration curves: hyperbolic curves, sigmoid curves, logit-log plots and the four-parameter logistic curve. Very often, the independent variable in RIA calibration curves is the concentration of unlabeled, standard antigen added whereas the dependent variable is B/B_0, i.e. the ratio of counts bound in the presence of unlabeled antigen to counts bound at zero dose of unlabeled antigen. Either arithmetic or logarithmic scales may be used, but logarithmic scales are preferred (especially for the x-axis) because they result in partial linearization and because they are useful for interpolation of values corresponding to low concentrations of antigen. Different kinds of curves are obtained depending on the scales used and on the mathematical nature of the dependent (response) variable. Thus, if B/B_0 is plotted versus the arithmetic or actual concentration [Ag], on linear paper, a *hyperbolic curve* is obtained (Figure 6-3) and if B/B_0 is plotted against ln [Ag], a *sigmoid curve* results which may have a region that is approximately linear (Figure 6-3).

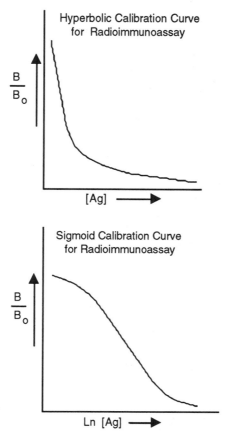

Figure 6-3.   Generalized hyperbolic and sigmoidal calibration curves for RIA

The sigmoidal shape of B/B$_0$ calibration curves suggested another way of plotting RIA data to pioneers in this field. They deduced that linearization may be achieved by *logit* transformation of the dependent variable (cf. Figure 6-4):

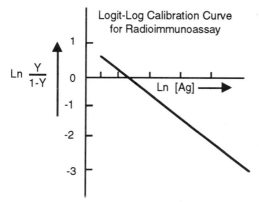

Figure 6-4.   Generalized logit-log curve

As a result, this method of transformation has become known as the logit-log method. Sometimes, similarities are pointed out between logit-log RIA calibration curves (Figure 6-4) and Hill plots (Figure 5-17), also known as Sips plots in the immunological literature. Indeed, comparisons of Hill plots with RIA calibration curves, and comparisons of the logit-log equation, i.e., equation (3) below, with equation (79) of Chapter 5, will reveal some similarities. However, the Hill equation and the RIA logit-log equation are not algebraically interconvertible. Mathematically, the logit-log method is often expressed in two-parameter form (slope m and intercept k):

$$Ln\left[\frac{Y}{1-Y}\right] \quad = \quad m \ln p + k, \tag{3}$$

where

$$Ln\left[\frac{Y}{1-Y}\right] \quad = \quad \text{logit } Y,$$

$$Y = \frac{B}{B_0} = \frac{\text{counts bound in the presence of p}}{\text{counts bound in the absence of p}}.$$

and where p = unlabeled antigen, m and k are constants and Y is a fraction or decimal, i.e. $0 < Y < 1$. Thus, in the *logit-log* method $\ln[(Y)/(1-Y)]$ is plotted versus $\ln p$ or $\ln[Ag]$ to obtain linear standard curves from which the concentration of unknown antigen present is estimated. Rodbard and colleagues, who have made extensive studies of mathematical aspects of competitive binding between labeled and unlabeled ligands, have indicated that the logit-log method is perhaps the method of choice for most routine processing of RIA data. We will, accordingly, use this method for RIA calculations and data processing in this chapter.

In using the logit-log method, N, the counts caused by *nonspecific binding*, (i.e. the radioactivity or counts bound with all components present except antibody) should be subtracted both from B and B_0 so that

$$Y = B/B_0 = \frac{B - N}{B_0 - N}$$

Accurate determinations of B_0 and N are essential, otherwise the logit-log curves may not be linear as predicted by equation (3). Because adequate curve fitting may be achieved with logit-log plots derived from properly conducted RIA experiments, and because unweighted linear least-squares regression is easier to use than the mathematically more rigorous weighted, nonlinear regression, the unweighted regression may be used to estimate levels of unlabeled samples present, but not necessarily to predict confidence limits. Estimates of how well a linear, logit-log curve is fitted may be made from parameters such as those described in sections 10-14 to 10-16 on linear regression.

If the two-parameter, logit-log transformation does not provide sufficiently accurate curves for the estimation of antigen concentrations by RIA, the two parameters (m and k), equation (3), and accurately determined measurements of B_0 and N, may be used as initial estimates for nonlinear curve fitting to the *four-parameter logistic equation*, equation (4):

$$B = \frac{a}{1 + bp^c} + d \tag{4}$$

The two variables B and p, and the initial estimates a, b, c and d of the corresponding parameters in equation (4), are then defined as:

$$
\begin{aligned}
B &= \text{counts bound expressed as counts per minute,} \\
p &= \text{the arithmetic or actual concentration of antigen,} \\
a &\equiv B_0 - N, \text{ the response when } p = 0, \\
b &\equiv \text{y-intercept of the logit-log curve,} \\
c &\equiv -1 \text{ times the slope of the logit-log curve,} \\
d &\equiv N, \text{ the counts due to nonspecific binding.}
\end{aligned}
$$

If B_0 and N are not measured experimentally, they may be estimated mathematically by extrapolating the logit-log data on the binding of standards both to zero and to infinite concentration to give initial estimates of a and d. Slope and intercept then may be obtained by linear least squares computation to estimate b and c. The main advantage of the four-parameter logistic equation is that it avoids the computational biases that are inherent in the hyperbolic, sigmoid and logit-log procedures. The main reason for using the logistic equation is to obtain the best fit of of a, b, c and d to equation (4) so that the most reliable calibration curve is provided. Reliable calibration curves lead, in turn, to increased accuracy in computing the unknown concentrations of antigens or other ligands. However, the four-parameter logistic approach is more complicated than the logit-log and other methods discussed above. For example, adjusting (refining) the initial estimates of the four parameters calls for the use of nonlinear regression to derive and solve a system of four linear equations with four variables. Solution of this system of four equations may be achieved by various mathematical methods, two of which, *Gauss-Jordan elimination* (also known as *Gauss-Jordan reduction*) and *matrix solution*, are used in this chapter. All of the equations required for transforming RIA data into the desired system of four linear equations, for minimizing the least-squares error function, and for computing adjustments to the four initial estimates, are listed in Ukraincik and Piknosh (1981) and demonstrated in Example 6-1. Steps for using the four-parameter logistic equation also are described in the reference by Ukraincik and Piknosh and may be summarized as follows:

1. Calculate logit Y; then plot the logit-log curve (linear regression), and determine the slope and y-intercept of the curve.
2. Calculate initial estimates of the parameters a, b, c, and d from the experimental data and from the logit-log curve.

3. Calculate the coefficients H_{ij} and the terms G_i by the Newton-Raphson procedure.

4. Calculate the correction factors Δa, Δb, Δc and Δd by Gauss-Jordan elimination.

5. Solve the system of four linear equations by successive substitutions using values obtained from steps 3 and 4 above.

6. Calculate the error function first using old values of a, b, c, and d and then using the new values corrected with Δa, Δb, Δc, and Δd.

7. Repeat the previous step until the smallest error function is obtained.

8. Use the corrected values of a, b, c, and d to estimate the values of B which will yield the best standard curve.

9. Calculate (for assessing the quality of the experimental binding data, B_i) the difference between $B_{estimated}$ and B_i as percent change.

10. Plot the calibration curve, ($B_{estimated}$) versus Ln [Ag] and estimate the concentrations of sample antigens from the curve.

As mentioned in sections 5-8 and 10-17, the use of increasingly more rigorous methods of data analysis cannot compensate for poor data. Therefore, it is important that experimental results be evaluated objectively to obtain measures of their quality. Among the tests recommended by workers who have studied methods of RIA data reduction are (1) calculation of the extent (large or small) to which the parameters a, b, c and d have to be adjusted; (2) measurement of the degree to which the concentrations of standards interpolated from calibration curves differ from the concentrations actually used; and (3) evaluation of measures of dispersion, e.g., standard deviation, associated with the usable range of calibration curves.

The reader would have noted that in the ten-step, Ukraincik and Piknosh algorithm outlined above, the coefficients rather than the variables are used to solve the system of four linear equations. Because matrix algebra is at the core of linear equations, and because the manual method of elimination (reduction) described in the algorithm may be simplified by using the built-in matrix capabilities of spreadsheets, a few comments about matrices and use of the built-in matrix capabilities of Excel and Lotus 1-2-3 are in order.

A *matrix* is a set of numbers arranged in columns and rows to form a rectangle. Individual numbers within a matrix are known as *elements*. Some of the basic operations of matrix algebra which may be of interest to the reader (addition and subtraction, multiplication by a scalar, multiplication of two matrices, inversion, transposition etc.) are described in the texts by Causton, and by Yaqub and Moore, listed in the references at the end of this chapter. Matrix algebra has numerous applications in statistics, business, mathematics, and the sciences. In section 6-14, we will show how Excel and Lotus 1-2-3 may be used in this context after we show solution by Gauss-Jordan elimination with Excel. Later, in sections 7-9 and 8-8, we will use the determinants of matrices to solve pairs of simultaneous equations with two unknowns.

6-11.   ELISA is run both competitively and noncompetitively

ELISA, enzyme-linked immunosorbent assay, is carried out on a solid phase, the "immunosorbent" material, unlike radioimmunoassay, which is carried out in solution. Several kinds of immunoadsorbents have been used for ELISA, including agarose, cellulose, polyacrylamide and various plastic surfaces made from polycarbonate, polypropylene, polystyrene, or polyvinyl. The important practical consequence of using such adsorbents is that they facilitate the separation of antigen-antibody complexes (the bound fraction) from free antigen or antibody. ELISA may be carried out either by competitive binding or by noncompetitive binding. However, regardless of the type of binding which occurs, all ELISA procedures involve common steps: *first*, antigen and antibody are allowed to react; *then*, the antigen-antibody complex is formed and separated from free antigen and antibody; *finally*, activity is determined in bound or free fractions as desired. In the following paragraphs, we will first discuss two variations of ELISA based on competitive binding; then we will discuss ELISA carried out by noncompetitive binding.

Competitive binding ELISA may be carried out either with enzyme-labeled antigen or with enzyme-labeled antibody. To carry out *competitive binding ELISA with enzyme-labeled antigen*, the main steps are first to attach the desired quantity of antibody to the solid phase by chemical or physical methods. Next, incubate the attached antibody with a fixed concentration of enzyme-labeled antigen in the presence or absence of standard or unknown antigen until equilibrium is attained. Then, remove unreacted materials and add substrate to determine the level of enzyme attached to the components on the solid phase. The *quantity of substrate used*, or of product formed, is *inversely proportional to the quantity of standard or unknown antigen.*

In *competitive binding ELISA with enzyme-labeled antibody*, the first step is to attach a suitable quantity of antigen to the solid phase. Next, the attached antigen is incubated with a fixed concentration of enzyme-labeled antibody in the presence or absence of standard or unknown antigen until equilibrium is attained. Unreacted materials are removed by washing, and substrate is added to the components on the solid phase. The components are then incubated to allow the enzyme reaction to take place and the level of enzyme attached to components on the solid phase is determined by an appropriate method. As before, *the substrate used is inversely proportional to the quantity of standard or unknown antigen.*

In *noncompetitive ELISA*, also known as the "sandwich" assay, key steps are first to attach antibody to the immunosorbent material. Then the attached antibody is incubated with a known quantity of standard or unknown antigen until equilibrium is attained. Unreacted antigen is removed by washing and enzyme-labeled antibody is added to bind to the remaining free antigenic sites on the antigen-antibody complex. Finally, unreacted materials are removed and substrate is added to determine the level of enzyme attached to the components on the solid phase. The quantity of *substrate used,* or of product formed, is *directly proportional to the quantity of standard or unknown antigen.*

Because of its ease of use, sensitivity, reproducibility, low cost and safety, ELISA has become one of the mainstays of the clinical laboratory in a very short time. Nevertheless, it should be borne in mind that ELISA does suffer from certain drawbacks. Among these are that competitive binding ELISA sometimes requires that enzyme-labeled antigens or antibodies be incubated directly with biological fluids or extracts. These fluids or extracts may contain proteolytic enzymes or enzyme inhibitors which may cause a loss of enzyme activity. Second, competitive binding ELISA with enzyme-labeled antigen requires substantial quantities of pure antigen. Obtaining such quantities of relatively pure antigen may be difficult or expensive or both. Noncompetitive binding ELISA does not, however, suffer from these two disadvantages because of the manner in which it is carried out. Like radioimmunoassay, plots of enzyme-linked immunoassay calibration data may yield sigmoid curves (Figure 6-5)

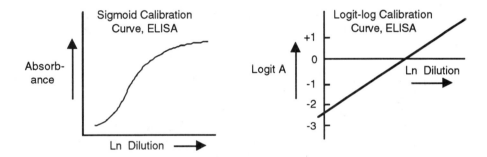

Figure 6-5. Generalized sigmoid and logit-Log calibration curves for ELISA

In ELISA, the products of the enzyme reactions usually are determined spectrophotometrically and absorbance is inversely proportional to concentration of antigen which is often expressed in terms of increasing dilution. Thus, the dependent variable is absorbance or some function of absorbance and the independent variable is dilution or Ln dilution as shown in Figure 6-5. To prepare an ELISA calibration curve from raw data, readings of absorbance may be transformed into logit units as described by Ritchie, Nickerson and Fuller (1983). In this transformation, Y is the ratio of the absorbance of a sample in the presence of added antigen, to the absorbance produced in the absence of added antigen. Or, expressed as a percentage:

$$Y = \frac{A_i}{A_{max}} \times 100, \qquad (5)$$

where A_i = absorbance of sample at dilution i,
and A_{max} = absorbance in the absence of added antigen.

Logit Y is then defined as

$$\text{logit } Y = \ln(Y/(100 - Y)) \qquad (6)$$

Or, rewriting equation (6) as a function of absorbance

$$f(A_i) \quad = \quad Ln \ (A_i/(A_{max} \ A_i)) \tag{7}$$
$$f(A_i) \quad = \quad b \ Ln \ dilution_i + a \tag{8}$$

where b is the slope and a the y-intercept of the straight line which results if experiment agrees with theory. As with RIA, unweighted linear least-squares regression may be used to estimate the concentrations of antigen present in unknowns and how well the data were fitted to the logit-log equation. Again, however, the unweighted regression should not be used for predicting confidence limits; and, if the logit-log transformation does not provide sufficiently accurate ELISA calibration curves, the four-parameter logistic may yield better results.

6-12. In EIA, antigens in gels react with immobilized antibodies

Electroimmunoassay (EIA) is an immunochemical method which may be used for the detection and quantitative estimation of proteins. In EIA, proteins to be studied are subjected to electrophoresis in agarose gels which contain large quantities of the related antibodies. The antigenic proteins react with antibodies to form antigen-antibody complexes, some of which are stable and insoluble. The resulting precipitates form rocket-shaped spots, the heights of which are directly proportional to the concentration of protein (antigen) present.

To carry out electroimmunoassay, plates are prepared containing thin layers of immunosorbent (usually agarose) into which antibody is uniformly incorporated. Small wells are punched into the layers, known quantities of standard or unknown antigen are added, and the plates are placed in a buffer solution at a pH corresponding to the isoelectric point of the antibody. An appropriate electric field then is applied, and, since the pH of the buffer corresponds to the isoelectric point of the antibody, the antibody does not migrate. However, the standards or unknowns do migrate, forming antigen-antibody complexes, some of which may initially be soluble. Eventually, when a complex with a critical ratio of antigen to antibody is formed, precipitation of the complex occurs; but, it should be pointed out, the principles which govern precipitation of antigen-antibody complexes in EIA are different from those governing the behavior of similar complexes in solution (cf. Figure 6-1). After electrophoresis, the plates are dried and the precipitates stained. As mentioned above, EIA produces rocket-shaped precipitates; consequently the method sometimes is referred to as "rocket" electroimmunoassay. In this immunochemical technique, the length of the rocket is proportional to the concentration of antigen, rocket length being measured from the center of the well to the tip of the rocket. Linear calibration curves usually are constructed from the lengths of the rockets versus the concentrations of the standards. However, to obtain reliable quantitative estimates of the amounts of antigen present in samples of unknowns during electroimmunoassay, it is important that interpolation from calibration curves be made only when standards and unknowns have given "rockets" of identical morphology. The sensitivity of electroimmunoassay generally is in the range of 10 to 50 nanograms of protein.

6-13. Worked examples

Example 6-1. The following data were obtained for the preparation of a radio-immunoassay calibration curve. Triplicate readings were made of the counts bound. The amounts of unlabeled antigen used varied between 1.25 and 100 picograms/50 μl of assay solution as indicated below.

B$_0$, cpm   N, cpm
Tube Number

1-3	4-6
4044	97
4212	152
4444	72

B, cpm
Tube Number

7-9	10-12	13-15	16-18	19-21	22-27	28-30	31-33
2710	2629	1878	1320	1177	731	436	243
2988	2571	2046	1776	1016	523	374	203
2462	2593	2149	1215	805	647	393	240

Concentration of Unlabeled Antigen, pg/50μl

Tube Number

7-9	10-12	13-15	16-18	19-21	22-27	28-30	31-33
1.25	1.56	3.12	6.25	12.50	25.00	50.00	100.00

a. Calculate, and then report in tabular form, the appropriate values for plotting the calibration curves as hyperbolic, sigmoid, and logit-log functions.

b. Plot the hyperbolic, sigmoid, and logit-log calibration curves.

c. Prepare the four-parameter logistic calibration curve.

d. Estimate from each of the four calibration curves the quantities of antigen present in samples which gave B/B$_0$ values of 0.17, 0.34, and 0.45.

SAMPLE RIA CALCULATIONS

To Calculate logit Y from raw data on B and B_o

1. Calculate the means of each set of B and B_o values. Then, as an example:

2. Let the means of B and B_o be 2720 cpm and 4233 cpm, respectively; N = 107 cpm.

$$
\begin{aligned}
3.\quad \text{Calculate} \qquad Y &= \frac{B\text{-}N}{B_o\text{-}N} &&= \frac{2613}{4126}, \\[2mm]
\therefore Y &= 0.63 \\[1mm]
\text{logit}\ Y &= \ln\frac{Y}{1\text{-}Y} &&= \ln\frac{0.63}{1 - 0.63}, \\[1mm]
\therefore \text{logit}\ Y &= 0.5322.
\end{aligned}
$$

4. Repeat the calculation of Logit Y for each mean B value..

5. Plot logit Y against ln [Ag] on linear paper, or, plot logit Y against [Ag] on semilog paper.

To calculate Y and B from Logit Y

$$
\begin{aligned}
\text{Let} \qquad \text{logit}\ Y &= \ln\frac{Y}{1\text{-}Y} &&= 0.42, \\[3mm]
\therefore \qquad e^x\ \text{logit}\ Y &= \frac{Y}{1\text{-}Y} &&= 1.522, \\[3mm]
\text{And} \qquad\quad Y &= 1.522 - 1.522Y, \\[1mm]
2.522\ \ Y &= 1.522 \\[1mm]
Y &= \frac{1.522}{2.522} &&= 0.6035, \\[2mm]
\text{Since} \qquad\quad Y &= \frac{B\text{-}N}{B_o\text{-}N}
\end{aligned}
$$

$$
\begin{aligned}
\text{Then if} \qquad B_o - N &= 4126\ \text{cpm}, \\[2mm]
B - N &= Y \times 4126\ \text{cpm} &&= 0.6035 \times 4126\ \text{cpm}, \\
&= 2490\ \text{cpm}. \\[2mm]
\therefore \qquad B &= 2490 + 107\ \text{cpm} &&= 2597\ \text{cpm}.
\end{aligned}
$$

6-14. Solution of example 6-1

	A	B	C	D	E	F	G	H	I	J	K
1	Raw data, B	Mean	Raw data, N		Mean,N		Bo-N				
2	4044	Bo	97								
3	4212		152								
4	4444	4233	72		107		4126				
5											
6	Raw data, B	Mean,	(B-N)/(Bo-N)	Ln[Ag]	Logit Y	Ln[Ag]	Y	Ag	Y	Ln[Ag]	B'/Bo
7	2710	B	0.63324986	0.223	0.5462	0.223	0.63	1.25	0.6	0.223	0.63
8	2988		0.60360288	0.446	0.4205	0.446	0.6	1.56	0.6	0.446	0.6
9	2462	2720	0.46465789	1.139	-0.142	1.139	0.46	3.13	0.5	1.139	0.46
10	2629		0.32232006	1.833	-0.743	1.833	0.32	6.25	0.3	1.833	0.32
11	2571		0.21625333	2.526	-1.288	2.526	0.22	12.5	0.2	2.526	0.22
12	2593	2598	0.12763551	3.219	-1.922	3.219	0.13	25	0.1	3.219	0.13
13	1878		0.0712497	3.912	-2.568	3.912	0.07	50	0.1	3.912	0.07
14	2046		0.02948542	4.605	-3.494	4.605	0.03	100	0	4.605	0.03
15	2149	2024									
16	1320										
17	1776										
18	1215	1437									
19	1177										
20	1016										
21	805	999.3									
22	731										
23	523										
24	647	633.7									
25	436										
26	374										
27	393	401									
28	243										
29	203										
30	240	228.7									

Hyperbolic RIA Calibration Curve

Sigmoid RIA Calibration Curve

Logit-Log RIA Calibration Curve

Four-parameter Logistic Curve

readsheet solutions Gauss-Jordan reduction, Excel

A	B	C	D	E	F	G	H	I	J	K	L
DATA ENTRY FOR CALIBRATION CURVES OF EXAMPLES 6-1(a) TO (d):											
Bo	Mean,E	N		Mean,N		Bo-N		B'/Bo' = Mean, B-N/Bo-N			
4044		97						(See transformations for logit etc. below).			
4212		152									
4444	4233	72		107		4126.3					

TRANSFORMATIONS FOR LOGIT, SIGMOID AND HYPERBOLIC CALIBRATION CURVES: | TRANSFORMATIONS FOR LOGISTIC:

B	Mean,B	B'/Bo'	Ln[Ag]	LogitY	Ln[Ag]	Y	Ag	Y	bp^c	x	y
2710		0.63	0.2231	0.55	0.22	0.63	1.25	0.63	0.52	105.34	0.66
2988		0.60	0.4460	0.42	0.45	0.60	1.56	0.60	0.63	36.93	0.61
2462	2720	0.46	1.1394	-0.14	1.14	0.46	3.13	0.46	1.18	-22.71	0.46
2629		0.32	1.8326	-0.74	1.83	0.32	6.25	0.32	2.19	-37.69	0.31
2571		0.22	2.5257	-1.29	2.53	0.22	12.50	0.22	4.08	-80.51	0.20
2593	2598	0.13	3.2189	-1.92	3.22	0.13	25.00	0.13	7.60	-46.92	0.12
1878		0.07	3.9120	-2.57	3.91	0.07	50.00	0.07	14.15	-21.66	0.07
2046		0.03	4.6052	-3.49	4.61	0.03	100.00	0.03	26.35	29.23	0.04
2149	2024	Slope		Intercept							
1320		(Logit-Log Curve):									
1776		-0.897	0.857938								
1215	1437	a	b	c	d						
1177		4126.3333	0.42403571	0.896658	107						

		NEWTON-RAPHSON EVALUATION OF COEFFICIENTS Hij AND TERMS Gi:									
1016	999	H11	H12	H13	H14	H22	H23	H24	H33	H34	H44
805		0.4339887	1496.9	141.6	0.66 5155773.7	436426.3	2187.4	41295.0	207.0	1	
731		0.3752216	1435.4	271.4	0.61 5489655.5	1000087.5	2309.5	189122.5	436.7	1	
523		0.2108233	1096.3	529.7	0.46 5699691.2	2816390.4	2416.5	1360769.5	1167.6	1	
647	634	0.0980862	636.4	494.6	0.31 4125825.7	3350653.5	2093.2	2603725.8	1626.6	1	
436		0.0387078	272.6	291.9	0.20 1895942.3	2343255.4	1537.9	2509624.5	1647.0	1	
374		0.0135174	104.9	143.1	0.12 804132.53	1248585.7	999.8	1704217.4	1364.7	1	
393	401	0.0043561	36.4	60.5	0.07 302610.25	552807.5	599.9	917017.8	995.1	1	
243		0.0013372	15.0	29.2	0.04 163066.55	272287.2	342.8	531710.6	669.5	1	
203		1.17604	5093.796	1962	2.46	2E+07	1.2E+07	12487.1	9857483	8114.2	8
240	229										G4

H11	H12	H13	H14	G1			69.39	230422.5	21802.8	105.3
1.176038343	-5093.79601	-1962.0425	2.45924996	-48.1343	Row1		22.62	85295.7	16129.9	36.9
H21	H22	H23	H24	G2			-10.43	-54871.9	-26512.0	-22.7
-5093.79601	23636697.8	12020493	-12487.0589	8270.944	Row2		-11.80	-78881.6	-61297.3	-37.7
H31	H32	H33	H34	G3			-15.84	-123806.9	-132597.1	-80.5
-1962.04245	12020493.4	9857483	-8114.16067	-248494	Row3		-5.46	-46914.0	-64033.8	-46.9
H41	H42	H43	H44	G4			-1.43	-12992.3	-21552.2	-21.7
2.459249959	-12487.0589	-8114.1607	8	37.98158	Row4		1.07	10019.6	19565.9	29.2
							48.1343	8270.94	-248494	-37.98

GAUSS-JORDAN REDUCTION OF THE SYSTEM OF EQUATIONS:										
1	-4331.32	-1668.35	2.091131	-40.93 Row1+A36 (1a)						
5093.796009	-22062850.2	-8498229.7	10651.7935	-208485 (1a)*-A38 (1b)						
0	1573847.57	3522263.7	-1835.26534	-200214 (1b)+Row2 (2a)						
1962.042452	-8498229.73	-3273371.7	4102.88733	-80304.8 (1a)*-A40 (1c)						
0	3522263.71	6584111.4	-4011.27334	-328799 (1c)+Row3 (3a)						
-2.45924996	10651.7935	4102.8873	-5.14261325	100.6551 (1a)*-A42 (1d)						
0	-1835.26534	-4011.2733	2.85738675	138.6367 (1d)+Row4 (4a)						
0	1	2.238	-0.00117	-0.127 (2a)+B47 (2b)						
0	-3522263.71	-7882810.2	4107.3155	448077.8 (2b)*-B49 (2c)						
0	0	-1298698.8	96.0421535	119279.2 (2c)+(3a) (3b)						
0	1835.26534	4107.3155	-2.14010489	-233.47 (2b)*-B51 (2d)						
0	0	96.042153	0.71728187	-94.833 (2d)+(4a) (4b)		Δ	Δ	Δ	Δd	
0	0	1	-7.4E-05	-0.092 (3b)+C54 (3c)		4126.3333	0.4240357	0.8966577	107	
0	0	-96.042153	0.00710257	8.82101 (3c)*-C56 (3d)		3990.5213	0.383563	0.7960316	-11.7381	
0	0	0	0.724384	-86.01 (3d)+(4b) (4c)						

Δa	Δb	Δc	Δd	Δe					
-135.811984	-0.04047268	-0.1006262	-118.738061						

Solution Of The System of Four Linear Equations:									
-159.7201	206.159553	197.43284	-292.006571	-48.1343 Equation 1	105.34	E	Er	Ln [Ag]	B calcd
691798.5398	-956640.4	-1209576.4	1482689.15	8270.944 Equation 2	36.93	11096.19	272428804	0.22	2725
266468.8772	-486501.531	-991920.87	963459.702	-248494 Equation 3	-22.71	1364.01	188409.72	0.45	2568
-333.995615	505.384683	816.497	-949.904485	37.98158 Equation 4	-37.69	515.60	14274.08	1.14	2035
ITERATION:		Δa'	Δb'	Δ'	Δd'	-80.51	1420.18	498434.68	1.83	1494
	Old	4126.3333	0.42403571	0.896658	107	-46.92	6481.25	1307969.5	2.53	1021
	delta	10038.509	-5.54457354	-11.0067	-12383.39	-21.66	2201.63	2278081.9	3.22	656
	New	14164.842	-5.12053784	-10.1101	-12276.39	29.23	469.09	3034558	3.91	402
							854.14	3664665.5	4.61	238
							24402.1	2.8E+08		

Spreadsheet formulas Gauss-Jordan reduction, Excel

	A	B	C	D	E
1	DATA ENTRY FOR CALIBRAT|				
2	Bo	Mean,Bo	N		Mean,N
3	4044		97		
4	4212		152		
5	4444	=AVERAGE(A3:A5)	72		=AVERAGE(C3:C5)
6					
7			TRANSFORM/ AND HYPERB(		
8	B	Mean,B	B'/Bo'	Ln[Ag]	LogitY
9	2710		=(B11-E5)/G5	=LN(1.25)	=LN(C9/(1-C9))
10	2988		=(B14-E5)/(G5)	=LN(1.562)	=LN(C10/(1-C10))
11	2462	=AVERAGE(A9:A11)	=(B17-E5)/(G5)	=LN(3.125)	=LN(C11/(1-C11))
12	2629		=(B20-E5)/(G5)	=LN(6.25)	=LN(C12/(1-C12))
13	2571		=(B23-E5)/G5	=LN(12.5)	=LN(C13/(1-C13))
14	2593	=AVERAGE(A12:A14)	=(B26-E5)/(G5)	=LN(25)	=LN(C14/(1-C14))
15	1878		=(B29-E5)/(G5)	=LN(50)	=LN(C15/(1-C15))
16	2046		=(B32-E5)/(G5)	=LN(100)	=LN(C16/(1-C16))
17	2149	=AVERAGE(A15:A17)	Slope		Intercept
18	1320		(Logit-Log Curve):		
19	1776		=LINEST(E9:E16,F9:F16)		=LINEST(E9:E16,F9:F16)
20	1215	=AVERAGE(A18:A20)	a	b	c
21	1177		=G5	=EXP(-E19)	=-D19
22	1016		NEWTON-RAP		
23	805	=AVERAGE(A21:A23)	H11	H12	H13
24	731		=L9^2*1	=L9^2*H9^E21*(C21*L9+K9)*1	=D21*H9^E21*LN(H9)*L9^2*(C21*L
25	523		=L10^2*1	=L10^2*H10^E21*(C21*L10+K10)	=D21*H10^E21*LN(H10)*L10^2*(C2
26	647	=AVERAGE(A24:A26)	=L11^2*1	=L11^2*H11^E21*(C21*L11+K11)	=D21*H11^E21*LN(H11)*L11^2*(C2
27	436		=L12^2*1	=L12^2*H12^E21*(C21*L12+K12)	=D21*H12^E21*LN(H12)*L12^2*(C2
28	374		=L13^2*1	=L13^2*H13^E21*(C21*L13+K13)	=D21*H13^E21*LN(H13)*L13^2*(C2
29	393	=AVERAGE(A27:A29)	=L14^2*1	=L14^2*H14^E21*(C21*L14+K14)	=D21*H14^E21*LN(H14)*L14^2*(C2
30	243		=L15^2*1	=L15^2*H15^E21*(C21*L15+K15)	=D21*H15^E21*LN(H15)*L15^2*(C2
31	203		=L16^2*1	=L16^2*H16^E21*(C21*L16+K16)	=D21*H16^E21*LN(H16)*L16^2*(C2
32	240	=AVERAGE(A30:A3	=SUM(C24:C3	=SUM(D24:D31)	=SUM(E24:E31)
33					
34	H11	H12	H13	H14	G1
35	=SUM(C24:C31)	=-SUM(D24:D31)	=-SUM(E24:E3	=SUM(F24:F31)	=-H42
36	H21	H22	H23	H24	G2
37	=B35	=SUM(G24:G31)	=SUM(H24:H31	=-SUM(I24:I31)	=I42
38	H31	H32	H33	H34	G3
39	=C35	=C37	=J32	=-SUM(K24:K31)	=J42
40	H41	H42	H43	H44	G4
41	=D35	=D37	=D39	=SUM(L24:L31)	=-K42
42					
43	GAUSS-JORDAN REDUCTION				
44	=A35/A35	=B35/A35	=C35/A35	=D35/A35	=E35/A35
45	=A44*-A37	=B44*-A37	=C44*-A37	=D44*-A37	=E44*-A37
46	=A45+A37	=B45+B37	=C45+C37	=D45+D37	=E45+E37
47	=A44*-A39	=B44*-A39	=C44*-A39	=D44*-A39	=E44*-A39
48	=A47+A39	=B47+B39	=C47+C39	=D47+D39	=E47+E39
49	=A44*-A41	=B44*-A41	=C44*-A41	=D44*-A41	=E44*-A41
50	=A49+A41	=B49+B41	=C49+C41	=D49+D41	=E49+E41
51	=A46/B46	=B46/B46	=C46/B46	=D46/B46	=E46/B46
52	=A51*-B48	=B51*-B48	=C51*-B48	=D51*-B48	=E51*-B48
53	=A52+A48	=B52+B48	=C52+C48	=D52+D48	=E52+E48
54	=A51*-B50	=B51*-B50	=C51*-B50	=D51*-B50	=E51*-B50
55	=A54+A50	=B54+B50	=C54+C50	=D54+D50	=E54+E50
56	=A53/C53	=B53/C53	=C53/C53	=D53/C53	=E53/C53
57	=A56*-C55	=B56*-C55	=C56*-C55	=D56*-C55	=E56*-C55
58	=A57+A55	=B57+B55	=C57+C55	=D57+D55	=E57+E55
59	Δ a	Δb	Δc	Δd	Δ e
60	=E44-(B44*B60+C44*C60+D44*I	=E51-(C51*C60+D51	=E56-(D56*D6(	=E58/D58	
61	Solution Of The System of F				
62	=A60*A35	=B60*B35	=C60*C35	=D60*D35	=A62+B62+C62+D62
63	=A60*A37	=B60*B37	=C60*C37	=D60*D37	=A63+B63+C63+D63
64	=A60*A39	=B60*B39	=C60*C39	=D60*D39	=A64+B64+C64+D64
65	=A60*A41	=B60*B41	=C60*C41	=D60*D41	=A65+B65+C65+D65
66	ITERATION:		Δa'	Δb'	Δc'
67		Old	=H56	=I56	=J56
68		delta	=C69+A60-C67	=D69+B60-D67	=E69+C60-E67
69		New	=C67+C68	=D67+D68	=E67+E68

eadsheet solutions Gauss-Jordan reduction, Excel

F	G	H
	Bo-N	(See transformations for logit etc. below).
	=B5-E5	
Ln[Ag]	**Y**	**Ag**
=LN(1.25)	=(B11-E5)/G5	1.25
=LN(1.562)	=(B14-E5)/(G5)	1.562
=LN(3.125)	=(B17-E5)/(G5)	3.125
=LN(6.25)	=(B20-E5)/(G5)	6.25
=LN(12.5)	=(B23-E5)/G5	12.5
=LN(25)	=(B26-E5)/(G5)	25
=LN(50)	=(B29-E5)/(G5)	50
=LN(100)	=(B32-E5)/(G5)	100
d		
=E5		
H14	**H22**	**H23**
=L9*1	=C21*(H9^2)^E21*L9^3*(C21*L9+2*K9)*	=C21*H9^E21*LN(H9)*L9^2*(C21*D21*H9^E21*L9^2+2*K9*L9*D21*H9^E21-K9)*1
=L10*1	=C21*H10^2^E21*L10^3*(C21*L10+2*K	=C21*H10^E21*LN(H10)*L10^2*(C21*D21*H10^E21*L10^2+2*K10*L10*D21*H10^E
=L11*1	=C21*H11^2^E21*L11^3*(C21*L11+2*K	=C21*H11^E21*LN(H11)*L11^2*(C21*D21*H11^E21*L11^2+2*K11*L11*D21*H11^E
=L12*1	=C21*H12^2^E21*L12^3*(C21*L12+2*K	=C21*H12^E21*LN(H12)*L12^2*(C21*D21*H12^E21*L12^2+2*K12*L12*D21*H12^E
=L13*1	=C21*H13^2^E21*L13^3*(C21*L13+2*K	=C21*H13^E21*LN(H13)*L13^2*(C21*D21*H13^E21*L13^2+2*K13*L13*D21*H13^E
=L14*1	=C21*H14^2^E21*L14^3*(C21*L14+2*K	=C21*H14^E21*LN(H14)*L14^2*(C21*D21*H14^E21*L14^2+2*K14*L14*D21*H14^E
=L15*1	=C21*H15^2^E21*L15^3*(C21*L15+2*K	=C21*H15^E21*LN(H15)*L15^2*(C21*D21*H15^E21*L15^2+2*K15*L15*D21*H15^E
=L16*1	=C21*H16^2^E21*L16^3*(C21*L16+2*K	=C21*H16^E21*LN(H16)*L16^2*(C21*D21*H16^E21*L16^2+2*K16*L16*D21*H16^E
=SUM(F24:F3	=SUM(G24:G31)	=SUM(H24:H31)
		G1
		=K9*L9*1
	Row1	=K10*L10*1
		=K11*L11*1
	Row2	=K12*L12*1
		=K13*L13*1
	Row3	=K14*L14*1
		=K15*L15*1
	Row4	=K16*L16*1
		=SUM(H34:H41)
Row1+A36	**(1a)**	
(1a)*-A38	(1b)	
(1b)+Row2	(2a)	
(1a)*-A40	(1c)	
(1c)+Row3	(3a)	
(1a)*-A42	(1d)	
(1d)+Row4	(4a)	
(2a)+B47	**(2b)**	
(2b)*-B49	(2c)	
(2c)+(3a)	(3b)	
(2b)*-B51	(2d)	
(2d)+(4a)	(4b)	**Δ a'**
(3b)+C54	**(3c)**	=C21
(3c)*-C56	(3d)	=a+A60
(3d)+(4b)	**(4c)**	
	x	**E old**
		=(C21/(1+D21*H9^E21)+F21-B11)^2
Equation 1	=C21/(1+D21*H9^E21)+F21-B11	=(C21/(1+D21*H10^E21)+F21-B14)^2
Equation 2	=C21/(1+D21*H10^E21)+F21-B14	=(C21/(1+D21*H11^E21)+F21-B17)^2
Equation 3	=C21/(1+D21*H11^E21)+F21-B17	=(C21/(1+D21*H12^E21)+F21-B20)^2
Equation 4	=C21/(1+D21*H12^E21)+F21-B20	=(C21/(1+D21*H13^E21)+F21-B23)^2
Δd'	=C21/(1+D21*H13^E21)+F21-B23	=(C21/(1+D21*H14^E21)+F21-B26)^2
=K56	=C21/(1+D21*H14^E21)+F21-B26	=(C21/(1+D21*H15^E21)+F21-B29)^2
=F69+D60-F6	=C21/(1+D21*H15^E21)+F21-B29	=(C21/(1+D21*H16^E21)+F21-B32)^2
=F67+F68	=C21/(1+D21*H16^E21)+F21-B32	=SUM(H61:H68)

Spreadsheet formulas Gauss-Jordan reduction, Excel

	I	J	K
1			
2	B'/Bo' = Mean, B-N/B₀		
3			
4			
5			
6		TRANSFORMATIONS	
7		FOR LOGISTIC:	
8	Y	bp^c	x
9	=(B11-E5)/G5	=D21*H9^E21	=C21/(1+J9)+F21-B11
10	=(B14-E5)/(G5)	=D21*H10^E21	=C21/(1+J10)+F21-B14
11	=(B17-E5)/(G5)	=D21*H11^E21	=C21/(1+J11)+F21-B17
12	=(B20-E5)/(G5)	=D21*H12^E21	=C21/(1+J12)+F21-B20
13	=(B23-E5)/G5	=D21*H13^E21	=C21/(1+J13)+F21-B23
14	=(B26-E5)/(G5)	=D21*H14^E21	=C21/(1+J14)+F21-B26
15	=(B29-E5)/(G5)	=D21*H15^E21	=C21/(1+J15)+F21-B29
16	=(B32-E5)/(G5)	=D21*H16^E21	=C21/(1+J16)+F21-B32
17			
18			
19			
20			
21			
22			
23	H24	H33	H34
24	=C21*H9^E21*L9^2*1	=C21*D21*(LN(H9)^2)*H9^E21*L9^2*(C21*J9*L9^2+2*K9*L9*D21*H9^E21-K9)*1	=C21*D21*H9^E21*LN(H9)*L9^2*1
25	=C21*H10^E21*L10^2	=C21*D21*(LN(H10)^2)*H10^E21*L10^2*(C21*J10*L10^2+2*K10*L10*D21*H10^E2	=C21*D21*H10^E21*LN(H10)*L10
26	=C21*H11^E21*L11^2	=C21*D21*(LN(H11)^2)*H11^E21*L11^2*(C21*J11*L11^2+2*D21*K11*L11*H11^E2	=C21*D21*H11^E21*LN(H11)*L11
27	=C21*H12^E21*L12^2	=C21*D21*(LN(H12)^2)*H12^E21*L12^2*(C21*J12*L12^2+2*D21*K12*L12*H12^E2	=C21*D21*H12^E21*LN(H12)*L12
28	=C21*H13^E21*L13^2	=C21*D21*(LN(H13)^2)*H13^E21*L13^2*(C21*J13*L13^2+2*D21*K13*L13*H13^E2	=C21*D21*H13^E21*LN(H13)*L13
29	=C21*H14^E21*L14^2	=C21*D21*(LN(H14)^2)*H14^E21*L14^2*(C21*J14*L14^2+2*D21*K14*L14*H14^E2	=C21*D21*H14^E21*LN(H14)*L14
30	=C21*H15^E21*L15^2	=C21*D21*(LN(H15)^2)*H15^E21*L15^2*(C21*J15*L15^2+2*D21*K15*L15*H15^E2	=C21*D21*H15^E21*LN(H15)*L15
31	=C21*H16^E21*L16^2	=C21*D21*(LN(H16)^2)*H16^E21*L16^2*(C21*J16*L16^2+2*K16*L16*D21*H16^E2	=C21*D21*H16^E21*LN(H16)*L16
32	=SUM(I24:I31)	=SUM(J24:J31)	=SUM(K24:K31)
33	G2	G3	G4
34	=C21*H9^E21*K9*L9^	=C21*D21*H9^E21*LN(H9)*K9*L9^2*1	=K9*1
35	=C21*H10^E21*K10*l	=C21*D21*H10^E21*LN(H10)*K10*L10^2*1	=K10*1
36	=C21*H11^E21*K11*l	=C21*D21*H11^E21*LN(H11)*K11*L11^2*1	=K11*1
37	=C21*H12^E21*K12*l	=C21*D21*H12^E21*LN(H12)*K12*L12^2*1	=K12*1
38	=C21*H13^E21*K13*l	=C21*D21*H13^E21*LN(H13)*K13*L13^2*1	=K13*1
39	=C21*H14^E21*K14*l	=C21*D21*H14^E21*LN(H14)*K14*L14^2*1	=K14*1
40	=C21*H15^E21*K15*l	=C21*D21*H15^E21*LN(H15)*K15*L15^2*1	=K15*1
41	=C21*H16^E21*K16*l	=C21*D21*H16^E21*LN(H16)*K16*L16^2*1	=K16*1
42	=SUM(I34:I41)	=SUM(J34:J41)	=SUM(K34:K41)
43			
44			
45			
46			
47			
48			
49			
50			
51			
52			
53			
54			
55	Δ b'	Δc'	Δd
56	=D21	=E21	=F21
57	=b+B60	=cc+C60	=d+D60
58			
59			
60	E new	Ln [Ag]	B calcd
61	=(C69/(1+D69*H9^E69	=D9	=(H57/(1+I57*H9^J57)+K57)
62	=(C69/(1+D69*H10^E6	=D10	=(H57/(1+I57*H10^J57)+K57)
63	=(C69/(1+D69*H11^E6	=D11	=(H57/(1+I57*H11^J57)+K57)
64	=(C69/(1+D69*H12^E6	=D12	=(H57/(1+I57*H12^J57)+K57)
65	=(C69/(1+D69*H13^E6	=D13	=(H57/(1+I57*H13^J57)+K57)
66	=(C69/(1+D69*H14^E6	=D14	=(H57/(1+I57*H14^J57)+K57)
67	=(C69/(1+D69*H15^E6	=D15	=(H57/(1+I57*H15^J57)+K57)
68	=(C69/(1+D69*H16^E6	=D16	=(H57/(1+I57*H16^J57)+K57)
69	=SUM(I61:I68)		

Use of matrix functions to solve example 6-1, Excel

	A	B	C	D	E	F
1		Solution of the Four Simultaneous Equations from Example 6-1				
2		Using the built-in Matrix Functions of Microsoft Excel				
3						
4						
5						
6	EQUATIONS:					
7	1.176x(1)	(-)5093.796x(2)	(-)1962.04x(3)	2.459x(4)	=	48.13
8	(-)5093.796x(1)	23636697.8x(2)	12020493x(3)	(-)12487.059x(4)	=	8270.9
9	(-)1962.04x(1)	12020493x(2)	9857483x(3)	(-)8114.1607x(4)	=	(-)248494
10	24592x(1)	(-)12487.0589x(2)	(-)8114.1607x(3)	8x(4)	=	37.982
11						COLUMN
12						MATRIX OF
13	COEFFICIENT MATRIX:					CONSTANTS
14	1.1760383	-5093.796	-1962.04	2.459		-48.1343
15	-5093.796	23636697.8	12020493	-12487.059		8270.944
16	-1962.04	12020493.4	9857483	-8114.1607		-248494
17	2.4592	-12487.0589	-8114.1607	8		37.9816
18						
19						
20	INVERSION OF COEFFICIENT MATRIX:					
21	-29.11848263	-0.007820015	0.006421058	3.256857551		
22	-0.007817075	-1.8423E-06	1.82561E-06	0.001378818		
23	0.006425854	1.82755E-06	-7.62229E-07	0.00010433		
24	3.267038391	0.001381881	0.000102622	1.381834283		
25						
26						
27	OUTPUT (Multiply inverted matrix and constants):					
28	x(1):	-134.9748125				
29	x(2):	-0.040251347				
30	x(3):	-0.10081641				
31	x(4):	-118.8437284				

Use of matrix functions to solve example 6-1, Excel

	A	B	C
1	**EXAMPLE 6-1, continued**		
2			
3			
4			
5			
6	**EQUATIONS:**		
7	1.176x(1)	(-)5093.796x(2)	(-)1962.0
8	(-)5093.796x(1)	23636697.8x(2)	1202049
9	(-)1962.04x(1)	12020493x(2)	985748
10	24592x(1)	(-)12487.0589x(2)	(-)8114.160
11			
12			
13	**COEFFICIENT MATRIX:**		
14	1.1760383	-5093.796	-1962.04
15	-5093.796	23636697.8	12020493
16	-1962.04	12020493.4	9857483
17	2.4592	-12487.0589	-8114.1607
18			
19			
20	**INVERSION OF COEFFICIENT MATRIX:**		
21	=MINVERSE(A14:D17)	=MINVERSE(A14:D17)	=MINVERSE(A14:D17)
22	=MINVERSE(A14:D17)	=MINVERSE(A14:D17)	=MINVERSE(A14:D17)
23	=MINVERSE(A14:D17)	=MINVERSE(A14:D17)	=MINVERSE(A14:D17)
24	=MINVERSE(A14:D17)	=MINVERSE(A14:D17)	=MINVERSE(A14:D17)
25			
26			
27	**OUTPUT (Multiply inverted ma**		
28	x(1):	=MMULT(A21:D24,F14:F17)	
29	x(2):	=MMULT(A21:D24,F14:F17)	
30	x(3):	=MMULT(A21:D24,F14:F17)	
31	x(4):	=MMULT(A21:D24,F14:F17)	

Use of matrix functions to solve example 6-1, Excel

	D	E	F
1	**EXAMPLE 6-1 continued**		
2			
3			
4			
5			
6			
7	2.459x(4)	=	48.13
8	(-)12487.059x(4)	=	8270.9
9	(-)8114.1607x(4)	=	(-)248494
10	8x(4)	=	37.982
11			**COLUMN**
12			**MATRIX OF**
13			**CONSTANTS**
14	2.459		-48.13
15	-12487.059		8270.9
16	-8114.1607		-248494
17	8		37.982
18			
19			
20			
21	=MINVERSE(A14:D17)		
22	=MINVERSE(A14:D17)		
23	=MINVERSE(A14:D17)		
24	=MINVERSE(A14:D17)		
25			
26			
27			
28			
29			
30			
31			

Use of the built-in matrix *command* to solve example 6-1, Lotus 1-2-3

A	A	B	C	D	E	F
1						
2	**Solution of the Four Simultaneous Equations from Example 6-1**					
3	**Using the built-in Data Menu Command "Matrix" of Lotus 1-2-3**					
4						
5	NOTE. Lotus 1-2-3 does not have separate matrix functions as does Excel.					
6	Instead, 1-2-3 has a "Matrix command" in the Data Menu to invert or multiply					
7	matrices. For this reason, we cannot show formulas for this worked example.					
8						
9	**EQUATIONS:**					
10	1.176x(1)	-5093.796x(2)	-1962.04x(3)	2.459x(4) =		48.13
11	-5093.796x(1)	23636697.8x(2)	12020493x(3)	-12487.059x(4) =		8270.9
12	-1962.04x(1)	12020493.4x(2)	9857483x(3)	-8114.1607x(4) =		-248494
13	2.4592x(1)	-12487.0589x(2)	-8114.1607x(3)	8x(4) =		37.982
14						**COLUMN**
15						**MATRIX OF**
16	**COEFFICIENT MATRIX:**					**CONSTANTS**
17	1.176	-5093.796	-1962.04	2.459		-48.13
18	-5093.796	23636697.8	12020493	-12487.059		8270.9
19	-1962.04	12020493.4	9857483	-8114.1607		-248494
20	2.4592	-12487.0589	-8114.1607	8		37.982
21						
22	**INVERSION OF COEFFICIENT MATRIX:**					
23	-2.908604E+01	-7.81130E-03	0.006413905	3.2532294262		
24	-7.808367E-03	-0.00000184	0.000001824	0.0013778445		
25	0.00641869567	0.0000018256	-7.6065E-07	0.0001051308		
26	3.26339892498	0.0013809039	0.000103424	1.3822413512		
27						
28	**OUTPUT (Multiply inverted matrix and constants):**					
29	x(1):	-1.34948E+02				
30	x(2):	-4.02439E-02				
31	x(3):	-0.100822029				
32	x(4):	-1.18846E+02				

6-15. Review questions

6-1.   Define the terms *antibody* and *antigen*.
6-2.   List the *major classes of immunoglobulins* and explain why, when an animal is injected with one antigen, a mixture containing more than one closely related antibody may be formed.
6-3.   What are the intermolecular forces holding antibody and antigen together in antigen-antibody complexes?
6-4.   Describe the shape of an antigen-antibody precipitation curve and explain this curve in terms of mass action relationships.
6-5.   Describe the primary structure of antibodies in general terms.
6-6.   What is the Dreyer-Bennett hypothesis of antibody diversity?
6-7.   Explain how the work of Susumi Tonegawa provided experimental support for the Dreyer-Bennett hypothesis.
6-8.   Distinguish between explanations of *antibody diversity* at the cellular and molecular levels.
6-9.   Define the term *major histocompatibility complex*.
6-10.   What is *hybridoma technology*, and how is it used to produce *monoclonal antibodies*?
6-11.   Why are monoclonal antibodies considered to be an important new tool both by academic and industrial scientists?
6-12.   Describe the *principle of radioimmunoassay*.
6-13.   Describe the experimental approach to the preparation of a radioimmunoassay calibration curve.
6-14.   What requirements should be met for successful use of the *logit-log* RIA calibration curve?
6-15.   List five common response variables for RIA calibration curves.
6-16.   When would you use the *four-parameter logistic equation* for the estimation of antibodies by RIA?
6-17.   Describe the major types of *enzyme-linked immunosorbent assay*.
6-18.   On what principle is ELISA based?
6-19.   Compare methods of plotting RIA and ELISA calibration curves.
6-20.   List some of the limitations of unweighted regression as applied to immunoassay calibration curves.
6-21.   In what three ways does the immune system protect mammalian cells from infections? Describe how each works.
6-22.   Describe the principle of electroimmunoassay, EIA.
6-23.   Outline the major similarities and differences between the *T cell receptor* and *antibodies*.
6-24.   Describe the nature and functions of the *major histocompatibility complex*.
6-25.   Define the term *cytokine* and explain why some scientists regard cytokines as potentially powerful medical tools.
6-26.   What are the abscissa and ordinate for each of the following *radioimmunoassay curves: hyperbolic, sigmoid* and *logit-log*?
6-27.   Explain how you would calculate logit Y from B, B_0 and N.
6-28.   Explain how you would calculate Y and B from logit Y.
6-29.   What are *macrophages*? How do they function in the immune system?
6-30.   Define the term stringent selection as applied to T cells.

6-31. Compare and contrast competitive binding ELISA and noncompetitive ELISA.
6-32. What have *Bennett, Dreyer, Köhler, Milstein* and *Tonegawa* contributed to modern immunology?
6-33. What roles do *Newton-Raphson evaluation* and *Gauss-Jordan reduction* play in calculation of the four-parameter logistic curve.
6-34. Describe the antigen binding sites of antibodies.
6-35. Describe the antigen binding protein of the T cell receptor.

Suggested Reading

Abraham, R.T., L.M. Karnitz, J.P. Secrist and P.J. Leibson 1992. Signal Transduction Through the T cell Antigen Receptor. *Trends. Biochem. Sci.* 17: 438-443.
Aloisi, R. M. 1979. Principles of Immunodiagnostics, C. V. Mosby, St Louis.
Alt, F.W., T.K. Blackwell and G.D. Yancopoulos 1987. Development of the Primary Antibody Repertoire, *Science* 238: 1079-1087.
Baxter, R.C. 1983. *Methods of Measuring Confidence Limits in Radioimmunoassay.* Methods in Enzymology, Academic Press, New York 92: 601-610.
Benkovic, S.J: *Catalytic Antibodies* 1992. Annu. Rev. Biochem. 61: 29-54.
Boulay, J-L and W.E. Paul 1992. The Interleukin-4-Related Lymphokines and their Binding to Hematopoietin Receptors. *J. Biol. Chem.* 267: 20525-20528.
Burston, S.G., R. Sleigh, D.L. Halsall, C.J. Smith, J.J. Holbrook and A.R. Clarke 1992. Enzyme Engineering XI. The Influence of Chaperonins on Protein Folding. *Ann. N.Y. Acad. Sci.* 672: 1-19.
Brown, M.S. and J.L 1983. Goldstein. *Lipoprotein Metabolism in the Macrophage: Implications for Cholesterol Deposition in Atherosclerosis.* Annu. Rev. Biochem. 52: 223-261.
Capra, J.D. and P.W. Tucker 1989. Human Immunoglobulin Heavy Chain Genes. *J. Biol. Chem.* 264: 12745-12748.
Carpenter, G. 1990. Epidermal Growth Factor. *J. Biol. Chem.* 265: 7709-7712.
Clausen, J 1969. *Immunochemical Techniques for the Identification and Estimation of Macromolecules,* Elsevier, NY.
Causton, D. 1983. *A Biologist's Basic Mathematics,* Arnold, Baltimore, Maryland.
Causton, D. R. 1987. *A Biologist's Advanced Mathematics,* Allen & Unwin, Boston.
Dahlquist, F. W. 1978. *The Meaning of Scatchard and Hill Plots, Methods in Enzymology,* Academic Press, New York, 48: 270-299.
Dinarello, C. 1988. Biology of Interleukin 1, *FASEB J.* 2: 108-115.
Engvall, E. 1980. *Enzyme Immunoassay ELISA and EMIT,* Methods in Enzymol, Academic Press, NY 70: 419-430.
Frisell, W. 1982. *Human Biochemistry,* Macmillan, New York.
Galfrè, G. and C. Milstein 1981. *Preparation of Monoclonal Antibodies: Strategies and Procedures,* Methods in Enzymology, Academic Press, New York 73: 3-46.
Honjo, T. and S. Habu 1985. *Origin of Immune Diversity: Genetic Variation and Selection,* Annu. Rev. Biochem. 54: 803-830.
Inagami, T. 1989. Atrial Natriuretic Factor. *J. Biol. Chem.* 264: 3043-3046.
Johnson, M.L. and S. G. Frasier 1985. *Nonlinear Least-Squares Analysis,* Methods in Enzymology, Academic Press, New York 117:301-342.
Kabat, E. A. 1980. *Basic Principles of Antigen-Antibody Reactions,* Methods in Enzymology, Academic Press, New York, 70: 3-49

Krieger, M., S. Acton, J. Ashkenas, A. Pearson, M. Penman and D. Resnick 1993 Molecular Flypaper, Host Defense, and Atherosclerosis. *J. Biol. Chem.* 268: 4569-4572.

Laurell, C.-B. and E. J. Mackay 1981. *Electroimmunoassay*, Methods in Enzymol., Academic Press, NY, 73: 339-369.

Lehninger, A. L 1975. *Biochemistry. 2d ed.*, Worth, New York.

Lehninger, A. L., D. J. Nelson and M. M. Cox 1992. *Principles of Biochemistry*, 2d. ed., Worth, New York.

Lerner, R.A., A. S. Kang, J.D. Bain, D.R. Burton and C.F. Barbass, III 1992. Antibodies Without Immunization. *Science* 258: 1313-1314.

Lindbladh, C., K. Mosbach and L. Bülow 1993. Use of Genetically Prepared Enzyme Conjugates in Enzyme Immunoassay. *Trends in Biochem. Sci.* 18: 279-283.

Marks, J.D., H.R. Hoogenboom, A.D. Griffiths and G. Winter 1992. Molecular Evolution of Proteins on Filamentous phage. Mimicking the Strategy of the Immune System. *J. Biol. Chem.* 267: 16007-16010.

Marrack, P. and J. Kappler 1987. The T Cell Receptor, *Science* 238: 1073-1079.

Marx, J. L. 1987. Antibody Research Garners Nobel Prize, *Science* 238: 484-485.

Motulsky, H. J. and L. A. Rasnas 1987. Fitting Curves to Data Using Nonlinear Regression: a Practical and Nonmathematical Review. *FASEB J. 1*: 365-374.

Munson, P.J. 1983. LIGAND: *A Computerized Analysis of Ligand Binding Data*. Methods in Enzymology, Academic Press, New York, 92: 543-576.

Nossal, G. J. V. 1988 Triumphs and Trials of Immunology in the 1980's. *Life Science Job Trends 2 No 1* Page 1, 19 October. NOTE. This article is an excerpt from a review that was first published in Immunology Today Vol. 9 No. 10, 286,1988.

Olson, L. and H. S. Kaplan 1983. *Human-Human Monoclonal Antibody-Producing Hybridomas: Technical Aspects*, Methods Enzymol., Academic Press, New York, 92: 3-16.

Oppenheim, J. J. 1985. *Antigen Nonspecific Lymphokines: An Overview*. Methods in Enzymology, Academic Press, New York, 116: 357-372.

Orth, D.N 1975. *General Considerations for Radioimmunoassay of Peptide Hormones*. Methods in Enzymology, Academic Press, New York, 37: 22-38.

Padlan, E.A. and E.A. Kabat 1991. *Modeling of Antibody Combining Sites*. Methods in Enzymology, Academic Press, New York, 203: 3-21.

Prescott, D.M. 1988. *Cells. Principles of Molecular Structure and Function*. Jones & Bartlett Publishers, Boston.

Prescott, S.M., G.A. Zimmerman and T.M. McIntyre 1990. Platelet-Activating Factor. *J. Biol. Chem.* 265: 17381-17384.

Retsky, K.L., M.W. Freeman and B. Frei 1993. Ascorbic Acid Oxidation Product(s) Protect Human Low Density Liproprotein Against Atherogernic Modification. *J. Biol. Chem.* 268:1304-1309.

Ritchie, D. G., J. M. Nickerson and G. M. Fuller 1983. *Two Simple Programs for the Analysis of Data from Enzyme-Linked Immunosorbent Assays (ELISA) on a Programmable Desk-Top Calculator*. Methods in Enzymology, Academic Press, New York, 92: 577-588.

Rodbard, D. and G. R. Frazier 1975. *Statistical Analysis of Radioligand Assay Data*, Methods in Enzymology, Academic Press, New York, 37: 3-22.

Samelson, L.E. and R.D.Klausner 1992. Tyrosine Kinases and Tyrosine-Based Activation Motifs. Current Research on Activation via the T cell Antigen Response. *J. Biol. Chem.* 267: 24913-24916.

Samuelson, B. and C.D. Funk 1989. Enzymes Involved in the Biosynthesis of Leukotriene B_4. *J. Biol. Chem.* 264: 19470-19472.

Sen, G.C. and P. Lengyel 1992. The interferon System. A Bird's Eye View of its Biochemistry. *J. Biol. Chem.* 267: 5017-5020.

Shokat K.M. and Schultz, P.G. 1991. *Catalytic Antibodies*. Methods in Enzymology, Academic Press, New York, 203: 327-351.

Stryer, L. 1981. *Biochemistry. 2d ed.*, Freeman, San Francisco.

Takahashi, M., S.A. Fuller and S. Winston 1991. *Design and Production of Bispecific Monoclonal Antibodies by Hybrid Bybridomas for Use in Immunoassay*. Methods in Enzymology, Academic Press, NY, 203: 312-327.

Tsomides, T.J. and H.N. Eisen 1991. Antigenic Structures Recognized by Cytotoxic T Lymphocytes. *J. Biol. Chem.* 264: 3357-3360.

Ukraincik, K and W. Piknosh 1981. *Microprocessor-Based Radioimmunoassay Data Analysis,* Methods in Enzymology, Academic Press, New York, 74: 497-508.

Van Vunakis, H. 1980. .*Radioimmunoassays,* Methods in Enzymology, Academic Press, NY, 70: 201-209.

Vilcek and T.H. Lee 1991. Tumor Necrosis Factor. New Insights into the Molecular Mechanisms of its Multiple Actions. *J. Biol. Chem.* 266: 7313-7316.

Waldman, T.A. 1991. The Interleukin-2-Receptor. *J. Biol. Chem.* 266: 2681-2684.

Yancopoulos, G. D. and F. W. Alt 1988. Reconstruction of an Immune System. *Science* 241: 1581-1583.

Yaqub, A. and H. G. Moore 1981. *Elementary Linear Algebra With Applications.* Addison-Wesley, Mass.

Yelton, D. E. and M. D. Scharff 1981. *Monoclonal Antibodies: A Powerful New Tool in Biology and Medicine,* Annu. Rev. Biochem., 50: 657-680.

PART III.

PHYSICOCHEMICAL APPROACHES

CHAPTERS 7 TO 9

Chapter 7
Biomolecular Spectroscopy

7-1. The electromagnetic spectrum defines the branches of spectroscopy

Spectroscopic methods have become the mainstay of major segments of the modern clinical and biomedical research enterprise. The underlying reason for this preeminence is that *molecular structure determines whether electromagnetic radiation is emitted or absorbed. Consequently, measurements of the emission or absorption of electromagnetic radiation can, in turn, provide information on molecular structure. Spectroscopic methods thus provide insights into the physical and chemical properties of biomolecules.* One index of the importance of spectroscopy is that even a brief review of pertinent physicochemical principles, as in this chapter, will include work for which several Nobel prizes were awarded between 1901 and 1991. These prizes and the 90-year time period remind us of two aphorisms: that physics was in a golden age during approximately the first third of the twentieth century; and that discovery, innovation, and the solution of major scientific problems often are gradual processes — not quick, quantum jumps.

Electromagnetic radiation is divided, for convenience, into regions of different wavelengths. These regions consist of radiation ranging from gamma rays which are of short wavelength to radiowaves which have the longest wavelengths. Figure 7-1 shows a diagram of the electromagnetic spectrum:

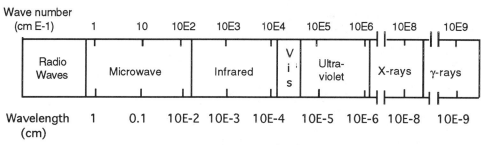

Figure 7-1. Regions of the electromagnetic epectrum

It may appear from Figure 7-1 and similar diagrams that each region or form of radiation is sharply distinct from the other; this, however, is not true. One region of the spectrum merges imperceptibly into the next. And happily, to simplify learning about them, not all regions of the electromagnetic spectrum are of equal interest to biomedical scientists. The spectroscopic regions of greatest biomedical interest are the ultraviolet, the visible, the infrared, and the magnetic resonance/radio frequency regions. Although all regions of the spectrum are not equally important to biomedical scientists, we will, nevertheless, undertake a brief discussion of the spectroscopic techniques associated with each spectroscopic region. The reason for this undertaking is to provide a broad understanding of the major techniques of biomolecular spectroscopy. Furthermore, because the emission or absorption of radiation depends on molecular structure, we will examine, at the outset, the physical nature of electromagnetic radiation and its interaction with matter.

7-2. Electromagnetic radiation has wave and particle properties

Classical physics, from the time of Isaac Newton (1642-1727) to the late nineteenth century, said that light existed as waves. But today, we know that properties of light such as absorption, emission, and scattering may best be explained if we assume that electromagnetic radiation has properties both of waves and particles.

In classical physics, five major properties of light supported the wave theory: energy, wavelength, frequency, velocity, and amplitude (i.e., the algebraic difference between peak and trough). For example, the transfer of energy from light to matter, the association of colors with specific wavelengths, and the dependence of intensity on amplitude were all consistent with the wave properties of light. Toward the end of the nineteenth century, however, it became apparent that the description of light exclusively as a wave could not explain certain other phenomena. One unexplained phenomenon was the wavelength distribution of light emitted from a hot body that did not reflect light, e.g. the red glow of an electric cooking range. In 1900, Max Planck explained this puzzling result (known as *blackbody radiation*), and, for the explanation, he was awarded the 1918 Nobel Prize in physics. Planck proposed in 1900 that *energy may be absorbed or emitted only in specific quantities which he referred to as quanta*, each single packet or *quantum* of energy having the size hv defined in the equation which bears Planck's name:

$$E \quad = \quad h\nu \tag{1}$$

where E =   a quantum of energy (in Joules),
                  h =   a quantity now called *Planck's constant,* and
                  ν =   frequency of radiation in Hz (with dimensions sec^{-1}).

The value of Planck's constant h is 6.626×10^{-34} Joules sec. In the derivation of equation (1), Planck suggested that *atoms and molecules may only occupy discrete sets of energy levels when raised to excited states* since radiation is absorbed or emitted discontinuously in integral multiples of the quantum E, so that

$$E = nh\nu \qquad (2)$$

where *n*, the *quantum number*, is a positive integer i.e. $n = 0,1,2,3$ ... but never 2.5, 3.7 etc. Thus, the lowest permitted energy level, the ground state, has the value zero, and allowed energy levels above the ground state (*excited states* or *quantum states*) have values $1,2,3...x\ h\nu$. According to Planck's quantum theory of radiation, energy levels different from the permitted values do not exist. Figure 7-2 illustrates the occurrence of discrete energy levels by showing energy on the ordinate and the permitted energy levels as horizontal lines:

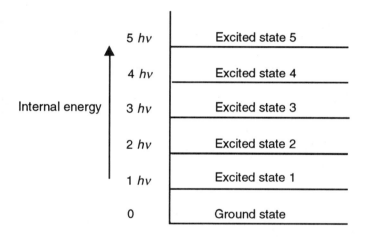

Figure 7-2. Representation of quantized (discrete) energy levels

Beginning in 1905, Albert Einstein (working as a clerk in the patent office in Bern, Switzerland, because he could not get a university job) made some important contributions to the development of quantum theory. Einstein was awarded the 1921 Nobel prize in physics for this work on the *photoelectric effect* - a phenomenon in which light, striking the surface of a metal, can cause the emission of electrons. Einstein suggested that light also consists of quanta, that each quantum or *photon* has a discrete amount of energy, and that when light strikes a particular metal, an electron either receives all of the energy of a given photon or none at all. In addition, Einstein proposed that, regardless of the intensity of a beam of light, it will cause a photochemical reaction to occur only if the frequency of the incident light is greater than a particular minimum or threshold value. Weak blue light could, for example, cause emission whereas strong red light would not. We explain the last statement as follows. In the wave theory of light, the frequency of electromagnetic radiation (ν) is related to *wavelength* (λ) by the relationship $\nu = c/\lambda$, where c is the velocity of light. Therefore, as shown by equation (3), the greater the frequency, or, the shorter the wavelength of radiation, the greater the energy:

$$E = h\nu = hc / \lambda, \tag{3}$$

where E = the energy of the photons in Joules,

h = Planck's constant in Joules-sec (energy * time)/quantum

ν = frequency in Hz, dimension sec^{-1},

c = velocity of light in a vacuum, 3.00×10^8 m s $^{-1}$,

λ = wavelength of the light in cm.

Another important spectroscopic milestone dates back to 1910 when New Zealand-born Ernest Rutherford, (Nobel Prize in chemistry 1908 for his pioneering work in nuclear physics; see section 8-1), carried out an experiment at Cambridge University in which he allowed α-particles (the nuclei of helium atoms) to pass through a thin sheet (film) of gold. This experiment underlined the close relation between spectroscopy and atomic and molecular structure. Since α-particles are ejected with high velocities and have great ability to penetrate matter, Rutherford anticipated that the α-particles would go right through the thin metallic film and would emerge on the other side of the film with somewhat reduced velocity. Instead, some particles were scattered at sharp angles, and some even were scattered backward. By shrewd analysis of this result, Rutherford became the first to deduce that atoms have dense, positively charged nuclei; but he suggested incorrectly that electrons rotate around the nuclei (as do planets about the sun) so that the inward attraction of the electrons to the nuclei is counterbalanced by the outward centrifugal force. To overcome the theoretical difficulties with Rutherford's model, Niels Bohr of Denmark (Nobel Prize in physics, 1922) proposed (at 28 years old in 1913) that electrons exist in quantized, fixed orbits (energy states) corresponding to the quantum number, n (equation 2). The result of Bohr's arbitrary combination of quantized and classical mechanics was that, it explained only the spectrum of the H atom. In 1923 a French graduate student, Prince Louis de Broglie, proposed in his doctoral dissertation that both matter and light have the properties of waves and particles. For this work, the Prince was awarded both the Nobel Prize in physics (1929) and his Ph.D. degree; but the latter was awarded with the lowest passing grade since his professors were displeased with his revolutionary views on particle-wave duality. Subsequently, de Broglie was further vindicated when his proposal that electrons can display wave character proved to be crucial to the development of the electron microscope. Then, in 1926, Werner Heisenberg (25 years old) worked out the mathematics for three-dimensional wave patterns of atoms (vibrating spheres), and Erwin Schrödinger (applying de Broglie's particle-wave ideas and Heisenberg's math) developed his *wave equation* which yields functions, $\psi_{(x,y,z)}$, called *wave functions or orbitals* representing the probability of finding an electron in a given region of space. Shortly after, in 1928, Paul Dirac developed a relativistic quantum theory of the electron. Their proposals earned Heisenberg, Schrödinger and Dirac the 1932 Nobel Prize in physics and established *quantum mechanics* (*wave mechanics*) as one of the most influential areas of modern science. The perspicacious American, Linus Pauling (Nobel Prize in chemistry 1954; for peace, 1962) quickly recognized the power of quantum mechanics for studying atomic and molecular structure. He deduced (in 1931) that the strength of covalent bonds depends on the degree of overlap of atomic orbitals, and he showed (in 1929, age 28) that whereas the bonds between the two silicon atoms and oxygen are stable in silicates, the analogous bonds in di- and tri-phosphates are not. This 1929 study led Fritz Lipmann, (Nobel Prize in medicine or physiology with Sir Hans Krebs, 1953), to propose in 1941 that ATP is a high-energy compound which plays a key role in energy metabolism.

To summarize: electromagnetic radiation is quantized and has the properties of waves and particles. Molecules occupy discrete levels when raised to excited states, and quantum mechanics is the basis of interpreting every spectroscopic technique now used in biomedical research. The application of quantum theory has been crucial for (a) developing electron microscopy, computed axial tomography (CAT), magnetic resonance imaging, telecommunications, laser printers, and compact discs; (b) modern car, aerospace, and computer manufacturing; and (c) understanding chemical bonding, molecular structure, molecular orbital theory, and properties of biomolecules as varied as ATP, heme compounds, and carcinogens.

7-3. The Boltzman law gives the most probable distribution of energies

Beginning at the end of the nineteenth century, the Austrian physical chemist Ludwig Boltzmann played a pivotal role in the development of *statistical mechanics,* which *spans the gap between thermodynamics and quantum mechanics.* Thermodynamics is the branch of physical chemistry which deals with the properties of matter in bulk, e.g., temperature, pressure, and concentration; quantum mechanics deals with the energetics and motions of individual particles in the submicroscopic state. In other words, statistical mechanics combines statistical theory with thermodynamics or energetics to describe matter in the macroscopic state consisting of large numbers of molecules (particles). Among the chief contributions of Boltzmann was the development of equations for calculating the probability of occupancy of energy states. These equations on the distribution of energies are based on several assumptions, among which are the following: that large numbers of particles (molecules) are present; that the particles are at or near equilibrium; that each individual particle has available to it a set of energy states; and that, the most probable distribution corresponds to the greatest number of arrangements (permutations) of molecules over the levels within each energy state. If we regard the valence bonds which join atoms in molecules as springs, the molecules which are formed with such 'springs' can undergo four types of motion: *translational, rotational, vibrational, and electronic* motions. The total molecular energy then would be the sum of the energy associated with each state:

$$E_{total} = E_{translational} + E_{rotational} + E_{vibrational} + E_{electronic}.$$

The energy spacings in translational states are very small. They are so small, in fact (cf. Figure 7-3), that we may regard them as being continuous or merged into each other. Accordingly, we may assume that translational motion is not quantized and rewrite the equation for the total energy of the molecules as:

$$E_{total} = E_{rotational} + E_{vibrational} + E_{electronic}.$$

A question of interest to us now would be, given a large number of molecules at *thermal equilibrium and near room temperature*, how would the individual molecules be distributed within a given energy state? *The laws of probability tell us that, in the most probable distribution, individual molecules will be distributed in the largest number of ways over the different states and that most of the molecules will occupy the lower states. Only a few molecules will occupy the highest energy states because occupancy at the highest states reduces the probability of occupancy at lower states for a given content of total energy.* Boltzmann showed that the ratio of molecules in two energy states, labeled ε_i and ε_o, may be given by equation (4). This equation shows quantitatively that occupancy of the highest energy states is unlikely since the molecular energy of these high energy states is much greater than the term kT, ≈ 0.6 kcal mole^{-1} (or $4*10^{-14}$ ergs molecule^{-1}) at room temperature:

$$\frac{n_i}{n_o} = e^{-(\varepsilon_i - \varepsilon_o)/kT} = e^{-\Delta\varepsilon_i/kT}, \tag{4}$$

where
n_i = the number of particles which have energy ε_i,

n_o = the number of particles which have energy ε_o,

k = the Boltzmann constant, (the gas constant R, divided by Avogadro's number),

T = temperature °K.

To express equation (4) in an alternative way, we can say that the *Boltzmann distribution law* shows that, *if the value of the exponent* Δe_i *is much greater than the value of* kT, *then the ratio* n_i/n_o *will be small (will be equal to a number with a large negative exponent).* Conversely, if the spacing between two states $\Delta\varepsilon_i$ is much less than kT (i.e. if $\Delta\varepsilon_i$ approaches 0), then n_i and n_o (the number of molecules in the two states) will be approximately equal since $x^0 = 1$ for any value of x except 0. According to equation (4) also, as *temperature is increased, the fraction of molecules in the higher energy states increases greatly and the average energy content of the molecules increases correspondingly.* We should note here that the term $e^{-\Delta\varepsilon_i/kT}$ is similar to the term $e^{-E_a/RT}$ in the Arrhenius equation, i.e. equation (32) in Chapter 5. Not surprisingly, then, equation (4) may be used not only to calculate which energy states are occupied but, also, to calculate the average energy of a set of particles. When large numbers of excited molecules in high energy states return to their ground states, the likelihood of emission of radiation increases.

Although we have, for the most part, referred to single, energy or quantum states in the foregoing, each state consists of several levels. In mathematical terms, quantum states are *degenerate* (as are the codons for amino acids in the genetic code). Therefore to calculate the distribution of an energy level (in contrast to calculating the distribution of an energy state) equation (4) should be weighted by a factor which takes this degeneracy into account. We introduce the degeneracy factor g to achieve this weighting in equation (5):

$$\frac{n_i}{n_o} = \left[\frac{g_i}{g_o}\right] e^{-(\varepsilon_i - \varepsilon_o)/kT} \tag{5}$$

where n_i and n_o are the numbers of particles in energy levels i and o respectively,

g_i and g_0 are corresponding weighting factors and the other terms are as defined previously. Equation (5) shows that *when the degeneracies of all states are equal, the lower energy states, those with small ε_i values, will be more highly populated than the higher energy states at any given temperature.* Although the Boltzmann distribution is the most probable distribution at equilibrium, it is not the only distribution that is possible and other distributions exist (even though for brief periods of time), both under equilibrium and nonequilibrium conditions - as we shall see in section 7-12 when we discuss lasers.

7-4.   The particle nature explains how light interacts with matter

A photon which strikes a sample may undergo several types of interaction with the sample: it may be absorbed by having its energy transferred to the sample; it may be scattered (deflected randomly) in various ways; it may interact with species in an excited state and cause emission of radiant energy as the excited species return to lower energy levels; or, if the incident radiation is plane-polarized light, the plane of polarization may be rotated by an optically active sample.

A biomolecule only absorbs electromagnetic radiation in quanta which have the appropriate energy values to raise that biomolecule from one energy level to another. As indicated in section 7-3 and in Figure 7-3, there are four types of processes by which quanta of energy may be absorbed. In order of increasing energy requirements, the four types of absorption processes are: *translational transition, rotational transition, vibrational transition, and electronic transition.*

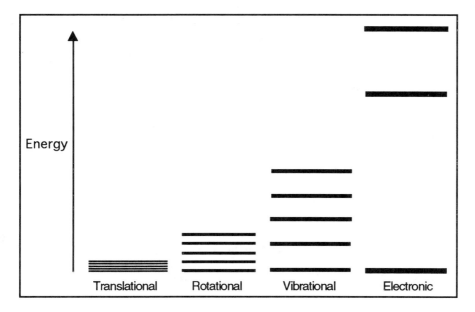

Figure 7-3. Diagram of energy spacings between levels in translational, rotational, vibrational, and electronic transition processes

Each of these induced transition processes occurs in quantized steps, as we have seen. But, in translational states, the spacings between energy levels are so close together (cf. Figure 7-3) that we may, for convenience, assume that they are continuous. Having made this assumption, we may then conclude that translational transitions are not quantized and proceed to consider rotational transitions. For rotational transitions, which have the next to lowest energy requirements of the four energy states, molecules may be raised to higher rotational energy levels if they interact with electromagnetic radiation in the far-infrared region, a relatively low-energy region of the spectrum. The energy which causes rotational transitions is inadequate for causing vibrational or electronic transitions. To raise molecules to vibrational energy levels, shorter, more energetic wavelengths (in the near infrared) are required than for rotational transitions; but, electromagnetic radiation in the near infrared which causes vibrational transitions also will cause rotational transitions. In an analogous fashion, absorption of still higher levels of electromagnetic radiation (in the ultraviolet or visible regions) raises the electrons of irradiated molecules to higher electronic transitions and causes, simultaneously, higher rotational and vibrational transitions.

Figure 7-4 depicts the relation among *rotational* (R), *vibrational* (V), and *electronic* (E) *levels of internal energy. This diagram shows that, as energy is increased (as wavelengths are decreased), vibrational transitions occur in addition to rotational transitions and, when electronic transitions occur, they are super-imposed on rotational and vibrational transitions. Transitions, therefore, occur in several levels of rotational, vibrational and electronic energy, corresponding to the absorption of radiation of different wavelengths.* The occurrence of several levels of three types of internal energy leads to the possibility of several different combinations of discrete transitions which may show up as a variety of absorption peaks. But, selection rules of quantum mechanics dictate that not all of these mathematically possible transitions will be allowed. Accordingly, each absorption peak will represent the actual physical transfer of energy from photons to absorbing biomolecules. In the laboratory, absorption of radiation corresponding to allowed

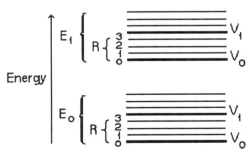

Figure 7-4. Relationships among rotational, vibrational and electronic Levels

transitions may be detected with spectrophotometers as peaks in recorded spectra. In the infrared and near infrared regions, these transitions yield spectra consisting of many peaks. In the ultraviolet and visible regions, the transitions are too close to each other to yield well-defined, sharp peaks in currently available instruments. Consequently, absorption peaks in the ultraviolet-visible region generally are broad.

7-5.   Spectroscopic techniques can detect absorbed radiation

In Figure 7-1, we classified electromagnetic radiation solely on the basis of wavelength. The electromagnetic spectrum additionally may be classified on the basis of the kinds of techniques used to detect radiation in each of the seven regions of the spectrum shown in Figure 7-1. We show such a grouping in Figure 7-5:

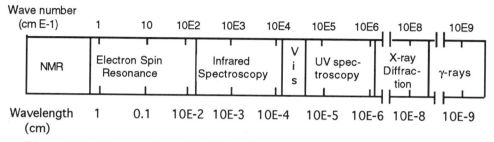

Figure 7-5. Frequencies of Radiation Detected By Spectroscopic Techniques

In the discussion of spectroscopic techniques which follows, we will use the grouping of Figure 7-5 and discuss the techniques in order of regions of increasing wavelength with one exception, that of γ-ray and X-ray spectroscopy. The latter technique will be discussed last so that the various spectroscopic methods will be discussed in the following sequence:

- Electronic spectroscopy (UV-visible, fluorescence, flame emission, atomic absorption and lasers).
- Raman scattering and infrared spectroscopy.
- Magnetic resonance spectroscopy and electron spin resonance.
- Optical rotatory dispersion and circular dichroism.
- X-ray spectroscopy.

7-6.   Ultraviolet-visible spectroscopy depends on electronic transitions

Whether or not electromagnetic radiation is absorbed in the UV and visible regions depends on the electronic structures of the biomolecules being investigated. Generally, molecular absorption in the UV and visible regions requires structures with at least one of the following features: (1) covalent groups such as C=C or C=O, (2) conjugated double-bond systems such as that in vitamin A, (3) aromatic structures such as those of purines, pyrimidines, and the amino acids phenylalanine, tyrosine, and tryptophan, or (4) structures with groups containing ions of transition metals (metals such as Mn, Fe, Co, Cu and Zn in period 4 and Mo in period 5 of the Periodic Table). All of these structures absorb radiation in the UV-visible range because they have energy levels with spacings in the UV-visible range.

As indicated above, spectroscopy in the ultraviolet and visible regions involves electronic transitions so we may ask, which electrons in biomolecules undergo transitions to higher energy levels when the molecules are subjected to radiation in the ultraviolet-visible region? According to molecular orbital theory, electrons are classified as σ, π or n depending on the orbitals (regions in space) in which there is a high probability that electrons will be found. Physicochemical considerations suggest that the only transitions allowed in common biomolecules are $\sigma \rightarrow \sigma^*$, $\pi \rightarrow \pi^*$, and $n \rightarrow \pi^*$. By convention, asterisks are used to indicate higher energy (excited state) arrangements of electrons. *Aliphatic compounds have tightly bound σ electrons; as a result, they require electromagnetic radiation of relatively high energy, in the near ultraviolet region, to undergo excitation.* Such aliphatic compounds therefore tend to require electromagnetic radiation in the region of 100 to 200 nm for excitation. In ethane, for example, the $\sigma \rightarrow \sigma^*$ transition occurs at 135 nm, and in proteins, $\sigma \rightarrow \sigma^*$ transitions of C-H and C-C groups occur at 125 and 135 nm respectively. *Biomolecules with aromatic rings have loosely bound π electrons and, as a result, require lower energies for excitation than aliphatic compounds. Aromatic compounds, therefore, often absorb at wavelengths longer than 250 nm. Unsaturated compounds which may have π electrons either more, or less, strongly held than those in aromatic compounds will require shorter, or longer, wavelengths for excitation, respectively.* Nonbonding electrons, such as those in aldehydes, (R-CH=Ö:) also can be excited to the π^* state ($\mathrm{RCH \dot= O:}$). In peptide groups, the $n \rightarrow \pi^*$ transition of the oxygen atom corresponds to absorption of radiation at 225 nm, indicating that lower quantities of energy are needed to achieve this transition than to cause the $\pi \rightarrow \pi^*$ transition in alkenes, which occurs, for example, at 165 nm in ethylene. *As conjugation increases, absorption occurs at progressively longer wavelengths than in simple alkenes since the excited state is stabilized by resonance to a greater extent than is the ground* state. β-carotene for example, has an orange color and absorbs radiation at 451 nm.

Much of biomolecular spectroscopy involves the use of absorption spectra. An *absorption spectrum* is a plot of the amount of light absorbed (or transmitted) by a sample versus wavelength of incident light. The chief characteristics of an absorption band are its location and intensity. Band location depends on whether the wavelength used for irradiation has quanta of energy equivalent to the energy required for an electronic transition to occur, i.e., to cause excitation from one level to another. Discussion of the factors on which the intensity of absorption depends is, however, beyond the scope of this book. At this point, let us note that absorption spectra are recorded routinely with spectrophotometers. Consequently, we will discuss this and other uses of spectrophotometers in the next section.

7-7. Spectrophotometers are used qualitatively and quantitatively

UV-visible spectrophotometers are built to perform two tasks: to produce a narrow band of light in the UV-visible region (200 to 800 nm) and to measure the fraction of that band of light which is absorbed or transmitted by molecules in solution. Qualitatively, spectrophotometers are used to identify biomolecules by their characteristic absorption spectra; quantitatively, they can be used either to measure the concentration of a single biomolecule in solution or to measure the concentrations of more than one absorbing biomolecule. The analysis of more than one component is based on two assumptions: (a) that each absorbing species in solution acts independently of the other and (b) that the fractions of light absorbed by two or more species are additive. In section 7-8, we will examine the mathematical basis of quantitative spectrophotometric analysis and the components of UV-visible spectrophotometers; then, in section 7-9, we will outline steps for the calculation of the concentrations of more than one biomolecule in multicomponent analysis.

7-8. Quantitative spectrophotometry depends on the Beer-Lambert law

The Beer-Lambert law, is a relationship which states that, at a specific wavelength, the fraction of light absorbed by a given biomolecule in solution is proportional to the concentration of the absorbing biomolecule (*Beer's law*) and to the length of the layer through which the light passes (*Lambert's law*). According to the combined Beer-Lambert law, if a narrow beam of light of intensity I passes through a very small length dl of an absorbing solution of concentration c, the decrease in intensity dl of the incident light, may be expressed mathematically as

$$dI/I \quad = \quad -kcdl. \tag{4}$$

Integration of equation (4) between 0 and 1, the entire length of the light path as shown in equation (5) gives equations (6) and (7):

$$\int_{I_o}^{I_t} \frac{dI}{I} \quad = \quad -kc \int_0^1 dl, \tag{5}$$

$$\ln I_t/I_o \quad = \quad -kcl, \tag{6}$$

$$\ln I_o/I_t \quad = \quad kcl, \tag{7}$$

where $\quad I_o \quad = \quad$ incident light,
and $\quad I_t \quad = \quad$ transmitted light.

Converting equation (6) to logarithms to the base 10

$$2.303 \log I_o/I_t \quad = \quad kcl, \tag{8}$$

and

$$\log I_o/I_t \quad = \quad (k/2.303)lc \quad = \quad \varepsilon lc. \tag{9}$$

When c is in molar units the term $k/2.303 = \varepsilon$ the *molar extinction coefficient*, known alternatively, as the *extinction coefficient*.

If we define the fraction of light transmitted through the sample as transmittance, T, and the fraction of light absorbed by the sample as *absorbance*, A then

$$I_t/I_o \quad = \quad T, \tag{10}$$
$$100 \times T \quad = \quad \% \, T, \tag{11}$$
$$\log I_o/I_t \quad = \quad A = - \log T, \tag{12}$$
$$A \quad = \quad 2 - \log \% T, \tag{13}$$
$$\% T \quad = \quad \text{antilog} \, (2 - A). \tag{14}$$

and

$$\boxed{A \qquad = \qquad \varepsilon l c} \tag{15}$$

According to equation (15), when the Beer-Lambert law is obeyed and absorbance is plotted against concentration, a straight line which passes through the origin should be obtained. In that straight line, as shown in Figure 7-6,

$$\text{slope} \quad = \quad \varepsilon l \tag{16}$$
$$\text{and} \quad \varepsilon \quad = \quad \text{slope}/l \tag{17}$$

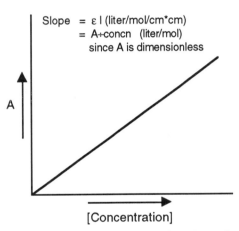

Slope = ε l (liter/mol/cm*cm)
      = A÷concn (liter/mol)
      since A is dimensionless

A

[Concentration]

Figure 7-6. Plot of A versus concentration when Beer-Lambert law is obeyed

According to equation (15) also, when A is determined spectrophotometrically and ε and l are known, concentration c, can be calculated from the relationship

$$c \quad = \quad A/\varepsilon l. \tag{18}$$
If $\quad l \quad = \quad$ 1 cm, as in standard spectrophotometric cells, then
$$c \quad = \quad A/\varepsilon. \tag{19}$$

In equation (15), A (i.e. the ratio $\log I_o/I_t$) is a dimensionless value; and since l is measured in cm and c has units of mol liter^{-1}, then

$$A \quad = \quad \varepsilon * (1 \text{ cm}) * (c, \text{ mol liter}^{-1}), \tag{20}$$
$$\varepsilon \quad = \quad A/lc \, (\text{liter mol}^{-1} \text{ cm}^{-1}). \tag{21}$$

∴ ε represents the absorbance of 1 liter of solution mol^{-1} cm^{-1} of light path.

Spectrophotometers (cf. section 7-7) are used for measuring the absorption or transmission of electromagnetic radiation according to the Beer-Lambert law. Spectrophotometers vary in design depending on the spectral region of interest; but usually (cf. Figure 7-7), they consist of (1) a source of electromagnetic radiation; (2) a monochromator; (3) a cell for holding the sample; (4) a detector; (5) an amplifier; and (6) a meter or recorder. Within a given spectral region, spectrophotometers may vary in sophistication. In general, however, the source of radiant energy sends polychromatic radiation (i.e. radiation with an entire range of wavelengths) to a *monochromator* which resolves the many wavelengths from the source into narrow bands of just a few wavelengths. UV-visible spectrophotometers are the most common type found in biomedical laboratories. They may use either one or two beams of radiation. In *single-beam instruments*, radiation from the source is sent through a monochromator to be resolved into narrow bands; the user then chooses a band with the desired wavelength, and this band is focused into a sample to determine how much radiation is absorbed. In *double-beam instruments*, the beam from the monochromator is split into two beams, one of which passes through the reference cell, the other through the sample; the detector then measures the ratio of radiation passing through sample and reference cells. Double-beam instruments may be used for recording absorption spectra. In both double- and single-beam operation, after the sample absorbs a fraction of the incident radiation, the remainder is transmitted to a detector which changes it into an electrical signal. The electrical signal is amplified and, finally, displayed on a meter or recorder. We show the components for performing this sequence of operations with a single-beam instrument in Figure 7-7.

Figure 7-7. Components of a single-beam, UV-visible spectrophotometer

Sources of visible radiation are tungsten filament lamps; sources of ultraviolet radiation are hydrogen or deuterium lamps; and sources of infrared radiation are heated silicon carbide (i.e. globar) rods, or Nernst glowers (i.e., hollow rods of zirconium and yttrium).

Usually, one places samples for UV-visible spectrophotometry in transparent cells (cuvettes). Quartz or fused silica cells are used for the UV region, and glass cells are used for the visible region. *In UV-visible spectrophotometers*, (cf. Figure 7-7), cuvettes are placed *after the monochromator; in IR spectrophotometers*, the cells are placed *before the monochromator.*

Several types of error may occur in using the Beer-Lambert law for spectrophotometric analysis. These include errors in reading absorbance or transmittance values; errors from dirty cuvettes; errors from using inappropriate blank solutions; errors causing nonlinearity of calibration curves; errors caused by instrumental abnormalities; and errors caused by chemical contamination. The reader should consult textbooks of quantitative chemical analysis for additional information.

7-9.   Spectrophotometers can measure more than component

If n noninteracting components are present in a solution and each component obeys the Beer-Lambert law, then the quantities of the individual components can be calculated from n absorbance readings at n wavelengths. Mathematically, the problem reduces to solving systems of n simultaneous equations for n variables. Because the determinant method of solving simultaneous equations is convenient, we will use it both for multicomponent spectrophotometric analyses and for double-label experiments (cf. section 8-8).

In the general form of simultaneous equations, each coefficient is given a double subscript:

$$a_{11}x_1 \quad + \quad a_{12}x_2 \quad = \quad k_1 \tag{22}$$
$$a_{21}x_1 \quad + \quad a_{22}x_2 \quad = \quad k_2 \tag{23}$$

The first subscript represents the row or number of the equation, the second represents the column or number of the variable. In order to solve the two equations for the two variables x_1 and x_2, only the coefficients and constants are used in the calculations. In the *determinant method* of solving for x_1 and x_2, the coefficients and constants are arranged in matrices according to the following rules:

$$x_1 = \frac{\det \begin{bmatrix} k_1 & a_{12} \\ k_2 & a_{22} \end{bmatrix}}{\det \begin{bmatrix} a_{11} & a_{12} \\ a_{21} & a_{22} \end{bmatrix}} \quad , \qquad x_2 = \frac{\det \begin{bmatrix} a_{11} & k_1 \\ a_{21} & k_2 \end{bmatrix}}{\det \begin{bmatrix} a_{11} & a_{12} \\ a_{21} & a_{22} \end{bmatrix}}$$

A rectangular array of numbers, such as each of the four above, is known as a *matrix* (i.e., there are four matrices above). As indicated by the arrows, the value of a determinant of a matrix is obtained by multiplying the top left by the bottom right element and subtracting from this the product of the bottom left and top right element. The determinant method is useful for solving systems of simultaneous equations which are found in biomedical research. Examples of such use of determinants are the calculations (outlined below) of concentrations of two biomolecules in multicomponent spectrophotometric analysis and the calculation of radioactivities in double label experiments (discussed in section 8-8).

We may rewrite equations (22) and (23) as shown in equations (24) and (25) for solving problems involving multicomponent spectrophotometric analysis:

$$\varepsilon_{11}C_1 \quad + \quad \varepsilon_{12}C_2 \quad = \quad A_1, \tag{24}$$

$$\varepsilon_{21}C_1 \quad + \quad \varepsilon_{22}C_2 \quad = \quad A_2, \tag{25}$$

where
$$
\begin{aligned}
\varepsilon &= \text{extinction coefficient,} \\
C_1 &= \text{Concentration of compound 1 (unknown),} \\
C_2 &= \text{Concentration of compound 2 (unknown),} \\
A_1 &= \text{absorbance of both compounds at wavelength 1,} \\
A_2 &= \text{absorbance of both compounds at wavelength 2,} \\
\text{1st number of double subscript} &= \text{wavelength (either wavelength 1 or 2),} \\
\text{2nd number of double subscript} &= \text{compound (either component 1 or 2).}
\end{aligned}
$$

Solution of equations (24) and (25) by determinants yields

$$C_1 \quad = \quad \left[\frac{A_1 - A_2 \left(\dfrac{\varepsilon_{12}}{\varepsilon_{22}} \right)}{\varepsilon_{11} - \varepsilon_{12} \left(\dfrac{\varepsilon_{21}}{\varepsilon_{22}} \right)} \right], \tag{26}$$

$$C_2 \quad = \quad \left[\frac{A_2 - A_1 \left(\dfrac{\varepsilon_{21}}{\varepsilon_{11}} \right)}{\varepsilon_{11} - \varepsilon_{12} \left(\dfrac{\varepsilon_{21}}{\varepsilon_{22}} \right)} \right]. \tag{27}$$

Equations (26) and (27) *or their rearranged equivalents* are used to determine the concentrations of two absorbing biomolecules in the computer programs of *Basic Biochemical Laboratory Procedures and Computing.*

7-10.   Fluorescence and phosphorescence occur with emission of light

As we saw in section 7-4, molecules which absorb electromagnetic radiation may be excited from the ground state to higher levels of internal energy. The excited molecules may return to the ground state, i.e., lose the absorbed energy, in a variety of ways. For example, they may undergo "radiationless" loss of energy by colliding with other molecules and undergoing a chemical reaction. Excited molecules also may convert absorbed energy to heat through the loss of vibrational or rotational motion. At other times, some of the absorbed energy of excited molecules may be emitted as photons of longer wavelength than the photons of incident light. When light is emitted rapidly (because all of the electron spins in the excited state are paired - a process of high probability), the emission is known as *fluorescence*. When emission of light from the excited state occurs slowly because two electrons in the excited state are unpaired (a process of low probability), the emission is known as *phosphorescence*.

Fluorescence is widely used in the biomedical sciences because of its sensitivity. Thus, methods of detection of biomolecules based on fluorescence are now used routinely in many chromatographic, immunochemical, and electrophoretic processes as well as in the detection and measurement of radioactivity by *liquid scintillation spectrometry* which we will discuss in section 8-5. Phosphorescence, on the other hand, is rarely used in the modern biomedical sciences and will not be discussed further. In UV-visible spectrophotometry, chemical groups which absorb radiation are known as *chromophores*. By analogy, *fluorophores* are the absorbing groups of fluorescent compounds. When fluorophores are excited by absorbed radiation, they may emit polarized light. *Fluorescence polarization* and *fluorescence anisotropy* both depend on two sequential measurements of the intensity of polarized emission. One measurement is made with a polarizer parallel to the exciting radiation, the second with the polarizer perpendicular to the exciting radiation. Fluorescence polarization, usually designated as p, and fluorescence anisotropy, designated r, are defined mathematically as follows:

$$p = \frac{I_{||} - I_{\perp}}{I_{||} + I_{\perp}} \qquad (28)$$

$$r = \frac{I_{||} - I_{\perp}}{I_{||} + 2I_{\perp}} \qquad (29)$$

where $I_{||}$ and $I_{\perp}$ are the emission intensities measured, respectively, at angles parallel, and perpendicular, to the exciting beam. Measurements of emission intensities may be made by excitation either with short pulses, or with continuous irradiation, from the exciting beam; these alternative approaches are known as *decay and steady-state* measurements, respectively.

An *anisotropic* biomolecule is one in which a given property has different values in different directions. Conversely, *isotropic* biomolecules are those in which a given property has the same value in all directions. Fluorophores may be *intrinsic* (i.e., may be covalently linked, innate parts of biomolecules). *Extrinsic* biomolecules are synthetic fluorescent probes that are inserted into biomolecules either covalently or noncovalently. Fluorescence polarization and anisotropy have their origins in the fact that fluorophores, at the ground state in solution, normally are randomly oriented. However, when excited (either by natural light or plane-polarized light) fluorophores may undergo molecular rotation; this rotation causes the direction of the emitted radiation to be different from that of the incident radiation. If polarized light is used for excitation and the fluorophore does not rotate, the direction of the emitted radiation will not be changed from that of the incident radiation. If, however, the irradiated fluorophore does rotate and, in turn, causes the axis (direction) of emission to be rotated, depolarization will occur, i.e., the measured polarization will be less than expected. Stated another way, if a fluorophore were placed in a rigid matrix, rotation would not take place and the limiting or maximum fluorescence anisotropy will be obtained. In a fluid or mobile medium, however, the fluorophore can rotate during the lifetime of excitation and this rotation will lead to a decrease in the limiting fluorescence anisotropy. Fluorescence polarization thus measures the extent to which axes of emission and excitation are or are not aligned. And, since the intensity and direction of emitted polarized light depend on molecular structure and molecular environment, measurements of intensity and polarization permit deductions to be made about changes in molecular structure and in the immediate molecular environment surrounding the irradiated fluorophore. In the laboratory, fluorescent polarization and anisotropy have been used to follow temperature-dependent phase transitions, and to measure the apparent microviscosity, of the hydrophobic regions of liposomes, lipoproteins, and biological membranes. Fluorescence polarization and anisotropy also have been used to follow the kinetics of binding and dissociation of ligand and receptor, of antigen and antibody and of low molecular weight fluorophores and macromolecules. In other studies, fluorescence polarization has been used to determine the rotational motion of macromolecules and to determine rates of conformational change. Yet again, the technique has been used to determine the relative orientations of transition dipoles in fluorophores and to estimate the effects of molecular environment on these orientations. Fluorescence polarization and anisotropy therefore provide information on the behavior and structure of macromolecules; and, because of its sensitivity to the immediate molecular environment, it is also useful for following the kinetics of assembly of macromolecules.

7-11. Flame emission spectroscopy and atomic absorption spectroscopy detect electronic transitions of atoms

Atoms raised to higher levels of internal energy do not undergo the rotational and vibrational motions that are observed when molecules are excited. But atoms, when excited, undergo quantized electronic transitions which yield characteristic line spectra. As indicated by its title, *flame emission spectroscopy (FES)*, previously known as flame photometry, is based on the *emission* of radiation by atoms raised to excited states in a flame. *Atomic absorption spectroscopy (AAS)*, on the other hand, is based on the *absorption* of radiation by atoms at their ground state levels in a flame. In both techniques, atoms of interest are introduced into a burner in solution and the solution is drawn into the flame by aspiration, thereby producing atoms in the gaseous state. The equilibrium ratio (distribution) of atoms in the excited state to the ground state is given (both in FES and AAS) by equation (5), section 7-3.

In *flame emission spectroscopy*, thermal energy from the flame excites a small fraction of atoms in the gaseous state and these atoms are raised to higher electronic energy levels. As the atoms in the excited state return to the ground state, they emit photons of light of characteristic wavelength. This emitted light is measured with the appropriate optical and electronic equipment. In flame emission spectroscopy, the intensity of emission is directly proportional to the concentration of the atom which is being measured so that the amounts of atoms in samples may be determined from calibration curves of intensity of emission versus concentration. We show the components which perform these tasks in Figure 7-8:

Figure 7-8. Components of a Flame Emission Spectrometer

FES is a useful technique for determinations of two alkali metals, sodium and potassium, and of two alkali earth metals, calcium and magnesium. In commercially available instruments, the technique usually is carried out with flames producing temperatures below 3 000°K and most often with flames giving temperatures in the range 2 000° to 3 000°K. Such flames are useful for the quantitative analysis of elements which meet three criteria: (1) the electronic transitions of the elements should require less than 5 eV of excitation energy to occur, (2) the elements should be capable of being atomized or converted to gaseous atoms in the flame, and (3) the elements must not undergo processes which decrease the concentration of gaseous atoms in the flame; examples of such processes are the formation of ions and of stable oxides in the flame. Some limitations of FES are that (a) easily produced, low cost flames with temperatures above 3 000°K are needed (but are not readily available in the biomedical laboratory) for detecting atoms whose excitation energies are greater than 5 eV; this limits the number of elements that can be analyzed by the technique. (b) At temperatures greater than 3 000 °K many elements form stable oxides and therefore do not undergo the emission process that is characteristic of atoms. (c) At some wavelengths, the flame itself interferes with the determination of some atoms by giving background radiation.

In *atomic absorption spectroscopy*, as in flame emission spectroscopy, most of the gaseous atoms in the flame remain in the ground state, i.e., only a small fraction of the atoms will be excited thermally to higher electronic levels. However, if the gaseous atoms in the ground state are irradiated with quanta of electromagnetic radiation from a special source which produces energy with spacings similar to those of the gaseous atoms in the ground state, then the gaseous atoms will absorb radiation from the special source and will be raised to higher levels of electronic energy. Absorption of radiation in atomic absorption spectroscopy thus is analogous to absorption in UV-visible spectrophotometry, and the block diagram of Figure 7-9 is thus similar to that of Figure 7-7. In AAS, the function of the chopper is to cut up the beam of radiation from the source to help in distinguishing source radiation from that of the flame.

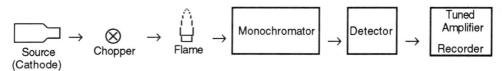

Figure 7-9. Components of an atomic absorption spectrophotometer

Because of the similarity of AAS to UV-visible spectrophorometry, the absorption of radiation by ground state atoms in atomic absorption spectroscopy obeys the Beer-Lambert law (section 7-8) and calibration curves are prepared as they would be in UV-visible spectrophotometry, i.e., absorption is plotted against concentration and, when the Beer-Lambert law is obeyed, a straight line which passes through the origin is obtained. As is to be expected from consideration of the Beer-Lambert law, the sensitivity of detection in atomic absorption spectroscopy can be improved by developing experimental conditions which increase the concentration of gaseous ground state atoms in the flame, or by increasing the path length over which absorption takes place in the burner. For inorganic samples, some advantages of

AAS are that it allows determination of more than 60 elements; that it has high specificity; and that it is very sensitive. For samples of biological origin, however, the situation is somewhat different because trace elements of interest may be trapped within, or associated tightly with, large biomolecules. In spite of these complications, however, AAS has had an enormous impact on knowledge of trace metals in biological tissues. Today, both the biomedical community and the general public are paying increasing attention to the importance of metal ions in human nutrition and disease.

7-12. Lasers are used both by research scientists and by surgeons

Charles H. Townes and colleagues at Columbia University announced in 1954 that they had developed a *maser*, an acronym for a device which achieved Microwave Amplification by Stimulated Emission of Radiation. Townes and the Russians Nikolai G. Basov and Aleksandr M. Prokhorov were awarded the 1964 Nobel Prize in physics for inventing masers and proposing the principle of lasers. We will discuss briefly the operation of lasers, an acronym for Light Amplification by Stimulated Emission of Radiation. Theodore H. Maiman developed the first laser in 1960; he used a laser medium consisting of ruby rod crystals, Al_2O_3 containing traces of Cr_2O_3. Since 1960, several other lasers have been introduced including gas lasers, liquid lasers, solid state lasers, metallic vapor lasers, glass rod lasers, injection lasers, plastic rod lasers, and free electron lasers. Lasers produce intense, coherent (i.e strongly directional), well-collimated (parallel), almost completely monochromatic radiation. To cause emission of radiation with this unique combination of optical properties, several methods (collectively referred to as pumping) are used to excite electrons of certain atoms. The excited atoms are raised to specific, available, higher energy levels, e.g., E_2 and E_3. Then, atoms in the excited states E_2 and E_3 decay by *spontaneous* transitions (by heat loss or by collision) to a lower, available, excited energy level E_1. The net effect of pumping is a greater population of atoms in E_1 than in E_0; i.e. the ratio of atoms in the activated state E_1, to atoms in the ground state E_0, greatly exceeds that expected from the Boltzmann distribution (section 7-3). This nonequilibrium condition is known as a *population inversion*. In lasers, the large number of atoms in E_1 which cannot return spontaneously to their ground state E_0, are *stimulated* to do so by irradiation with photons of frequency = $(E_1 - E_0)/h$; this is the same frequency (cf. equation 1, section 7-2) which induces *absorption*. That is, irradiation with photons which *induce absorption* also will *induce emission* from E_1 to E_0 in a population inversion since, as first shown by Albert Einstein, *the probability of induced transition is the same* for absorption and emission. When one irradiating photon strikes an excited atom in E_1 two photons are emitted: one from the incoming photon, the other from the excited atom. *This emission is amplified* by optical resonance. *Optical resonance* is attained by positioning the laser medium between two highly reflective mirrors with the distance between the mirrors equal to integral multiples of the desired wavelength. Huge amplification results as photons flash back and forth repeatedly (oscillate) between the mirrors. Since one of the mirrors is partially transparent, it allows some amplified light (the laser beam) to leave the chamber.

Thus, laser operation entails *four major, sequential events*: (a) *pumping* (b) *population inversion* (c) *stimulated emission* and (d) *optical resonance*. Most existing lasers emit narrow ranges of fixed wavelengths in the UV, visible or infrared spectral regions, depending on the space between E_1-E_0; newer lasers can be "tuned" to emit light from γ-rays to radiowaves. Industry uses lasers in telecommunications and to cut, drill and weld. Physicians use lasers in opthalmology and to excise polyps, tumors and obstructions, a major advantage of such surgery being that laser heat cauterizes excisions thus causing less bleeding than traditional surgery. Scientists use lasers to study cellular functions, to conduct fusion research and relaxation experiments, and to provide a light source for Raman spectroscopy, our next topic.

7-13. Raman spectroscopy complements studies in the infrared

Raman scattering, discovered in 1928, is named for its discoverer, the Indian scientist Sir Chandrasekhar Venkata Raman (Nobel Prize in physics, 1930). Raman found that a small fraction of light scattered by molecules in a sample consisted of frequencies that were much different from the frequency of the incident light. In Raman spectroscopy, monochromatic light of high intensity is used to irradiate a sample. The photons used are very much larger than the vibrational energy spacings of the sample. Absorption therefore will *not* occur. The photons can, however, interact with molecules in the sample in another way. Some energy of the photons can induce oscillations of electronic charge in molecules of the sample and the oscillating charges can scatter the remainder of the photon in many directions. If such interactions occur with molecules in the ground state, the unused fraction of the photon will be scattered and will be reemitted as light of lower frequencies in the UV-visible region. Alternatively, if a photon which is far from the absorption band gains energy from molecules in the excited state, scattered light of higher frequency will be emitted.

In Raman spectroscopy, the differences between the frequencies of scattered and incident light are proportional to transition energies of vibrational energy levels. When the reemitted light has *frequencies lower* than those of the incident light, the lines are called *Stokes lines*. When the reemitted light has *frequencies higher* than the incident light, the lines are known as *anti-Stokes* lines. The intensity of the scattered light in Raman spectroscopy is low. Therefore, an intense light (e.g. lasers section 7-12) is needed to produce scattered light of detectable intensity. As we shall see in the following section, Raman and infrared spectroscopy both detect vibrations of atoms in biomolecules; so why are two techniques used which measure the same property? The answer: Raman peaks depend on a change in polarizability as the atoms in a molecule vibrate; but the appearance of infrared peaks depends on a change of dipole moment when vibration of atoms occurs. Thus, vibrations which are observed by one technique may not necessarily be observed by the other, and, when both techniques are used to characterize a biomolecule, a more comprehensive description of its molecular vibrations may be obtained than if only one of the two techniques were used. Also important is that useful Raman spectra may be obtained in aqueous solution, in biological tissues or in a wide range of physical states. Infrared spectra, however, usually cannot be made in the presence of water because water has very strong infrared absorption bands which can obscure the absorption peaks in the biomolecules of interest. Until recently, infrared spectroscopy was, by far, more useful than Raman spectroscopy. However, the availability of lasers (section 7-12) for producing high-intensity light, and improvements in the sensitivity and stability of electronic instruments, have recently made Raman spectroscopy an easier, more useful technique than it was not long ago. Today, Raman spectroscopy can produce useful information on the secondary and tertiary structure of proteins and on the conformation of nucleic acids and biological membranes.

7-14. Infrared spectroscopy identifies molecules and functional groups

The *infrared (IR) spectroscopy* of biomolecules generally is carried out in the region 4,000 cm^{-1} to 666 cm^{-1} (i.e., 2.5 to 15.0 μm). Absorption of this radiation by a biomolecule causes *quantized excitation* of its atoms to higher vibrational levels. Atoms of molecules can vibrate in several different ways among which are *stretching, bending, wagging, rocking, twisting, and scissoring* defined by Pecsok et al., 1976. In addition to these classes of vibration, various combinations of vibrations also may occur. Consequently, infrared spectra of biomolecules are complex and can provide a large amount of information. This complexity makes it statistically probable that, when two or more compounds yield identical IR spectra, they have the same chemical structure.

Vibrations induced by infrared radiation are infrared active (give infrared absorption bands) only if the electric dipole moment changes during excitation. Very often, the absorption of infrared radiation by a molecule causes all of its atoms to vibrate; therefore, in this case, when absorption bands appear, they are characteristic of the entire molecule. At other times, modes of vibration arise only from specific groups and this gives rise to absorption bands that are characterisitc of those functional groups. The frequencies at which absorption bands of functional groups appear are almost, but not entirely, independent of the structure of the compound in which the groups are found. In fact, very useful tables have been published listing the frequencies at which infrared absorption bands are located in many biomolecules. These tables also list the changes in band position which structural changes such as hydrogen bonding, the introduction of double bonds, conjugation, and London-van der Waals interactions induce.

Infrared spectra of samples are obtained with infrared spectrophotometers. Today, these spectrophotometers are of two broad types:

- traditional, dispersive instruments and
- Fourier-transform instruments.

Dispersive infrared spectrophotometers contain two key components: monochromators to produce wavelengths accurately, and *prisms or diffraction gratings* to separate (disperse) wavelengths of radiation onto a slit. The slit permits only a narrow range of wavelengths to reach the sample and detector. Samples of biological origin examined by conventional or dispersive, infrared spectroscopy may be prepared as solutions or solids. But, as mentioned in section 7-13, water is not a suitable solvent because of its strong absorption in the infrared. Another disadvantage of water as a solvent in traditional infrared spectroscopy is that the spectroscopic cells used for holding the samples are often made of sodium chloride. Among suitable solvents for conventional IR spectroscopy are carbon disulfide and carbon tetrachloride; in general, however, samples examined in the solid state yield more detailed spectra than samples examined in solution. When the purpose of an investigation is to identify an unknown sample, identification can be made with some certainty by infrared spectroscopy if the unknown, upon comparison with a pure standard, yields identical infrared absorption bands. Identical infrared spectra will be found only if the unknown and standard both absorb light at each of several different frequencies, implying that the unknown and standard have the same physical properties.

Fourier-transform infrared spectroscopy (FTIR) is carried out with non-dispersive instruments. These instruments depend on a *Michelson interferometer* which manipulates the incident beam of radiation so that eventually, the component frequencies become discrete in a recombined beam which is measured at precise intervals by a detector. Fourier transformation of the signal yields the infrared spectrum of the sample. Among the advantages of Fourier-transform infrared spectroscopy compared with traditional dispersive methods are improved sensitivity; improved signal-to-noise ratio; improved resolution of absorption bands; improved accuracy of generating desired wavelengths of incident radiation; and the possibility of obtaining spectra in aqueous solutions (cf. section 7-16). Spectra of biomolecules in aqueous solution are obtained readily by FTIR because computerization makes it possible to subtract the absorption peak assigned to water. Improved resolution of absorption bands also is attained in FTIR by mathematical procedures, two of which are *Fourier self-deconvolution* and *derivative spectroscopy*. Both procedures use mathematical methods to decrease the widths of absorption bands.

Both in dispersive and FTIR, the positions of IR absorption band are determined by bond strength and by the masses of the bonded atoms. If a bond is strong and the masses of the bonded atoms are small, more energy is needed to cause vibration than if the bond is weak and the masses of the bonded atoms are large. Differences in bond strength and atomic mass also work independently to influence band position.

According to Taillandier and Liquier cited in the references, some advantages of infrared spectroscopy for studying nucleic acids are that it is a nondestructive method which provides the opportunity of obtaining results on

- many physical states such as solutions, gels, hydrated fibers, and crystals;
- small as well as long polynucleotides;
- small amounts of samples such as 2 OD units of material;
- comparisons with data from x-ray diffraction and NMR solution studies.

Regarding the study of proteins by FTIR, Haris and Chapman (1992) point out that its ease of application, its capability of measuring the secondary structure of peptides, and its ability to probe structural changes in relation to temperature, light, pH or other experimental variables, are important practical considerations. We present additional information on FTIR in section 7-16.

7-15. NMR can elucidate chemical structure, metabolism, and disease

Nuclear magnetic resonance (NMR) depends on the fact that all nuclei carry a charge; but, the nuclei of certain atoms (approximately one-half of the known nuclei), behave as though they are able to spin about the nuclear axis. The spinning charge in nuclei generates a magnetic field with a magnetic moment directed along the axis of spin. When a spinning nucleus, e.g., a proton, is placed in an external magnetic field, the nuclear magnet is not allowed to point in all directions. It may have only certain orientations. The energy of a magnetic moment in a magnetic field depends on these orientations. Magnetic nuclei interact with the external field and assume orientations with quantized energy levels. Protons have quantized energy levels only in two directions: aligned with the applied field (lower energy) or aligned against the applied field (higher energy). Since the spinning nucleus has quantized energy levels, transitions may be induced between them if electromagnetic radiation of the appropriate frequency is applied. Calculations show that the energy difference between the number of protons in the higher and lower energy levels corresponds to radiant energy in the radiofrequency range. Application of radiant energy of this frequency thus may cause protons in the lower energy level to absorb the radiant energy and jump to the higher level. Such absorption of energy by magnetic nuclei is known as *nuclear magnetic resonance*, NMR; it was discovered in 1946 by Edward M. Purcell (Nobel Prize in physics, 1952, with Felix Bloch). Some of the magnetic nuclei commonly used in the biomedical sciences are ^1H, ^2H, ^{13}C, ^{19}F, and ^{31}P.

For experimental convenience, the radiofrequency in earlier models of NMR instruments was generally held constant whereas the strength of the magnetic field was varied slowly until each proton undergoes resonance. In general, when the energy of the applied magnetic field (in gauss) is sufficient to cause transition from the lower to the upper level, the magnetic field matches the frequency of the radio waves and absorption occurs. When absorption occurs, a signal is recorded and the spectrum which results is the NMR spectrum. If all protons were affected only by the external magnetic field, NMR spectra would consist of a single peak that would not be of any diagnostic value. In actual experiments, however, the magnetic properties of nuclei are affected by their surroundings, not simply by the applied magnetic field. In other words, the magnetic field experienced by a nucleus depends not only on the external field, but also on the fields produced by electrons and by other nuclei in a given molecule. Protons in different molecular environments thus need somewhat different magnetic fields to achieve resonance. The difference in frequency between the absorption position of a proton in a given molecular environment and the position of absorption of a proton in a reference compound added to the sample is referred to as the *chemical shift*, δ. Chemical shifts provide sensitive and reliable indices of proton environments. The chemical shift is calculated as a relative rather than an absolute value, is multiplied by 10^6 for mathematical convenience and is expressed in parts per million, ppm:

$$\delta = \frac{H_O \text{ (ref)} - H_O \text{ (s)}}{H_O \text{ (ref)}} * 10^6 \text{ ppm,}$$

where $\quad H_O \quad =$ intensity of the applied magnetic field

and $\quad$ ref and s $\quad =$ reference and sample respectively.

From the foregoing considerations, it is not surprising to learn that NMR is a powerful tool for the determination of chemical structure; that the number of absorption peaks in an NMR spectrum indicates how many different kinds of magnetic nuclei are present; that the positions (chemical shifts) of the peaks are a reflection of the various environments in which the magnetic nuclei are found; and that the intensity of a peak (its signal area) is proportional to the numbers of magnetic nuclei that are present in that peak. The proportionality between peak intensity and numbers of magnetic nuclei permits estimation of the total amount of the detectable species present in a sample.

In the biomedical sciences, NMR has become an important method for studying protein structure, when necessary, at atomic resolution. Indeed, it is often regarded as the best available method for investigating protein structure in solution. NMR also is used for studying the metabolism of cells, tissues and organs. A major advantage of the technique is that it is noninvasive. Therefore, NMR permits direct observation of cellular phenomena without disturbing cellular structure or function. As a tool in the biomedical sciences, NMR has been used for the identification of biomolecules; for the measurement of transport processes, bioenergetics and the distribution of pH and $[H^+]$; for performing pH titration of individual amino acid chains at the active sites of enzymes and for studying protein conformation. A recent application that has elicited much public interest is *magnetic resonance imaging* (MRI) of human subjects which permits scientists to evaluate the metabolic state of, and clinicians to visualize, soft tissues within patients. The principle of magnetic resonance imaging is as follows. Cells contain approximately 70 to 80% water on a weight basis. Consequently, the single absorption peak of water is the major NMR signal from biological tissues. As explained above, the magnetic field experienced by a proton depends both on the externally applied field and on the molecular environment in which the proton is found. Consequently, when a magnetic field is applied in a gradient across the various tissues of a patient, a spectrum will be obtained in which proton signals of varying intensities occur exclusively at specific locations in various tissues. Scans of the patient taken at different projections and different magnetic field gradients, are combined to give a composite image or composite spectrum of the tissue that was examined (cf. two-dimensional NMR, see section 7-16). The noninvasive nature of this clinical application of NMR, the absence of harmful radiation associated with it, and the clarity of the pictures obtained, make it an attractive tool for the diagnosis of tumors and other diseases of soft tissues in patients.

7-16.   Fourier transformation increases the sensitivity of NMR

Many quantities and phenomena - both in nature and in everyday life - occur in a cyclic or periodic manner. When these periodic processes are repeated indefinitely, over and over again, each cyclic process constitutes a series which may be approximated by a periodic function. The period of a periodic function is that segment along the horizontal axis which determines the periodic behavior. The *Fourier series* is a trigonometric function which recurs every 2π units along the horizontal axis. This series has become an important tool for analyzing periodic functions in the biomedical and physical sciences. As the name implies, the Fourier series was named after its developer, Jean Baptiste Joseph Fourier, a French mathematician, civil servant and one-time officer in Napoleon's army. Examples of commonly occurring cyclic processes are the cyclic levels of hormones in animals, the menstrual cycle, the patterns of electrocardiograms, lunar behavior, tidal behavior and the swings of a pendulum. One area of the biomedical sciences in which Fourier transformation has become very important is in spectroscopy, and this is particularly so in the case of NMR spectroscopy. Much of the credit for this development goes to Richard R. Ernst and colleagues in Switzerland who introduced Fourier transform NMR (FT-NMR) spectroscopy and two-dimensional NMR. For these and other achievements, Ernst was awarded the 1991 Nobel Prize in Chemistry. In FT-NMR spectroscopy, data from each sample are collected at many, equally spaced intervals of time (the "time domain"). These data then are averaged and expressed in terms of discrete frequencies (the "frequency domain"). Experimentally, FT-NMR spectroscopy depends on the fact that a short, intense pulse of an electric field produces broad-band, electromagnetic radiation over a range of frequencies. This range of frequencies is inversely proportional to the duration of the pulse. Therefore, using very short, intense pulses of an electric field is an effective way to excite several spectral lines simultaneously.

In FT-NMR spectroscopy, the time-domain data from repeated scans of the same spectrum are averaged, then stored in a computer and converted to the traditional representation of spectral lines. To achieve this result, the operator subjects a sample in a magnetic field to numerous pulses of radiofrequency energy. In response, the sample emits characteristic signals. These signals arise because, when a given single oscillator (e.g., a spinning nucleus) is subjected to a pulse of irradiation at its resonant frequency, the amplitude of its oscillation will increase. However, when the pulse is over, the nucleus will continue to oscillate but with an amplitude which decreases exponentially with time. The signal elicited by each pulse (known as the *free induction signal*) is the Fourier transform of the conventional continuous-wave display. The free oscillation induces a decaying sinusoidal voltage (*free induction decay* or FID) in a sensitive coil of wire around the sample. The FID and the continuous-wave display are a Fourier transform pair; so, one member of the pair can be converted to the other mathematically. Once the data have been obtained and stored (in computer memory or on a disk), appropriate software is used to convert one (the FID) to the other, the continuous-wave display of the NMR spectrum.

The popularity of Fourier transformation in NMR spectroscopy arises from its speed and sensitivity relative to continuous-wave spectroscopy in which the spectrometer makes single and slow scans of frequency (the spectrum) as a function of time. As an example of such slow scans, suppose that we are planning to scan a spectral band which is 6 kHz wide. Then, suppose we make two additional assumptions: (a) that linewidths are approximately 1 Hz, and (b) that the 6 kHz region is to be scanned at 1 Hz sec^{-1}. At this rate, the time required for the scan would be greater than $1\frac{1}{2}$ hr; more specifically, at least 100 min would be required for performing this scan. Thus, not only can continuous-wave NMR be time-consuming, but, in addition, only a small range of frequencies can be examined at any given time. This slow scanning is overcome in FT-NMR spectroscopy. Fourier transformation provides the additional advantage that the signal-to-noise ratio increases as the square root of the number of scans. The reason for this increase is that the intensity of a signal is directly proportional to the number of scans, N, whereas random noise only increases as $\sqrt{N}$. Therefore the signal-to-noise ratio (the sensitivity) increases by a factor of $N/\sqrt{N} = \sqrt{N}$.

Fourier transformation, thus, is a mathematical operation which converts data collected quickly and repetitively in the time domain (e.g. a single frequency that might be represented as a cosine wave) to the equivalent in the frequency domain (a single line). Fourier transformation increases the speed and enhances the signal-to-noise ratio (the sensitivity) of modern spectroscopic analyses.

Two-dimensional NMR is an outgrowth of FT-NMR. Here, instead of sequences of single pulses, the operator uses sequences of paired radiofrequency pulses. After the first pulse (a period known as the "preparation period"), a delay is introduced and the sample is allowed to "evolve." The operator then subjects the sample to the second pulse and the reemitted radiofrequencies again are allowed to evolve. The paired sequences are repeated many times; but with each repeat, the duration of the first evolution period is increased. The sequential data are stored as they are collected and, finally, the operator carries out Fourier transformation with respect to the two time periods to obtain the composite spectrum. Two-dimensional NMR has been particularly useful in studies of proteins in solution because it helps investigators both to resolve signals from individual functional groups and to assign the signals to specific atoms or groups of equivalent atoms. Two-dimensional NMR, with its ability to derive composite spectra, is the foundation on which magnetic resonance imaging (MRI), (section 7-15) is based.

Fourier transform is also used in infrared spectroscopy (section 7-14). The methods of mathematical treatment in FT-IR are the same as in FT-NMR spectroscopy but a different experimental system is used. Specifically, the equivalent of multiple, fixed-frequency channels is achieved with a piece of equipment called a *Michelson interferometer*. The function of the latter is first to split the incident beam of radiation from the sample into two beams. Then, one half of the split beam is passed to a fixed mirror and the other to a moving mirror. The paths of the split beams are made progressively different through reflection to the fixed and movable mirrors. The net result is that the component frequencies become discrete in a recombined beam which is measured at precise intervals by a detector. Fourier transformation of the signal, known as an *interferogram*, yields the infrared spectrum of the sample.

To conclude this section, we note that in addition to NMR and infrared spectroscopy, Fourier transform is used to analyze data from x-ray diffraction and mass spectrometery (*ion cyclotron resonance spectrometers*). In x-ray diffraction, Fourier transform is used to deduce molecular structure from electron densities (section 7-19).

7-17. Electron spin resonance detects unpaired electrons

Electron spin resonance (ESR) is based on the presence of unpaired electrons in molecules. Like spinning nuclei (section 7-15), the spins of electrons are quantized and their energies depend on their orientations in a magnetic field. Electrons, like protons, can assume only two energy levels (section 7-15), and, resonance between the two states, may be induced by applying electromagnetic radiation of the appropriate frequency. The correct frequency for inducing transitions between these two states corresponds to microwave radiation. As with NMR, since the spacing of energy levels in ESR depends on the external field, it is convenient to keep the microwave radiation constant and vary the magnetic field (in gauss). When magnetic resonance is achieved, it is indicated by the occurrence of resonance absorption peaks. Biomolecules containing unpaired electrons are rare. For this reason, ESR is usually carried out with synthetic probes designed to provide information on such phenomena as the functions of biological membranes and the mechanisms of oxidation-reduction enzymes. The sensitivity of ESR is very high, extending to the detection of picomole quantities of substances with unpaired electrons.

7-18. ORD and CD show the conformation of macromolecules

Optical rotatory dispersion (ORD) and *circular dichroism (CD)* can be used to probe and measure the helical coiling (the secondary structure) of biomolecules in solution. ORD measures how optical rotation changes with the wavelength of incident light. CD measures the differential absorption of left- and right-circularly polarized light by optically active biomolecules.

Visible light, as we saw in Figure 7-1, is a form of electromagnetic radiation and, as the words "electromagnetic radiation" suggest, visible light consists of electric and magnetic fields. These fields, which are perpendicular to each other, can rotate about the axis of propagation, but always remain mutually perpendicular. In an ordinary (unpolarized) beam of light, the electric component consists of waves oscillating in several directions. In *plane polarized light*, however, all of the waves point in a single direction. Plane polarized light is produced by passing ordinary light through a lens made of crystalline calcium carbonate (calcite) or through a polaroid material. *Circularly polarized light* has its electric field rotating about the direction of propagation and it is produced by passing plane polarized light through a device known as a quartz-wave plate. Circularly polarized light may be either right- or left-handed, depending on whether its sense of rotation with respect to the sense of propagation is clockwise or anticlockwise.

A substance which can rotate the plane of polarized light is said to be optically active. The *specific rotation* of that substance [α] at a fixed wavelength is defined by the equation

$$[\alpha] \;=\; \frac{\alpha}{1 * c'} \tag{30}$$

where   α = angle of rotation in degrees,
and   l = the path length in decimeters,
        c = concentration in g/100ml.

In measuring optical rotatory power, it is desirable to express optical activity in molar quantities; but, for a variety of reasons, the quantity most frequently used in working with macromolecules is the *reduced residue rotation* which is defined as

$$[m'] \;=\; [\alpha]\,\frac{3}{n^2 + 2} * \frac{MRW}{100}, \tag{31}$$

where n is the refractive index of the solution and MRW is the *mean residue weight*, i.e., the average weight of the residues in the biomolecule. For polypeptides of unknown composition, the value of MRW is taken as 115, because the calculated values for several proteins cluster about this value. Measurements of [m'] are made with *spectropolarimeters* and analysis of the results is made with two equations which we will not discuss: the Drude equation and the Moffitt equation .

Circular dichroism is detected at wavelengths where optically active compounds absorb the radiation. Circular dichroism is measured by passing circularly polarized light of specified wavelength through solutions of optically compounds. It usually is found that some optically active compounds absorb left-circularly more strongly than right-circularly polarized light (or vice versa). Thus, the circular dichroism at the specified wavelength of absorption, is defined as the difference between the extinction coefficients of left- and right-circularly polarized light. Individual absorption bands are described by a quantity known as the *rotational strength*. It is obtained by integrating the extinction coefficient over the absorption band.

7-19. X-ray diffraction maps the locations of atoms in molecules

X-rays have wavelengths of 0.1 to 25 ångstrom ($1\text{Å} = 0.1\text{nm} = 10^{-10}$ m). X-rays are produced by bombarding a metal target (e.g. copper, molybdenum or tungsten) with a beam of high energy electrons in an evacuated tube. X-rays were discovered accidentally in 1895 by Wilhelm Konrad Röntgen who called them x-rays because, at the time, he did not understand them. From CAT scans *(computed axial tomography, a method which reconstructs cross-sectional images of biological structures by recording and summing several x-ray scans)*, to the detection of bone-fractures, the discovery of x-rays has had enormous impact on medical diagnosis and other aspects of modern life. For his discovery, Röntgen was awarded the first Nobel prize in physics, in 1901.

Diffraction, for our purposes, is the change in direction of propagation of electromagnetic radiation caused by interaction of the radiation with electrons of atoms (or ions) in crystals or fibers.

We can think of a crystal as being composed of layers of reflecting planes. Incident radiation is reflected (scattered) from these planes. When the waves are scattered so that they cross the axis of propagation at the same time and in the same direction, they reinforce each other. Under these circumstances, the amplitude of the resultant wave is equal to the sum of the component waves and *constructive interference* is said to have occurred. When the waves are scattered in such a way that addition of the waves tends to cancel out each other *destructive interference* is said to have occurred.

The general experimental approach for conducting x-ray diffraction of biomolecules is as follows. The sample is placed in the path of a beam of x-ray radiation. These rays are diffracted and thus emerge at various angles and various intensities. The diffracted rays are recorded on photographic film. Depending on the physical state of the sample, angles and intensities may appear either as spots ("reflections") or as a series of concentric arcs where the x-rays intersect the film. There are two general types of method in use for obtaining diffraction patterns:

- the powder method and
- single crystal methods.

The powder method

The powdered sample, placed in a small tube at the center of the x-ray camera, is irradiated with monochromatic x-rays. The camera usually has the shape of a cylinder. As the rays are diffracted and emanate from the sample, they assume a conical shape and form a circular spot where they intercept the film. The circular spot is formed because the powder consists of several randomly oriented, small crystals or "crystallites," and as a result, the diffraction spots produced are smeared into a continuous circle about the origin. Portions of the diffracted beams are intercepted by a photographic film placed around the inside of the cylindrical camera. The photographs thus formed will consist of a series of slightly curved lines. Information on the arrangements of atoms in the biomolecule may be deduced from the positions of the lines on the photograph in relation to the location of the incident beam.

Single crystal methods

• **Rotating crystals.** A single crystal, of side 0.2 mm or greater, is mounted in a device known as a *goniometer*. The latter turns and tilts the crystal in various different directions at various angular displacements. A beam of x-rays of known wavelength then is passed into the crystal with the result that the x-rays are scattered by interactions with electrons of atoms in the biomolecule. The diffracted rays emerge from the crystal at different angles and intensities. Each ray makes a spot when it intersects the x-ray film and diffraction patterns are obtained for many different orientations. The arrangements of atoms and molecules in the crystals are deduced from the spacings of spots ("reflections") in the pattern.

• **Laue method.** This method is named after Max von Laue who received the Nobel prize in physics in 1914 first for suggesting (at a scientific meeting in Munich), and then for demonstrating almost immediately with Friedrich and Knipping in 1912, that crystals can be used as three-dimensional gratings for the diffraction of x-rays. In the Laue method, the crystal is kept in a stationary state and a parallel beam of x-rays is passed through it. The photographic film is placed perpendicular to the incident radiation, and the diffracted rays are recorded as a collection of spots at various distances from the center of the beam.

• **Fibrous macromolecules.** Many macromolecules, e.g. proteins and DNA, form fibers. Such fibers have great regularity of structure. For example, in fibrous proteins, polypeptide chains are coiled in a single direction and often are arranged in bundles which are parallel to the fiber axis. Consequently, diffraction patterns may be obtained using an experimental approach similar to that described for rotating crystals. The resulting diffraction patterns appear in various forms such as spots, rings or arcs depending on the arrangement of fibers in the macro-molecule. From these patterns, the electron density of one unit cell of the crystal can be deduced and then correlated with the sequence of monomers in the macro-molecule. The unit cell is defined as the smallest part of a crystal which, when repeated many times, will yield the entire structure of the macromolecule.

The Braggs made major contributions in x-ray crystallography

An important method of deriving structural information from x-ray diffraction patterns was proposed by a graduate student named William Lawrence Bragg (later, Sir Lawrence, the son of Sir William Henry Bragg). Father and son shared various honors including the 1915 Nobel Prize in physics for their pioneering work in x-ray crystallography. And, as mentioned by Eisenberg and Crothers in their Physical Chemistry textbook, Lawrence Bragg had the unique experience of hearing that he and his father had been awarded the Nobel Prize while he was serving at the front during World War 1.

Since x-rays have great penetrating power, they can penetrate different layers of crystals. Some rays will be scattered from the first layer, some will be scattered from the second, some from the third and so forth. Lawrence Bragg noted that one may think of x-rays as being "reflected" from the various planes in a crystal. But, in contrast to ordinary reflection, x-rays are reflected at specific angles which depend on the wavelength of the x-rays and the spacing between the planes in the crystal. Moreover, according to Lawrence Bragg, if constructive interference occurs, x-rays reflected from adjacent layers of a crystal will recombine to yield a resultant which will have the maximum amplitude. Lawrence Bragg showed that this requirement is expressed by the equation now known as *Bragg's reflection law*:

$$n\lambda = 2d \sin \theta \tag{32}$$

where n = 1,2,3...so that reflections from parallel planes reinforce each other;

λ = wavelength; d = distance between layers in the crystal;

θ = the angle of incidence.

Later, William Bragg suggested that since the electron density function in a crystal repeats itself many times (i.e. is a periodic function which may be expressed as a sum of sine and cosine terms), Fourier analysis may be used to deduce crystal structure from electron density. As we have seen, Fourier transform is now a common technique for analyzing other periodic functions that are measured and stored with modern scientific instruments (cf. section 7-16).

7-20.   Worked examples

Example 7-1.   Predict the absorbance and percent transmittance, in a 1 cm cuvette, of a solution which contains 0.5 mg ml^{-1} of cytochrome c. The molecular weight of cytochrome c is 13 370 and its molar absorbancy index, $E_{553 \ nm \ red}$ = 15.3 * 10^3 M^{-1}cm^{-1}.

Solution.

From equation 7-15

$$A \quad = \quad \varepsilon l c,$$

$$= \quad 15 \ 300 \ l \ mol^{-1} \ cm^{-1} * 1.0 \ cm * \frac{0.5 \ g \ liter^{-1}}{13 \ 370 \ g \ mol^{-1}},$$

$$= \quad 0.57.$$

From equation 7-14

$$\% \ T \quad = \quad antilog \ (2-A),$$
$$= \quad 10^{(2 \ - \ 0.57)},$$
$$= \quad 27 \ \% \ T.$$

Example 7-2. A biomolecule B has a molar absorbancy index of 6.00 * 10^4 at 620 nm in a 1.0 cm cuvette. Show that the amount of B, as n mol in a final volume of 3 ml, is expressed by the relation:

$$c \quad = \quad \frac{10^{-9} \ mol \ B}{3 * 10^{-3} \ l} \quad = \quad 50 * Absorbance.$$

Solution.

From equation 7-18

$$c \quad = \quad A/\varepsilon l,$$

$$c \quad = \quad \frac{A * 3 * 10^{-3} \ l}{6.00 * 10^4 \ l \ mol^{-1} \ cm^{-1} * 1 \ cm * 10^{-9} \ mol},$$

$$c \quad = \quad \frac{3 * 10^{-3} \ A}{6.00 * 10^{-5}} \quad = \quad \frac{3 \ A}{0.06},$$

$$c \quad = \quad 50 * A.$$

Example 7-3.   Calculate the molar extinction coefficient of a solution containing 5 * 10^{-4} g liter^{-1} of a biomolecule, molecular weight 275 g mol^{-1}, and absorbance 0.75 in a 1.2 cm cuvette.

Solution.

From equation 7-15

$$\varepsilon = A/lc,$$
$$= \frac{0.75}{1.2 \text{ cm} * \left[\dfrac{5 * 10^{-4} \text{ g liter}^{-1}}{275 \text{ g mol}^{-1}}\right]},$$
$$= 343\ 750\ \text{l cm}^{-1}\ \text{mol}^{-1}.$$

Example 7-4. A $3.00 * 10^{-6}$ M solution in a 1.0 cm cuvette read 16 % T at 620 nm. What were the absorbance and the molar absorbancy index of the solution?

Solution.

From equation 7-13

$$A = 2 - \log \% \text{ T},$$
$$= 2 - \log 16 = 2 - 1.2041,$$
$$= 0.7959,$$
$$A = 0.80.$$

From equation 7-15

$$\varepsilon = A/lc,$$
$$= \frac{0.80}{1.0 \text{ cm} * 3 * 10^{-6} \text{ mol l}^{-1}},$$
$$= 266\ 667\ \text{l cm}^{-1}\ \text{mol}^{-1}.$$

Example 7-5. A standard solution of bovine serum albumin was used to construct a standard curve by the method of Lowry et al. with the following results:

BSA (µg/µl)	0	20	40	60	80	100
Absorbance	0	0.05	0.10	0.15	0.20	0.23

What is the concentration of a protein solution which gave an absorbance of 0.12 under the same conditions that were used for the standards?

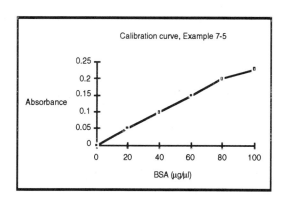

Calibration curve, Example 7-5

The calibration curve shows that the solution contained 49 µg of protein per µl.

Example 7-6. At pH 7.0, the molar extinction coefficients of AMP are 15 400 $M^{-1}cm^{-1}$ at 260 nm and 2 460 M^{-1} cm^{-1} at 280 nm. Also at pH 7.0, the molar extinction coefficients of GMP are 13 700 M^{-1} cm^{-1} at 260 nm and 9 040 M^{-1} cm^{-1} at 280 nm. If a solution containing both nucleotides gave absorbance readings of 0.65 at 260 nm and 0.23 at 280 nm, calculate the concentrations of AMP and GMP in that solution.

Solution.

From equations (7-26) and (7-27)

$$C_1 = \frac{\left[A_1 - A_2\left(\dfrac{\varepsilon_{12}}{\varepsilon_{22}}\right) \right]}{\left[\varepsilon_{11} - \varepsilon_{12}\left(\dfrac{\varepsilon_{21}}{\varepsilon_{22}}\right) \right]}; \quad C_2 = \frac{\left[A_2 - A_1\left(\dfrac{\varepsilon_{21}}{\varepsilon_{11}}\right) \right]}{\left[\varepsilon_{11} - \varepsilon_{12}\left(\dfrac{\varepsilon_{21}}{\varepsilon_{22}}\right) \right]}.$$

Concentration of compound 1:

$$C_1 = \frac{0.65 - 0.23\,(13\,700/9\,040)}{15\,400 - 13\,700\,(2\,460/9\,040)},$$

$$C_1 = \frac{(0.65 * 9{,}040) - (0.23 * 13\,700)}{(15\,400 * 9\,040) - (2\,460 * 13\,700)},$$

$$C_1 = (5876 - 3151)/(139\,216\,000 - 33\,702\,000),$$

$$C_1 = (2725/105\,514\,000),$$

$$C_1 = 2.58 * 10^{-5}\ M.$$

Concentration of compound 2:

$$C_2 = \frac{0.23 - 0.65\,(2\,460/15\,400)}{15\,400 - 13\,700\,(2\,460/9\,040)},$$

$$C_2 = \frac{(0.23 * 15{,}400) - (0.65 * 2{,}460)}{(15\,400 * 9\,040) - (2\,460 * 13\,700)},$$

$$C_2 = (3542 - 1599)/(139{,}216{,}000 - 33{,}702{,}000),$$

$$C_2 = (1943/105{,}514{,}000),$$

$$C_2 = 1.84 * 10^{-5}\ M.$$

7-21. Spreadsheet solutions

	A	B	C	D	E	F	G
1	EXAMPLE 7-1				EXAMPLE 7-6		
2		INPUT	OUTPUT				
3	Molar absorbancy index	15300				INPUT	OUTPUT
4	Length of light path	1			A1	0.65	
5	Weight of analyte	0.5			A2	0.23	
6	Mol. wt. of analyte	13370			e12	13700	
7	Absorbance		0.57		e22	9040	
8	% Transmittance		27		e11	15400	
9					e21	2460	
10	EXAMPLE 7-2				SOLUTION:		
11		INPUT	OUTPUT		CALCN 1:		
12	Vol. of solution	0.003			Numerator		2725
13	Molar absorbancy index	60000			Denominator		105514000
14	Length of light path	1			[Compound 1]		2.583E-05
15	Number of mols.	1E-09			CALCN 2:		
16	Concn., (nmol/3ml)		3.333E-07		Numerator		1943
17	Conversion factor		50		Denominator		105514000
18					[Compound 2]		1.841E-05
19	EXAMPLE 7-3						
20		INPUT	OUTPUT				
21	Absorbance	0.75					
22	Length of light path	1.2					
23	Concn. of solution	0.0005					
24	Molecular weight	275					
25	Molar extinction coeff.		343750				
26							
27	EXAMPLE 7-4						
28		INPUT	OUTPUT				
29	% Transmittance	16					
30	Absorbance		0.79588				
31							
32	Absorbance	0.79588					
33	Length of light path	1					
34	Concn. of solution	0.000003					
35	Molar absorbancy index		265293.34				

Spreadsheet formulas

	A	B	C	D	E	F	G
1	EXAMPLE 7-1				EXAMPLE 7		
2		INPUT	OUTPUT				
3	Molar absorbancy index	15300				INPUT	OUTPUT
4	Length of light path	1			A1	0.65	
5	Weight of analyte	0.5			A2	0.23	
6	Mol. Wt. of analyte	13370			e12	13700	
7	Absorbance		=(B3*B4*B5)/B6		e22	9040	
8	% Transmission		=10^(2-0.57)		e11	15400	
9					e21	2460	
10	EXAMPLE 7-2				SOLUTION:		
11		INPUT	OUTPUT		CALCN 1:		
12	Vol. of solution	0.003			Numerator		=((F4*F7)-(F5*F
13	Molar absorbancy index	60000			Denominator		=((F8*F7)-(F9*F
14	Length of light path	1			[Compound 1]		=G12/G13
15	Number of mols.	0.000000001			CALCN 2:		
16	Concn., (nmol/3 ml)		=B15/B12		Numerator		=((F5*F8)-(F4*F
17	Conversion factor		=((1/C16)/B13)		Denominator		=((F8*F7)-(F9*F
18					[Compound 2]		=G16/G17
19	EXAMPLE 7-3						
20		INPUT	OUTPUT				
21	Absorbance	0.75					
22	Length of light path	1.2					
23	Concn. of solution	0.0005					
24	Molecular weight	275					
25	Molar extinction coeff.		=B21/((B22*B23)/B24)				
26							
27	EXAMPLE 7-4						
28		INPUT	OUTPUT				
29	% Transmittance	16					
30	Absorbance		=2-LOG10(B29)				
31							
32	Absorbance	=C30					
33	Length of light path	1					
34	Concn. of solution	0.000003					
35	Molar absorbancy index		=B32/(B33*B34)				

Spreadsheet solutions

	A	B	C	D	E	F	G	H	I	J
1	EXAMPLE 7-5									
2										
3										
4	INPUT		INPUT							OUTPUT
5	BSA (μg/μl)	x	A	y= A	x^2	x*y	y^2			
6	0	0	0	0	0	0	0		SLOPE	0.0023571
7	20	20	0.05	0.05	400	1	0.0025		INTCP.	0.0038095
8	40	40	0.1	0.1	1600	4	0.01		CORREL	0.9975602
9	60	60	0.15	0.15	3600	9	0.0225			
10	80	80	0.2	0.2	6400	16	0.04		A of Sample:	
11	100	100	0.23	0.23	10000	23	0.0529		0.12	
12										
13	Count	6							Concn. of	
14	Σ	300	0.73	0.73	22000	53	0.1279		Sample	49
15	(Σx OR Σy)^2	90000		0.5329						
16	Correl	99	204.9	0.4843	99.24					
17										
18										
19										
20										
21										
22										
23										
24										
25										
26										
27										
28										
29										
30										
31										
32										
33										
34										
35										

Calibration curve, Example 7-5

Absorbance

BSA (μg/μl)

Spreadsheet formulas

	A	B	C	D
1	EXAMPLE 7-5			
2				
3				
4	INPUT		INPUT	
5	BSA (µg/µl)	x	A	y= A
6	0	=A6	0	0
7	20	=A7	0.05	0.05
8	40	=A8	0.1	0.1
9	60	=A9	0.15	0.15
10	80	=A10	0.2	0.2
11	100	=A11	0.23	0.23
12				
13	Count	=COUNT(B6:B11)		
14	Σ	=SUM(B6:B11)	=SUM(C6:C11)	=SUM(D6:D11)
15	(Σx OR Σy)^2	=B14*B14		=D14*D14
16	Correl. Calcn.	=B13*F14-B14*D14	=SQRT(B13*E14-B15)	=SQRT(B13*G14-D1
17				
18				

	E	F	G	H	I	J
1						
2						
3						
4						OUTPUT
5	x^2	x*y	y^2			
6	=B6*B6	=B6*D6	=D6*D6		SLOPE	=(B13*F14-B14*D14)/(B13*E14-
7	=B7*B7	=B7*D7	=D7*D7		INTCP.	=(D14*E14-B14*F14)/(B13*E14-
8	=B8*B8	=B8*D8	=D8*D8		CORREL	=B16/E16
9	=B9*B9	=B9*D9	=D9*D9			
10	=B10*B10	=B10*D10	=D10*D10		A of Sample	
11	=B11*B11	=B11*D11	=D11*D11		=0.12	
12						
13					Concn. of	
14	=SUM(E6:E11)	=SUM(F6:F11)	=SUM(G6:G11)		Sample	=(I11-J7)/J6
15						
16	=C16*D16					
17						
18						

7-22. Review questions

7-1. Define the terms *transmittance, percent transmittance, extinction coefficient, absorbance,* and *absorption spectrum.*

7-2. Explain why *absorbance* is a dimensionless value.

7-3. What are the dimensions of the *molar extinction coefficient?*

7-4. In general, how are spectrophotometers used qualitatively?

7-5. In general, how are spectrophotometers used quantitatively?

7-6. On what law is quantitative spectrophotometry based?

7-7. Write and explain the equation for the law of question 7-6.

7-8. How is a spectrophotometric calibration curve prepared?

7-9. On what assumptions is *multicomponent spectrophotometric analysis* based?

7-10. What property of an absorbing biomolecule in solution may be obtained from the slope of its calibration curve?

7-11. Define the terms *constructive* and *destructive interference.*

7-12. List the major classes of transition processes by which quanta of energy may be absorbed and explain why the possibility of several different combinations of discrete transitions may show up as many absorption peaks.

7-13. What are some limitations of *flame emission spectroscopy?*

7-14. Why is absorption of radiation in *atomic absorption spectroscopy* analogous to absorption in UV-visible spectrophotometry?

7-15. Explain how flame emission spectroscopy is used for the quantitative determination of atoms.

7-16. Explain how Boltzmann influenced the development of statistical mechanics.

7-17. Explain why quantum mechanics is considered by some to be the basis of interpreting data from modern spectroscopic techniques.

7-18. Distinguish between the Rutherford and Bohr models of the structure of the atom. What were the deficiencies of these models?

7-20. What is Raman scattering, and why is Raman spectroscopy increasingly useful in modern biomedical research?

7-21. Why do biomedical scientists use both infrared and Raman spectroscopy even though both measure the same molecular property? What is that property?

7-22. Describe the principle of NMR.

7-23. List five magnetic nuclei that are commonly used by biomedical scientists.

7-24. What requirements should be met for an atom to exhibit nuclear magnetic resonance?

7-25. Define the term *chemical shift.*

7-26. How do you explain the fact that the signal-to-noise ratio is increased in FT-NMR? What is the practical consequence of this increase?

7-27. Describe the contributions of Planck, Pauling, and Ernst to modern spectroscopy.

7-28. On what principle is ESR based?

7-29. Define the terms fluorescence and phosphorescence.

7-30. On what electronic structures of biomolecules does absorption in the ultraviolet region of the electromagnetic spectrum depend?

7-31. For what does the acronym laser stand? Outline briefly the general operation, and some biomedical applications, of lasers.

7-32. Describe the principle of *magnetic resonance imaging*.

7-33. Describe how the contributions of Planck and Einstein to quantum theory both depended on and supported each other.

7-34. Describe the nature and uses of two-dimensional NMR.

7-35. Define the term Fourier transformation and explain why the Fourier series, discovered in the time of Napoleon, is so important today.

7-36. What properties of biomolecules may be studied by ORD and CD?

7-37. List the major regions of the electromagnetic spectrum.

7-38. What techniques are associated with the major regions of the electromagnetic spectrum?

7-39. What is the general experimental approach for conducting x-ray diffraction of biomolecules?

7-40. Define the term *diffraction*.

7-41. Explain the importance of the *free induction decay* in NMR.

7-42. Draw an energy diagram which shows the relationships among rotational, vibrational and electronic levels of internal energy.

7-43. Explain why fluorescence polarization and anisotropy can provide information on the behavior and structure of macromolecules.

7-44. Define the terms *anisotropic* and *isotropic* biomolecules.

7-45. Define the terms *extrinsic* and *intrinsic* biomolecules.

7-46. Why are the names Heisenberg and Schrödinger important in spectroscopy?

7-47. Why is the name Bragg important in x-ray crystallography?

7-23. Review problems

7-48. Convert the following values of percent transmittance to absorbance: (a) 92; (b) 75; (c) 60; (d) 27; (e) 10.

7-49. Convert the following values of absorbance to percent transmittance: (a) 2.00; (b) 1.8; (c) 1.2; (d) 0.8; (e) 0.31; (f) 0.15; (g) 0.05.

7-50. Calculate the molar extinction coefficient of a biomolecule of molecular weight 200 if a 5.0 mM solution of the compound in a cuvette of 1.2 cm path length has an absorbance of 0.75 at 660 nm.

7-51. What is the molar extinction coefficient of a biomolecule which has an absorbance of 0.70 in a cuvette of path length 1.0 cm, at 260 nm? Assume that the biomolecule has a molar extinction coefficient of 12 500 at the same wavelength.

7-24. Additional problems

7-52. The slope of a calibration curve (equal to εl) was 25 476 liters mol^{-1} and the width of the cuvette was 1.15 cm. Calculate the extinction coefficient.

7-53. A solute with extinction coefficient of 22 150 liter mol^{-1} cm^{-1} gave an absorbance of 0.72 in a 1.15 cm cuvette. What was the concentration of solute?

7-54. Convert 60 %T to absorbance and convert an absorbance of 0.47 to %T.

7-55. The following values were obtained for a spectrophotometric standard curve:

Concentration, μg ml^{-1}	1.0	2.0	3.0	4.0	5.0
Absorbance	0.29	0.41	0.52	0.64	0.75

If two samples carried through the same procedure as the standards gave absorbance readings of 0.35 and 0.70, respectively, what concentrations of ester did they contain? How good was your calibration curve? Give reasons to support your answer.

7-56. Predict A and % T, in a 1 cm cuvette, of 0.5 mg ml^{-1} of cytochrome c (molecular weight 13 370 and molar absorbancy index, E_{553} nm reduced, $15.3*10^3$ M^{-1} cm^{-1}).

7-57. What is the molar extinction coefficient of a $5*10^{-4}$ g liter^{-1} solution of a compound, molecular weight 275 g mol^{-1}, and absorbance 0.675, in a 1.2 cm cuvette?

7-58. A solution at a concentration of $3.00*10^{-6}$ mol liter^{-1} in a cuvette of 1.0 cm path length gave a percent transmittance of 16 %T at 620 nm. What were the absorbance and the molar absorbancy index of the solution?

7-59. Aliquots of a standard solution of bovine serum albumin were used to construct a standard curve by the method of Lowry with the following results:

[Protein] , μg ml^{-1}	20	40	60	80	100
Absorbance	0.05	0.09	0.14	0.19	0.22

What is the concentration of a protein which gave an absorbance reading of 0.12 under the same conditions used for the standards?

7-60. At pH 7.0, the molar extinction coefficients of AMP are 15 400 M^{-1} cm^{-1} at 260 nm and 2 460 M^{-1} cm^{-1} at 280 nm. Also at pH 7.0, the molar extinction coefficients of GMP are 13 700 M^{-1} cm^{-1} at 260 nm and 9 040 M^{-1} cm^{-1} at 280 nm. If a solution containing both nucleotides gave absorbance readings of 0.65 at 260 nm and 0.23 at 280 nm, calculate the concentrations of AMP and GMP in that solution.

Suggested Reading

Amato, I. 1991. Nobel Prizes '91. Chemistry: A Certain Resonance. *Science* 254: 518-519.

Anderson, S.R. 1991. Time-Resolved Fluorescence Spectroscopy. Applications to Calmodulin. *J. Biol. Chem.* 266:11405-11408.

Appendix, 1989. *Computer Programs Related to Nuclear Magnetic Resonsnce: Availability, Summaries, and Critiques.* Methods in Enzymology, Academic Press, New York, 176: 477-489.

Adler, A. J., N. J. Greenfield and G. D. Fasman 1973. *Circular Dichroism and Optical Rotatory Dispersion of Proteins and Polypeptides.* Methods in Enzymology, Academic Press, New York, 27: 675-735.

Bajzer, Z. and F.G. Prendergast 1992. *Maximum Likelihood Analysis of F,luorescence Data.* Methods in Enzymology, Academic Press, New York 210: 200-237.

Barrow, G. M. 1964. The Structure of Molecules. Benjamin, New York.

Barrow, G. M. 1974. *Physical Chemistry for the Life Sciences. 2d ed.*, McGraw-Hill, New York.

Bax, A. and L. Lerner 1986. Two-dimensional Nuclear Magnetic Resonance Spectroscopy. *Science* 232: 960-967.

Brand, L. and B. Witholt 1967. *Fluorescence Measurements.* Methods in Enzymology, Academic Press, New York, 11: 776-856.

Causton, D. R. 1983. *A Biologist's Basic Mathematics*, Arnold, Baltimore, Maryland.

Causton, D. R. 1987. *A Biologist's Advanced Mathematics*, Allen & Unwin, Boston.

Chang, R. 1981. *Physical Chemistry With Applications to Biological Systems.* 2d. ed., Macmillan, New York.

Chapman, D. 1965. *The Structure of Lipids by Spectroscopic and X-Ray Techniques*, Wiley, London.

Clegg, R.M. 1992. *Fluorescence Resonance Energy Transfer and Nucleic Acids.* Methods in Enzymology, Academic Press, New York, 211: 353-388.

Cornish-Bowden, A. 1981 *Basic Mathematics for Biochemists*, Chapman & Hall, London.

Christian, G.D. 1980. *Analytical Chemistry, 3d ed.*, Wiley, New York.

Dandliker, W.B., J. Dandliker, S. A. Levison, R. J. Kelly, A. N. Hicks, and J. U. White 1978. *Fluorescence Methods for Measuring Reaction Equilibria and Kinetics.* Methods in Enzymology, Academic Press, New York, 48: 380-415.

Dickerson, R. E., H. B. Gray, M. Y. Darensbourg and D. J. Darensbourg 1984. *Chemical Principles, 4th ed.*, Benjamin/Cummings, Reading, Massaachusetts.

Eisenberg, D. and D. M. Crothers 1979. *Physical Chemistry With Applications to the Life Sciences.* Benjamin/Cummings, Massachusetts.

Fasman, G. D. 1963. *Optical Rotatory Dispersion.* Methods in Enzymology, Academic Press, NY, 6: 928-957.

Feigon, J., V. Skelnár, E. Wang, D.E. Gilbert, R.F.Macaya and P. Schultze 1992. ^1H *NMR Spectroscopy of DNA.* Methods in Enzymology, Academic Press, New York 211: 235-253.

Gadian, D. G. and G. K. Radda 1981. *NMR Studies of Tissue Metabolism.* Annu. Rev. Biochem, 50: 69-83.

Gibson, Q.H. 1989. Hemoproteins, Ligands, and Quanta. *J. Biol. Chem.* 264: 20155-20158.

Gorenstein, D.G. 1992. ^{31}P *NMR of DNA.* Methods in Enzymology, Academic Press, New York 211: 254-286.

Gray, D.M., R.L. Ratliff and M.R.Vaughan 1992. Circular Dichroism Spectroscopy of DNA Methods in Enzymology, Academic Press, New York 211: 389-405.

Haris, P. I. and D. Chapman. 1992. Does Fourier-Transform Infrared Spectroscopy Provide Useful Information on Protein Structures. *Trends in Biochem. Sci.*, 1 328-333.

Hileman, B. 1993. Health Effects of Electromagnetic Fields Remain Unresolved. *Chem. & Engineering News.* Pages 15-29, November 8.

Hudson, Bruce and L. Mayne 1986. *Ultraviolet Resonance Raman Spectroscopy of Biopolymers*. Academic Press, Methods in Enzymology, Academic Press, New York, 130: 331-409.

Johnson, W.C. Jr 1992. *Analysis of Circular Dicuroism Spectra*.Methods in Enzymology, Academic Press, New York 210: 426-447.

Klotz, I. M. 1967. *Introduction to Biomolecular Energetics*. Academic Press, New York.

Lehninger, A. L. 1975. *Biochemistry*. 2d ed., Worth, New York.

Lehninger, A. L., D. J. Nelson and M. M. Cox 1992. *Principles of Biochemistry*, 2d. ed., Worth, New York.

Litman, B. J. and Y. Barenholz 1982. *Fluorescence Probe: Diphenylhexatriene*. Methods in Enzymology, Academic Press, New York, 81: 678-685.

Markley, J. L. 1989. *Two-Dimensional Nuclear Magnetic Resonance of Proteins: An Overview*. Methods in Enzymology, Academic Press, New York, 176: 12-64.

Mislow, K. 1966. *Introduction to Stereochemistry*, Benjamin, New York.

Montgomery, R., and C. A. Swenson 1976. *Quantitative Problems in the Biochemical Sciences*, 2d ed., Freeman, San Francisco.

Moffatt, D.J. and H.M. Mantsch 1992. *Fourier Resolution Enhancement of Infrared Spectral Data*. Methods in Enzymology, Academic Press, New York 210: 192-200.

Morris, J. G. 1974. *A Biologist's Physical Chemistry, 2d ed.*, Arnold, Baltimore, Maryland.

Morrison, R. T. and R. N. Boyd 1983. *Organic Chemistry, 4th ed.*, Allyn and Bacon, Boston.

Pauling, L. 1976. The birth of quantum mechanics. *Trends in Biochem. Sci.*, 1: 214-215.

Pauling, L., 1960. *The Nature of the Chemical Bond, 3d ed.*, Cornell Univ. Press.

Pecsok, R. L., L. D. Shields, T. Cairns and I. G. McWilliam 1976. *Modern Methods of Chemical Analysis, 2d ed.*, Wiley, New York.

Peticolas, W.L. and E. Evertsz 1992. *Conformation of DNA in Vitro and in Vivo from Laser Raman Scattering*. Methods in Enzymology, Academic Press, New York 211: 335-352.

Radda, G. K. 1986. The Use of NMR Spectroscopy for the Understanding of Disease. *Science* 233: 640-645.

Shinitzky, M. and Y. Barenholz 1978. Fluidity Parameters of Lipid Regions Determined by Fluorescence Polarization. *Biochim. et Biophys. Acta*. 515: 367-394.

Silverstein, R. M. and G. C. Bassler 1967. *Spectrometric Identification of Organic Compounds, 2d ed.*, Wiley, New York.

Spiro, T. G. and B. P. Gaber 1977. *Laser Raman Scattering as a Probe of Protein Structure*. Annu. Rev. Biochem. 46: 553-572.

Stinson, S. 1991. Nobel Chemistry Prize: Ernst honored for NMR achievements. *Chem. Eng. News*, Page 4, Oct. 21.

Susi, H. and D. M. Byler 1986. *Resolution-Enhanced Fourier Transform Infrared Spectroscopy of Enzymes*. Methods in Enzymology, Academic Press, New York 130: 290-311.

Taillander, E. and J. Liquier 1992. *Infrared Spectroscopy of DNA*. Methods in Enzymology, Academic Press, New York 211: 307-335.

Van Holde, K. E. 1971. *Physical Biochemistry*. Prentice-Hall, New Jersey.

Veillon, C. and B. L. Vallee 1978. *Atomic Absorption in Metal Analysis of Enzymes and Other Biological Materials*. Methods in Enzymology, Academic Press, New York, 44: 446-484. Note. This article contains information on flame emission spectrometry.

Willard, H. H. 1981. *Instrumental Methods of Analysis*. Wadsworth, Belmont, California.

Wüthrich, K. 1990. Protein Structure Determination in Solution by NMR Spectroscopy. *J. Biol. Chem.* 265: 22059-22062.

Yaqub, A. and H. G. Moore 1981. *Elementary Linear Algebra With Applications*. Addison-Wesley, Mass.

Chapter 8

Radioactivity

8-1.　Radioactive isotopes emit characteristic types of radiation

Radioisotopes (or radionuclides) are atoms whose nuclei disintegrate spontaneously and continuously yielding, in the process, several types of radiation. Ernest Rutherford (section 7-2) who, in 1918, carried out the first artificial *transmutation* or conversion of one element into another by bombarding nuclei with high-speed, radioactive particles, also was the first (in 1908) to characterize the types of radiation emitted by naturally occurring radionuclides: *helium nuclei (alpha particles), electrons (beta particles), and high-energy photons (gamma rays).* The latter are similar to, but are of greater energy than, x-rays (sections 7-1 and 7-5). In the life sciences, ^3H, ^{14}C, ^{32}P, and ^{35}S are the radioactive isotopes used most frequently. They are all beta emitters and they decay by ejecting an electron and a neutrino. The varying energies with which their electrons are emitted permit beta emitters to be distinguished from each other. Beta emitters such as ^{32}P which eject electrons with high energy can be detected with a Geiger counter because their electrons can pass through the window of the Geiger tube, can then ionize the gas contained in the tube and can, as a result, cause a flow of electric current to occur. Low energy beta emitters such as ^3H cannot, though, be detected by Geiger counters, because their ejected electrons are not sufficiently energetic to pass through the window of a Geiger tube. All beta emitters can be detected and measured by liquid scintillation counting (section 8-5). Two gamma emitters, ^{125}I and ^{131}I also are used in biomedical research. They can be detected by Geiger counters but usually are measured in gamma counters.

In this chapter, we will provide computations for the efficiency of counting radioisotopes, for quench correction, for the radioactivity of samples in double label experiments, and for the half-lives of radioisotopes.

8-2.　Radioisotopes decay at mathematically predictable rates

Radioisotopes decay to more stable forms at rates which are dependent on the concentrations of radioactive atoms in the radioisotopes. Accordingly, the rate of decay may be expressed in terms of the derivative or the rate of change of radioactive atoms with time:

$$\frac{-dN}{dt} = \lambda N \quad or \quad \frac{dN}{N} = -\lambda t, \tag{1}$$

where $-dN/dt$ = the number of atoms decaying with time,

dN/N = fraction of original number of radioactive atoms undergoing decay,

and λ = a disintegration constant that is characteristic of each radioisotope.

On integrating equation (1) we get:

$$\int_{N_0}^{N} \frac{dN}{N} = \lambda \int_{0}^{t} dt, \tag{2}$$

$$\ln \frac{N}{N_0} = -\lambda t, \tag{3}$$

$$\ln \frac{N_0}{N} = \lambda t, \tag{4}$$

where N_0 = the number of radioactive atoms at a time selected as zero time,

N = the number of radioactive atoms after time t has elapsed,

and N_0/N = ratio of radioactive atoms present at the two times.

In the laboratory, radioactivity is measured as the number of counts or disintegrations per minute and not as a ratio of radioactive atoms. Therefore, to be consistent with laboratory practice, we rewrite equation (4) as

$$\ln \frac{A_0}{A} = \lambda t, \tag{5}$$

where A_0 = the number of disintegrations (activity) at a time selected as zero time,

and A = the number of disintegrations after time t has elapsed.

Taking antilogarithms of equation (5) we get

$$A = A_0 e^{-\lambda t}. \tag{6}$$

8-3.   Measured radioactivity should be corrected for counting errors

The counts of radioactivity in a sample, whether made in a liquid scintillation spectrometer (section 8-5), a gamma counter or a Geiger counter, should be corrected for background counts, statistical variability, coincidence losses, dead time and quenching. The latter will be discussed in the next section. Here, we will discuss elementary aspects of each of the other factors in turn.

Background radiation may arise from several sources including cosmic radiation, ^{40}K in glass vials, thermal noise, naturally occurring radioisotopes in a laboratory, and contamination of a liquid scintillation counter or counting vials. Such background counts should be measured in the absence of radioactive samples and subtracted from the counts obtained for each sample to give the *net counts per minute (net cpm)* in the samples. Subtraction of background counts is particularly important when counts from the samples are not much higher than the background.

Statistical variation in measuring radioactivity results from the fact that the disintegration of radioisotopes occurs in a random manner in time and space. This means that, if sources of experimental error such as instrument malfunction are avoided, the number of disintegrations should follow the *Poisson distribution* (described in section 10-11). It means also that the number of disintegrations in any interval of time should be independent of the disintegrations in any previous or subsequent interval. Thus, if a radioactive sample is counted ten times for one minute each time, ten different counts will be obtained. Approaching this problem in terms of ensuring reliable counting reduces to getting a good estimate of the true mean counts in the sample and estimating the degree to which the measured counts are scattered around the mean. We will see in section 10-4 that when values are scattered normally around a mean, 68% of the values will fall within ± 1 standard deviation, 95% within ± 1.96 standard deviations and 99.5% within ± 3.31 standard deviations. If N, the mean number of counts > 100, then the standard deviation is $\sqrt{N}$ since the *variance of a Poisson process is equal to the mean;* i.e., $\sigma^2 = \mu$. As the rate of nuclear disintegrations increases, the Poisson distribution approaches the normal distribution (see section 10-4). Usually, better results are obtained by making one ten-minute count than by taking ten one-minute counts. As N increases, the absolute value of the standard deviation increases but the relative standard deviation; i.e., $1/\sqrt{N}$ and the percent relative standard deviation, $100/\sqrt{N}$, both decrease.

Coincidence losses occur when samples of very high radioactivity are counted. In such highly radioactive samples, the measuring instrument is unable to count all of the disintegrations or counts. One solution to this problem is to dilute the samples or to calculate the coincidence losses by methods which will not be discussed here.

Dead time is related to coincidence losses. Dead time is the period of time during which a counter is unable to record the disintegrations of highly radioactive samples. It is usually measured in microseconds.

8-4. The measurement of radioactivity is expressed in several ways

The *curie,* abbreviated Ci, is the basic unit of radioactivity. Today, 1 Ci is defined as $3.700 * 10^{10}$ disintegrations per second, dps. For many reasons (including those discussed in sections 8-3 and 8-5), when a radioactive sample is counted, the radioactivity that is measured and recorded by an instrument is less than the total number of disintegrations produced by the sample. The measured or observed radioactivity is reported as *net counts per minute, net cpm* (section 8-3) in contrast to the total number of disintegrations per minute or dpm actually produced. The *dpm of a radioactive sample* is obtained from the net cpm of that sample by simple division using the relationships:

$$E = \frac{\text{net cpm of standard}}{\text{dpm of standard}} \tag{7}$$

where   E   =   counting efficiency so that

$$\text{dpm of sample} \quad = \quad \frac{\text{net cpm of sample}}{E} \tag{8}$$

The counting efficiency usually is determined for a given radioisotope from counting the radioactivity of standards containing known levels of that radioisotope. Frequently, net cpm and dpm are reported as *specific activity*, defined as radioactivity per unit weight e.g. net cpm/mg, or net cpm/mol of sample.

The rad is another unit of radioactivity. It is the amount of radiation absorbed per unit time by a given mass of material. Mathematically, this relationship may be expressed as follows:

$$\text{number of rads} \quad = \quad \frac{dN}{dt} \; * \; E_{avg} \; * \; \frac{t}{mass} \tag{9}$$

where dN/dt = number of atoms decaying with time

E_{avg} = average energy of particles emitted, measured usually in millions of electron-volts (1 MeV = $1.6 * 10^{-6}$ ergs)

and t = time of exposure.

By definition 1 rad equals the absorption of radiation corresponding to 100 ergs g^{-1}.

The *rem*, another unit of radioactivity, is used to express levels of radiation allowed for human subjects. The rem is equal to rads multiplied by a quality factor Q. Q has different values for different types of radiation. For example, Q can vary from 1 to 20, with the lower value being assigned to beta particles, gamma rays and X-rays and the higher values to alpha particles, protons and neutrons. Currently, the maximum dose of radioactivity allowed for pregnant women is 0.5 rem per year. It has been estimated that background radiation exposes human subjects to approximately 0.1 rem per year.

8-5.   Quenching reduces counts in liquid scintillation spectrometry

Liquid scintillation counters (LSC) measure the radioactivity of beta emitters in solutions containing scintillator molecules (molecules which undergo fluorescence, cf. section 7-10). When a beta emitter ejects a beta particle in a scintillation solution or "cocktail", the ejected particle excites solvent molecules. The latter then transfer their excitation energy to the scintillator molecules by collision. An activated scintillator molecule emits a flash of light (a *photon*, section 7-2) as it returns to its ground state, the total number of photons produced depending on the initial energies of the beta particles. The photon yield of a scintillator molecule is measured as an anode pulse, the height of which depends both on the energy of the beta particle, and on the detection efficiency of the liquid scintillation spectrometer.

Scintillation spectrometers usually are set to count pulses above or below certain energy levels. The device which is adjusted to eliminate the pulses below a specific energy level is known as the lower discriminator. The device which is adjusted to eliminate pulses above a given energy level is the upper discriminator. The region between the two discriminator settings is known as a *window* or *channel*. Modern liquid scintillation spectrometers usually have two counting channels. Pulses of higher energy levels are counted in the upper channel, those of lower energy levels in the lower channel.

Quenching is a process which reduces the number of photons produced by beta particles during liquid scintillation counting. Thus, given two samples with equal activities of the same radioisotopes, if one were quenched and the other were not, the quenched sample will yield a lower count rate than the unquenched. Quenching also decreases the pulse height spectra of beta emitters and shifts pulses from the upper to the lower channel.

There are two broad types of quenching: chemical quenching and optical quenching. *Chemical quenching* occurs when solvents or solutes, added with the radioisotopic sample to a scintillation solution, interfere with the transfer of energy. Chemical interference with energy transfer may occur in three ways: from the beta particle to the solvent, from the solvent to solvent or from the solvent to scintillator molecules. Among well-known chemical quenching agents are chloroform, ethanol and methanol. *Optical quenching* occurs when properties or components associated with the scintillation system interfere with the transfer of photons. Causes of optical quenching include colored or turbid solutions, both of which reduce the number of scintillations (photons) recorded by the liquid scintillation spectrometer.

Because quenching reduces the number of photons and decreases pulse height, it also decreases counting efficiency in liquid scintillation spectrometry. Consequently, it is necessary to correct the measured count rate of the quenched samples (in net cpm) for this decreased efficiency. We outline how this may be done in the next section.

8-6.   There are three methods for making quench corrections

Three methods generally used for quench corrections in liquid scintillation spectrometry are: the *internal standard*, the *sample channels ratio* and the *external standard*.

Internal standard

The labeled sample is counted alone and background counts are subtracted. We show these corrected counts, given by the sample alone, as cpm_0 in equation (10). Next, a quantity of an appropriate standard of known radioactivity (dpm) is added to the sample. The sample, with the internal standard, is recounted to obtain the *additional counts* recorded. These additional counts are an estimate of the radioactivity of the internal standard and are represented as $cpm_1 - cpm_0$ in equation (10). *It is assumed in this case that the sample alone, and the sample plus standard, are each counted with the same counting efficiency.* This efficiency, E, is obtained from the equation

$$E \quad = \quad \frac{cpm_1 - cpm_0}{dpm_{standard}} \quad = \quad \frac{\text{additional counts}}{dpm_{standard}} \quad\quad (10)$$

Therefore, as shown in equation (8), the *dpm of the sample* is given by

$$\text{dpm}_{\text{sample}} \ = \ \frac{\text{cpm}_0}{E} \tag{11}$$

Sample channels ratio

The *sample channels ratio, SCR*, is defined as the ratio of counts from a given sample in the two counting channels of a liquid scintillation spectrometer. Generally

$$\text{SCR} \ = \ \frac{\text{net cpm in upper channel}}{\text{net cpm in lower channel}} . \tag{12}$$

Quench correction by the sample channels ratio depends on the fact that counts are shifted from the upper to the lower channel in quenched samples. Modern liquid scintillation spectrometers compute SCR automatically. To correct for quenching by the sample channels ratio, one prepares and counts a set of quenched standards. Then, the counting efficiency of each standard (net cpm/dpm) is obtained from the count rate in one of the two channels. The *efficiency of counting each standard* is next plotted against its SCR to give a calibration curve. The *counting efficiency of each sample* is then obtained by reading from the curve the efficiency values corresponding to the computed SCR. Finally, the dpm of the sample is computed from its net cpm and efficiency with equation (8).

External standard ratio

Quench correction by the *external standard ratio* (abbreviated ESR by some instrument manufacturers) calls for a set of quenched standards, as does quench correction by the sample channels ratio. The *external standard* is a source of gamma radiation. When the source is placed next to the counting vials it produces *Compton electrons* in the scintillation solution. Compton electrons behave similarly to beta particles; consequently, both the number produced, and their counting efficiencies in scintillation solutions, are reduced by quenching.

Steps for measuring ESR. Four measurements are made automatically on the quenched standards for applying quench correction by the external standard ratio. They are

• Count the *combined radioactivities* of the standards and the gamma source in each of the two *external standards channels*. We refer to this value as *count 1*. Next,
• count the radioactivities of the *quenched standards alone* in the two *external standards channels*. We refer to this measurement as *count 2*. Then,
• calculate the *external standard ratio*; {(count 1 - count 2)/dpm of the source}. Finally,
• count the radioactivities of the *quenched standards* (net cpm) in the *two counting channels* to obtain the counting efficiencies of the standards.

Calibration curves. One prepares calibration curves (see, for instance, worked example number 8-7 below) by plotting the efficiency of counting each standard (net cpm/dpm) against the corresponding ESR. As with the SCR method, the counting efficiency of each unknown sample is obtained by reading from the appropriate calibration curve the efficiency values corresponding to the ESR for that unknown.

Radioactivities of unknowns. One calculates the radioactivities of the unknowns, as dpm, by dividing net cpm of each unknown by its counting efficiency.

Each of the quench correction methods outlined above has advantages and disadvantages, details of which are beyond the scope of this book.

8-7. Scintillation spectrometers can count the cpm of radionuclide pairs

Beta particles are emitted at several different speeds. More precisely, beta particles are emitted with a continuous distribution of kinetic energy from zero up to some maximum value E_{max}. When the number of electrons ejected is plotted against energy interval, the resulting curve is the *beta spectrum*. Because the E_{max} of each beta particle occurs as a sharp cutoff point, usually in the range 15 keV to 15 MeV, beta emitters can be distinguished from each other on the basis of the characteristic maximum and average speeds of their beta particles. Since beta particles are ejected with a continuous distribution of energy, however, some overlap occurs between spectra of radioisotopes with different E_{max} values.

Liquid scintillation spectrometers can be used to count one beta emitter in the presence of another if the instrument is set properly and if E_{max} of the higher energy radioisotope is at least 3 times greater than E_{max} of the lower. As examples, the E_{max} values of ^3H (18 keV) and ^{14}C (156 keV), as well as those of ^{35}S (168 keV) and ^{32}P (1.71 MeV), are sufficiently far apart that spectral overlap can be minimized in a properly adjusted liquid scintillation spectrometer. Two counting channels are required for double label experiments with such pairs of radioisotopes; one for counting the radioisotope with the greater E_{max}, the other for counting the radioisotope with the smaller E_{max}. Usually, the smallest counting errors are obtained if

• A fraction of the counts from the radioisotope with the greater E_{max} is allowed to spill into the channel for counting the radioisotope with the smaller E_{max} (to allow efficient counting of the radioisotope with the greater E_{max}) but
• The radioisotope with the smaller E_{max} is virtually excluded from the channel with the radioisotope of greater E_{max}.

Generally, the isotope of greater E_{max} is counted in the upper window or channel, that of the smaller E_{max} in the lower window or channel.

8-8.   Simultaneous equations yield the DPM of radionuclide pairs

Although E_{max} values of the radioisotopes in double label experiments usually are widely separated, radioactivity from both radioisotopes, as net cpm, is counted in each channel. The reason is spectral overlap. Computation of the dpm of each radioisotope in a sample therefore involves two simultaneous equations. In these equations, the dpm values of the two radioisotopes are the unknowns, the net cpm values are the constants and the two sets of efficiencies (total of four efficiencies) are the coefficients. Thus, if we let the lower and upper channels be A and B, respectively, the two simultaneous equations can be written as follows for ^3H and ^{14}C:

$$EHA * {}^3Hdpm + ECA * {}^{14}Cdpm = \text{cpm in A.} \tag{13}$$
$$EHB * {}^3Hdpm + ECB * {}^{14}Cdpm = \text{cpm in B.} \tag{14}$$

In these equations,

EHA	=	efficiency of ^3H in the lower channel, A,
ECA	=	efficiency of ^{14}C in the lower channel, A,
EHB	=	efficiency of ^3H in the upper channel, B,
ECB	=	efficiency of ^{14}C in the upper channel, B,
cpm in A	=	total cpm in the lower channel - background counts,
cpm in B	=	total cpm in the upper channel - background counts,
^3Hdpm	=	dpm of ^3H in the sample,
$^{14}Cdpm$	=	dpm of ^{14}C in the sample.

Solution of equations(13) and (14) by determinants (cf. section 7-9) yields

$$^3Hdpm = \left[\frac{cpm\ A - cpm\ B \left(\dfrac{ECA}{ECB} \right)}{EHA - EHB \left(\dfrac{ECA}{ECB} \right)} \right], \tag{15}$$

$$^{14}Cdpm = \left[\frac{cpm\ B - cpm\ A \left(\dfrac{EHB}{EHA} \right)}{ECB - EHB \left(\dfrac{ECA}{EHA} \right)} \right], \tag{16}$$

Equation (15) is equivalent to equation (13) rearranged to

$$^3Hdpm = \left[\frac{cpm\ A - (ECA * {}^{14}Cdpm)}{EHA} \right]. \tag{17}$$

If the product in the numerator of the right hand side of equation (17) is large (if there is a high proportion of radioactivity from the higher energy radioisotope in the lower channel), then the computed value of ^3H dpm will be small.

8-9.   The rate of radioisotope decay may be given by the half-life

The half-life ($t\frac{1}{2}$) of a radioisotope is the time required for one-half of the original number of radioactive atoms to undergo decay. A useful equation for calculating parameters related to $t\frac{1}{2}$ (section 5-12) may be developed from equations (3) and (4):

$$\text{At } t\tfrac{1}{2}, \qquad \ln N_0/N \quad = \quad \ln 1/0.5; \tag{18}$$

Substituting (18) in equation (4);

$$\ln 1/0.5 \quad = \quad \lambda t\tfrac{1}{2}, \tag{19}$$

$$\ln 2 \quad = \quad \lambda t\tfrac{1}{2}, \tag{20}$$

$$0.693 \quad = \quad \lambda t\tfrac{1}{2}, \tag{21}$$

$$\lambda \quad = \quad \frac{0.693}{t\tfrac{1}{2}}. \tag{22}$$

Substituting (22) in equation (6),

$$A \quad = \quad A_0 e^{-(0.693\,/\,t\frac{1}{2})\,*\,t}, \tag{23}$$

$$A \quad = \quad A_0 2^{-t/t\frac{1}{2}}, \tag{24}$$

$$A/A_0 \quad = \quad 2^{-t/t\frac{1}{2}}, \tag{25}$$

$$\quad = \quad 0.5^{\,t/t\frac{1}{2}}, \tag{26}$$

$$\text{Let} \qquad t/t\tfrac{1}{2} \quad = \quad N,$$

$$A_0 \quad = \quad A/0.5^N. \tag{27}$$

where
A	=	radioactivity at time t,
A_0	=	radioactivity at zero time,
A/A_0	=	fraction of radioactivity remaining at time t,
$t/t\frac{1}{2}$ or N	=	the number of half-lives between t_0 and t.

In practice, A/A_0 can be obtained in two ways:

*   by direct calculation given $t/t\frac{1}{2}$, or
*   by interpolation from a graph of A/A_0 versus.

We use direct calculation to calculate A/A_0 in *Basic Biochemical Laboratory Procedures andComputing*.

8-10.   Worked examples

Example 8-1. An aliquot of a radioactive triacylglycerol equivalent to 5 mg of the lipid gave 12 000 dpm when counted in a liquid scintillation spectrometer. What was the specific radioactivity of the labeled triacylglycerol?

Solution.

$$\text{Specific activity} \quad = \quad \text{radioactivity per unit weight,}$$

$$= \quad 12\,000 \text{ dpm/ 5 mg,}$$
$$= \quad 2\,400 \text{ dpm mg}^{-1}.$$

Example 8-2. A commercially prepared solution of adenosine $[2\text{-}^3\text{H}]\text{-}5'\text{-}$ monophosphate contained 20 Ci m mol^{-1}. If you wanted to use 250 μ Ci of the labeled compound in an experiment, how many m mol would you be using?

Solution.

$$20 \text{ Ci} \quad \equiv \quad 20 * 10^6 \,\mu\text{Ci} \quad \equiv \quad 1 \text{ m mol,}$$
$$1 \,\mu \text{ Ci} \quad \equiv \quad 1/20 \text{ m mol} \quad = \quad 0.05 \text{ m mol,}$$
$$250 \,\mu \text{ Ci} \quad = \quad 0.05 * 250 \text{ m mol,}$$
$$= \quad 12.5 \text{ m mol.}$$

Example 8-3. If a carbon-14 standard of 4×10^4 dpm read 27 000 net cpm in a liquid scintillation spectrometer and a tritium standard of 1×10^5 dpm read 41 000 net cpm, what were the counting efficiencies of the two standards?

Solution.

$$\text{Efficiency, E} \quad = \quad \text{net cpm standard/dpm standard,}$$

$$\text{For } {}^{14}\text{C, E} \quad = \quad 27\,000/40\,000 \quad = \quad 0.68,$$
$$\text{For } {}^{3}\text{H, E} \quad = \quad 41\,000/100\,000 \quad = \quad 0.41.$$

Example 8-4. (a) If a sample of L-methionine-$[^{35}\text{S}]$ gave 122 000 net cpm at the beginning of an experiment, what would be its radioactivity 190 days later? The half-life of sulfur-35 is 87.2 days. (b) Phosphorus-32 has a half-life of 14.3 days. How much radioactivity would be lost from a sample containing 800 becquerels after thirty-two days? What fraction of the radioactivity remains?

Solution.

From equation (24):

(a) $\quad A = A_0 2^{-t/t_{\frac{1}{2}}}$,

$\quad = 122\ 000 * 2^{-190/87.2}$ net cpm,

$\quad = 122\ 000 * 2^{-2.18}$ net cpm,

$\quad = 26\ 922$ net cpm.

(b) $\quad A = A_0 2^{-t/t_{\frac{1}{2}}}$,

	$= 800 * 2^{-32/14.3}$	$=$	169 Bq,
Amount lost	$= 800 - 169$	$=$	631 Bq,
Fraction remaining	$= 169/800$	$=$	0.21.

Example 8-5. Given that 1 Ci $= 3.7 * 10^{10}$ Bq and that 1 μCi of a radionuclide was counted with a counting efficiency of 70%, calculate the observed radioactivity of the radionuclide, expressing your answer in net cpm and becquerels. (1 Bq = 60 dpm).

Solution.

$$\text{Efficiency} \quad = \frac{\text{net cpm of radionuclide}}{\text{dpm of radionuclide}},$$

$\qquad$ 1 Ci $\quad = (3.7 * 10^{10}) * 60$ dpm,

$\qquad\qquad\quad = 2.22 * 10^{12}$ dpm.

When counted with 70% efficiency,

$\qquad$ 1 Ci $\quad = (2.22 * 10^{12}) * 0.70$ net cpm,

$\qquad\qquad\quad = 1.554 * 10^{12}$ net cpm,

$\qquad\qquad\quad = 2.59 * 10^{10}$ Bq.

Example 8-6. A radioactive compound, counted in a liquid scintillation spectrometer, registered 750 net cpm. A quantity of an appropriate standard of known radioactivity (2000 dpm) was added to the sample and together the sample + standard registered 2175 net cpm. Calculate the radioactivity of the sample as dpm.

Solution.

From equation (10),

$$\text{Efficiency,} \quad E \quad = \quad \frac{cpm_1 - cpm_0}{dpm_{standard}} \quad = \quad \frac{\text{additional counts}}{dpm_{standard}},$$

$$E \quad = \quad \frac{2175 - 750}{2000} \quad = \quad 0.7125,$$

$$\text{and} \quad dpm \quad = \quad 750/0.7125 \quad = \quad 1053 \text{ dpm.}$$

Example 8-7. Six commercially available quenched standards, each containing $1.45 * 10^5$ dpm (0.0516 µCi) of tritium; $3.60 * 10^4$ dpm (0.0162 µ Ci) of carbon-14; and varying levels of a quenching agent, were counted in a liquid scintillation spectrometer equipped with an external standard. Each quenched standard was counted for 10 minutes at an appropriate instrument setting with the following results:

ESR	Sample	Time (min)	Channel A (Lower)	Channel B (Upper)	EHA or ECA	EHB(*100) or ECB
TRITIUM	**STANDARDS**					
0.9226	A	10	540 670	2695	0.472	0.24
0.7880	B	10	471 284	663	0.412	0.05
0.7084	C	10	422 983	363	0.369	0.03
0.6067	D	10	344 981	269	0.301	0.023
0.4639	E	10	234 150	229	0.204	0.02
0.2522	F	10	105 141	173	0.092	0.015
CARBON -14	**STANDARDS**					
0.9286	A	10	66 981	261 799	0.19	0.73
0.7010	B	10	103 324	210 041	0.29	0.58
0.5639	C	10	145 816	151 465	0.41	0.42
0.3677	D	10	204 430	73 531	0.57	0.20
0.2797	E	10	233 267	33 128	0.65	0.09
0.2163	F	10	246 696	11 106	0.69	0.03

Calculate the radioactivity (as dpm) of a ^{14}C- and ^{3}H-labeled sample which had an ESR of 0.90 and activities of 16 000 net cpm and 21 000 net cpm in channels A and B respectively.

Solution.

Plots of efficiency against the external standard ratio for EHA, EHB, ECA and ECB, yield the four calibration curves shown below.

At ESR = 0.90 the four calibration curves show that EHA = 0.47; EHB = 0.002; ECA = 0.20 and ECB = 0.73.

Substitution of these efficiencies into equations (15) and (16) yields the answers: ^{3}H = 21 827 dpm; ^{14}C = 28 707 dpm as detailed in the calculations on the following page.

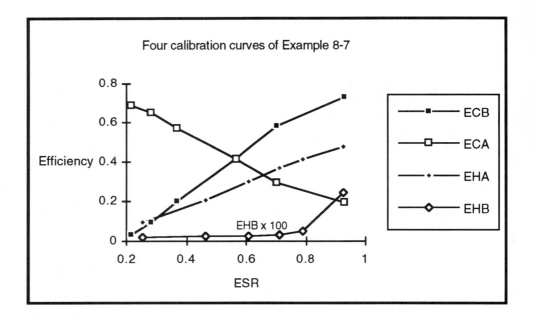

From equation (15)

$$^3\text{Hdpm} = \left[\frac{\text{cpm A} - \text{cpm B}\left(\dfrac{\text{ECA}}{\text{ECB}}\right)}{\text{EHA} - \text{EHB}\left(\dfrac{\text{ECA}}{\text{ECB}}\right)} \right] ,$$

$$= \frac{(16\,000 * 0.73) - (21\,000 * 0.20)}{(0.47 * 0.73) - (0.002 * 0.20)} , \qquad = \frac{7\,480}{0.3427}$$

$$= 21\,827 \text{ dpm.}$$

From equation (16)

$$^{14}\text{Cdpm} = \left[\frac{\text{cpm B} - \text{cpm A}\left(\dfrac{\text{EHB}}{\text{EHA}}\right)}{\text{ECB} - \text{EHB}\left(\dfrac{\text{ECA}}{\text{EHA}}\right)} \right] ,$$

$$= \frac{(21\,000 * 0.47) - (16\,000 * 0.002)}{(0.73 * 0.47) - (0.002 * 0.20)} , \qquad = \frac{9838}{0.3427}$$

$$= 28\,707 \text{ dpm.}$$

8-11. Spreadsheet solutions

	A	B	C	D	E	F	G
1	EXAMPLE 8-1				EXAMPLE 8-5		
2		INPUT	OUTPUT			INPUT	OUTPUT
3	Radioactivity				Dpm in 1 Ci		2.22E+12
4	of lipid, dpm	12000			70% Efficiency		1.55E+12
5	Wt. of lipid	5			Convert to Bq		2.59E+10
6	Specific activity		2400				
7					EXAMPLE 8-6		
8	EXAMPLE 8-2					INPUT	OUTPUT
9		INPUT	OUTPUT		cpm2	2175	
10	Specific activity				cpm1	750	
11	(Ci/mmol)	20			dpm added	2000	
12	mmol in 250μCi		12.5		Efficiency		0.7125
13					dpm of compound		1053
14	EXAMPLE 8-3						
15		INPUT	OUTPUT				
16	14-C:				EXAMPLE 8-7		
17	Observed dpm	27000					
18	Theoretical dpm	40000				INPUT	OUTPUT
19	Efficiency		0.675		cpm(A)	16000	
20					cpm(B)	21000	
21	3-H:				EHA	0.47	
22	Observed dpm	41000			EHB	0.002	
23	Theoretical dpm	100000			ECA	0.2	
24	Efficiency		0.41		ECB	0.73	
25							
26	EXAMPLE 8-4				SOLUTION:		
27		INPUT	OUTPUT		Calculate 3-H:		
28	Ao	122000			Numerator		7480
29	t	190			Denominator		0.3427
30	t1/2	87.2			DPM 3-H		21827
31	cpm after 190 d		26943				
32	Ao	800			Calculate 14-C:		
33	t	32			Numerator		9838
34	t1/2	14.3			Denominator		0.3427
35	Amount lost		630		DPM 14-C		28707
36							

Spreadsheet formulas

	A	B	C	D	E	F	G
1	EXAMPLE 8-1				EXAMPLE 8-5		
2		INPUT	OUTPUT			INPUT	OUTPUT
3	Radioactivity				Dpm in 1 Ci		=37000000000*60
4	of lipid, dpm	12000			70% Efficiency		=G3*0.7
5	Wt. of lipid	5			Convert to Bq		=G4/60
6	Specific activity		=B4/B5				
7					EXAMPLE 8-6		
8	EXAMPLE 8-2					INPUT	OUTPUT
9		INPUT	OUTPUT		cpm2	2175	
10	Specific activity				cpm1	750	
11	(Ci/mmol)	20			dpm added	2000	
12	mmol in 250µCi		=(1/B11)*250		Efficiency		=(F9-F10)/F11
13					dpm of compound		=F10/G12
14	EXAMPLE 8-3						
15		INPUT	OUTPUT				
16	14-C:				EXAMPLE 8-7		
17	Observed dpm	27000					
18	Theoretical dpm	40000				INPUT	OUTPUT
19	Efficiency		=B17/B18		cpm(A)	16000	
20					cpm(B)	21000	
21	3-H:				EHA	0.47	
22	Observed dpm	41000			EHB	0.002	
23	Theoretical dpm	100000			ECA	0.2	
24	Efficiency		=B22/B23		ECB	0.73	
25							
26	EXAMPLE 8-4				SOLUTION:		
27		INPUT	OUTPUT		Calculate 3-H:		
28	Ao	122000			Numerator		=(F19*F24)-(F20*F23)
29	t	190			Denominator		=(F21*F24)-(F22*F23)
30	t1/2	87.2			DPM 3-H		=G28/G29
31	cpm after 190 d		=B28*2^-(B29/B30)				
32	Ao	800			Calculate 14-C:		
33	t	32			Numerator		=(F20*F21)-(F19*F22)
34	t1/2	14.3			Denominator		=(F24*F21)-(F22*F23)
35	Amount lost		=(B32-(B32*2^(-B33/B34)))		DPM 14-C		=G33/G34
36							

8-12. Review questions

8-1. What is *quenching* in liquid scintillation spectrometry?
8-2. Define the terms *alpha particle, beta particle* and *gamma ray.*
8-3. What is a *rem?*
8-4. What is a *rad?*
8-5. Explain the difference between *cpm* and *dpm.*
8-6. Explain the terms *lower discriminator* and *upper discriminator.*
8-7. What is *pulse height?*
8-8. Define the term *Compton electron.*
8-9. Explain briefly the liquid scintillation process.
8-10. What is the *disintegration constant* of a radioisotope?
8-11. Define the term *half-life* of a radioisotope.
8-12. Write an equation which expresses the rate of radioisotopic decay.
8-13. Define *specific radioactivity.*
8-14. List three chemical quenching agents.
8-15. Name three methods of quench correction.
8-16. Explain how a *beta spectrum* is obtained.
8-17. Define E_{max}.
8-18. Explain the term *counting efficiency.*
8-19. How is *counting efficiency* determined?
8-20. What is the *base unit of radioactivity?*

CIRCLE THE APPROPRIATE UPPER CASE LETTERS (A, B, C, D, or E) TO INDICATE WHICH PARTS OF QUESTIONS 8-21 TO 8-30 ARE CORRECT.

8-21. The following radioisotopes can be detected by gamma scintillation spectrometry: A. ^{32}P; B. ^{14}C; C. ^{35}S; D. ^{131}I; E. ^{3}H (A, B, C, D, E).

8-22. The following radioisotopes can be detected by Geiger counters: A. ^{32}P; B. ^{3}H; C. ^{35}S; D. ^{131}I; E. ^{14}C (A, B, C, D, E).

8-23. The following radioisotopes usually are counted in liquid scintillation spectrometers: A. ^{3}H; B. ^{14}C; C. ^{32}P; D. ^{131}I; E. ^{35}S (A, B, C, D, E).

8-24. Chemical quenching of beta emitters involves the participation of : A. scintillator molecules; B. solvent molecules; C. the beta particles; D. the scintillation vials; E. Compton electrons (A, B, C, D, E).

8-25. Chemical quenching of external standards in liquid scintillation spectrometry involves: A. Compton electrons; B. turbidity of the sample; C. beta particles of the samples; D. solvent molecules; E. scintillator molecules (A, B, C, D,E).

8-26. Quench corrections are made by A. SCR; B. internal standardization; C. ESR; D. specific radioactivity of samples; E. dpm of the standard (A, B, C, D, E).

8-27. To calculate counting efficiencies, one needs: A. the number of atoms decaying with time; B. E_{max} of the radioisotope; C. net cpm of a standard; D. specific radioactivity of the sample; E. dpm of the standard (A, B,C, D, E).

8-28. To calculate the number of rads, one needs: A. time of exposure; B. E_{avg}; C. counting efficiency; D. mass of material absorbing radiation; E. logarithm of the ratio of radioactive atoms present at zero time and time t (A, B, C, D, E).

8-29. Optical quenching involves: A. colored samples; B. turbid samples; C. Compton electrons; D. solvent molecules; E. fingerprints and other factors which decrease the transparency of scintillation vials (A, B, C, D, E).

8-30. The following are units of radioactivity: A. specific activity; B. Rem; C. Rad; D. the Curie; E. keV (A, B, C, D,E).

8-13. Review problems

8-31. Calculate the dose in rads absorbed by 5 grams of tissue exposed to 5 mCi of ^{32}P for 30 minutes. Assume that all of the radioactivity was absorbed.

8-32. Calculate the specific radioactivity (as dpm mol^{-1}) of a solution of NaH_2PO_4 which contains 50 mCi of phosphorus-32 per mmol.

8-33. If the solution in question 8-32 contained 25 mCi in a total of 1 ml of water solution, what was the total amount of ^{32}P present as dpm and how much Na_2HPO_4 was present (as mmols)?

8-34. What fraction of the total radioactivity in question 8-33 would have remained after 22 days? How long would it take for the solution to lose 80% of its initial radioactivity?

8-35. Calculate the half-life of a radioisotope which gives 7 000 net cpm at zero time and 4 000 net cpm at 145 days later.

8-36. A standard with 10 ml of ^3H contained $2 * 10^6$ dpm ml^{-1} and gave $1.22 * 10^7$ net cpm when counted on a liquid scintillation spectrometer. What was the counting efficiency?

8-37. Calculate λ of ^{32}P (in units sec^{-1}) if the half-life of ^{32}P is 14.3 days.

8-38. Calculate λ of ^3H (in units sec^{-1}) if the $t_{\frac{1}{2}}$ of ^3H is 12.26 years. From question 8-37 and this question, what do you conclude about the fraction of ^{32}P radioactive atoms decaying per second compared with the fraction of ^3H which decays per unit time?

8-39. How many mCi of ^3H would remain after 4.5 years if there were 0.80 mCi present initially? The $t_{\frac{1}{2}}$ of tritium is 12.26 years.

8-40. If a standard of ^{35}S gave 88 000 net cpm after 60 days, what was its initial radioactivity (as net cpm)? The $t_{\frac{1}{2}}$ of ^{35}S is 87.1 days.

8-14. Additional problems

8-41. You plan to carry out two parallel experiments with ^{32}P ($t_{\frac{1}{2}}$ = 14.3 days). If the radioactive precursor which will be used in the two experiments contains 372 000 net cpm ml^{-1} at the beginning of the first experiment, what will be its radioactivity (net cpm ml^{-1}) 37 days later at the beginning of the second experiment?

8-42. The half-life of ^{35}S is 87.2 days. What levels of radioactivity will be present in a sample containing 5 000 net cpm of this radionuclide after numbers of the half-lives corresponding to 0.01, 0.1, 0.5, 0.75, 1.0 and 7.0 have elapsed?

8-43. What are the efficiencies of counting the following standards:

1.	200 000 dpm of	^3H	yielding	40 000	net	cpm
2.	40 000 dpm of	^{14}C	yielding	20 000	net	cpm
3.	40 000 dpm of	^{14}C	yielding	32 000	net	cpm
4.	58 000 dpm of	^3H	yielding	39 000	net	cpm
5.	110 000 dpm of	^{14}C	yielding	18 000	net	cpm
6.	58 000 dpm of	^3H	yielding	14 000	net	cpm
7.	110 000 dpm of	^{14}C	yielding	56 000	net	cpm

8-44. Rather fortuitously, an experiment in which uniformly labeled ^3H-glucose and ^{14}C-glucose were precursors gave three samples, each of which read 10 000 net cpm in channel A, the lower channel of a liquid scintillation spectrometer, and 33 000 net cpm in channel B, the upper channel. Because of quenching, however, the efficiencies of counting the three samples in the two counting channels varied as follows:

	Sample 1	Sample 2	Sample 3
ELA	0.1	0.67	0.25
ELB	0.0001	0.0005	0.0005
EHA	0.1	0.2	0.18
EHB	0.41	0.82	0.70

In these data
ELA	=	Efficiency of LOWER ENERGY radioisotope in channel A
ELB	=	Efficiency of LOWER ENERGY radioisotope in channel B
EHA	=	Efficiency of HIGHER ENERGY radioisotope in channel A
EHB	=	Efficiency of HIGHER ENERGY radioisotope in channel B

Calculate the ^3H and ^{14}C radioactivities of the three samples as dpm.

8-45.   You would like to dispose of a phospholipid sample containing 150 000 cpm of ^{32}P. The commercial company that disposes of such wastes for your laboratory says that it will not dispose of radioactive samples with greater than 50 000 cpm. How long would you have to wait before the company will accept the waste phospholipid? The half-life of ^{32}P is 14.3 days.

8-46.   Phosphorus-32 has a half-life of 14.3 days. What was the original radioactivity of a sample containing 500 net cpm assuming that two half-lives have elapsed?

8-47.   Phosphorus-32 has a half-life of 14.3 days. What would be the radioactivity of a sample containing 2000 net cpm after two half-lives have elapsed?

Suggested Reading

Bransome, E. D. 1975. *The Design of Double Label Experiments*. Methods in Enzymology, Academic Press, New York, 40: 293-302.

Causton, D. R. 1983. *A Biologist's Basic Mathematics*, Arnold, Baltimore, Maryland.

Causton, D. R. 1987. *A Biologist's Advanced Mathematics*, Allen & Unwin, Boston.

Christian, G.D. 1980. *Analytical Chemistry*, 3d ed., Wiley, New York.

Cornish-Bowden, A. 1981. *Basic Mathematics for Biochemists*, Chapman and Hall, London.

Dickerson, R. E., H. B. Gray, M. Y. Darensbourg and D. J. Darensbourg 1984. *Chemical Principles, 4th ed.*, Benjamin/Cummings, Reading, Massachusetts.

Eisenberg, D. and D. M. Crothers 1979. *Physical Chemistry With Applications to the Life Sciences*. Benjamin/Cummings, Massachusetts.

Fox, B. W. 1976. *Techniques of Sample Preparation for Liquid Scintillation Counting*. Elsevier, New York.

Montgomery, R., and C. A. Swenson 1976. *Quantitative Problems in the Biochemical Sciences, 2d ed.*, Freeman, San Francisco.

Morris, J. G. 1974. *A Biologist's Physical Chemistry, 2d ed.*, Arnold, Baltimore, Maryland..

Pecsok, R. L., L. D. Shields, T. Cairns and I. G. McWilliam 1976. *Modern Methods of Chemical Analysis, 2d ed.*, Wiley, New York.

Rapkin, E. October 1966. *Determination of Counting Efficiency for Liquid Scintillation Spectrometry*. Laboratory Scintillator, Technical Bull., Picker Nuclear, White Plains, New York, 11: No. 1L.

Rapkin, E., November 1966. *Liquid Scintillation Counting Instrumentation Logic,*Technical Bulletin, Picker Nuclear, White Plains, New York, 11: No. 1L.

Yaqub, A. and H. G. Moore 1981. *Elementary Linear Algebra With Applications*. Addison-Wesley, Massachusetts.

Chapter 9

Oxidation-Reduction

9-1. Introduction

Most nonphotosynthetic organisms carry out biosynthetic, transport, and other energy-requiring processes with the energy released during the oxidation of reduced biomolecules. In metabolic terms, energy released from the oxidation of substrates during catabolism is coupled to the energy demands of anabolism and transport. Therefore, to understand how metabolic energy is produced and used, it is important to understand the nature of biological oxidation-reduction systems.

We will present some principles of oxidation-reduction systems in a three-step sequence. *First*, we will describe some basic facts of oxidation-reduction processes. *Then* we will derive the Nernst equation which is as important to oxidation-reduction reactions as the Henderson-Hasselbalch equation is to acid-base chemistry. *Finally*, we will outline the experimental approach to redox potentiometric titrations and indicate how the midpoint potential may be calculated from the resulting experimental data.

9-2. Oxidation is the loss, reduction is the gain, of electrons

A *reducing biomolecule* is one with a strong tendency to *donate electrons* and become oxidized (lose electrons). Conversely, an *oxidizing biomolecule* has a strong tendency to *accept electrons* and to become reduced (gain electrons). Consequently, an *oxidation-reduction* reaction or *redox reaction* is one in which electrons are transferred from a donor or reducing component to an acceptor or oxidizing component.

When a reducing agent is oxidized, it is converted by reversible loss of electrons to its conjugate oxidized form. Together, the reducing component and its conjugate oxidized form constitute, and function as, a redox couple. A generalized example of such a couple is

$$A_{reduced} \quad \rightleftharpoons \quad A_{oxidized} + n\varepsilon. \tag{1}$$

When an oxidizing agent is reduced it is transformed, by reversible gain of electrons to its conjugate reduced form and together, the oxidizing component and its conjugate reduced form also constitute, and function as, a redox couple. Equation (2) is a generalized example

$$B_{oxidized} + n\varepsilon \quad \rightleftharpoons \quad B_{reduced}. \tag{2}$$

In equations (1) and (2), n is the number of electrons transferred and ε denotes an electron. Since every reducing or oxidizing agent functions as a conjugate couple, a complete oxidation-reduction or redox reaction requires the participation of two redox couples, each of which differs in its affinities for electrons. For example, if in equations (1) and (2), $A_{reduced}$ and $A_{oxidized}$ together with $B_{oxidized}$ and $B_{reduced}$ were, respectively, the donor and acceptor couples of an oxidation-reduction reaction, the complete reaction would be given by the sum of the two equations:

$$A_{reduced} + B_{oxidized} \quad \rightleftharpoons \quad A_{oxidized} + B_{reduced} \tag{3}$$

Because each redox reaction consists of two redox couples, each couple is referred to as a *half-cell* or *half-reaction*. The tendency of a reducing compound to donate electrons, or of an oxidizing compound to accept electrons, is given by the *standard oxidation-reduction potential. This potential is measured in volts and is defined as the relative tendency of a redox couple to give electrons to, or accept them from, a suitable reference couple.* By international agreement, the standard electron affinities of redox compounds (their standard redox potentials) are obtained by reference to the standard hydrogen half-cell. This half-cell consists of hydrogen gas at standard atmospheric pressure (1.01325×10^5 pascal) in equilibrium with H^+ (H_3O^+) ions in solution at 1.0 M concentration (pH = 0). The standard redox potential for the standard hydrogen half-cell is given the value 0.0 volt at all temperatures. Thus, with the zero point established, a hydrogen scale of potentials with values above or below the zero point could be devised. Again, by international agreement, a redox couple which undergoes reduction and gains electrons from the standard hydrogen half-cell is given a positive redox potential. Conversely, a redox couple which undergoes oxidation and donates electrons to the standard hydrogen electrode will have a negative potential.

The relation between a reducing biomolecule and its conjugate oxidant expressed by equation (1) is analogous to that between an acid and its conjugate base expressed by the Lowry-Brønsted definition of acids and bases (chapter 2, section 2-5). Therefore, since the Henderson-Hasselbalch equation relates pH to pK, one would expect that a similar equation should play an analogous role in oxidation-reduction. This is indeed true and the equation for oxidation-reduction is the Nernst equation. Its derivation follows.

9-3. The Nernst equation is derived from expressions for ΔG

We begin by making two observations about ΔG, the *free energy change* of a reaction. The *first* is that ΔG is the maximum amount of useful work which can be obtained from a reaction. The *second* observation is that ΔG is related to the equilibrium constant, to the reaction quotient, and to the oxidation-reduction potential. These observations are expressed algebraically in equations (4) to (7):

$$\Delta G = -W_{max} \tag{4}$$
$$\Delta G^\circ = -RT \ln K \tag{5}$$
$$\Delta G = \Delta G^\circ + RT \ln Q \tag{6}$$
$$\Delta G^\circ = -n\mathcal{F}E^\circ \tag{7}$$

In these four equations, ΔG is the change in free energy; W_{max} = the maximum amount of work; ΔG° = the standard free energy change; R = the gas constant; T = the absolute temperature in degrees kelvin; K = [products]/[reactants] at equilibrium; Q, the reaction quotient = [products]/ [reactants] when the reaction is not at equilibrium; n = the number of electrons transferred in the reaction; $\mathcal{F}$, the faraday constant, i.e., the charge of one mole of electrons, = 96, 484.6 coulombs mol^{-1} and E° = the standard electrode potential.

If, in equation (8) below, the redox couple A_{red}/A_{ox} is the standard hydrogen half-cell, and if the tendency of a biological redox couple B_{ox}/B_{red} (at any concentration) to accept electrons is being measured by comparison with the standard hydrogen electrode, then according to equation (6), ΔG will depend on the standard free energy, ΔG°, and on the reaction quotient. That is, for the generalized equation

$$A_{red} + B_{ox} = A_{ox} + B_{red} \tag{8}$$

$$Q = \ln \frac{[A_{ox}][B_{red}]}{[A_{red}][B_{ox}]} \tag{9}$$

So that
$$\Delta G = \Delta G^\circ + RT \ln Q \tag{10}$$

Hence
$$-n\mathcal{F} E = -n\mathcal{F}E^\circ + RT \ln \frac{[A_{ox}][B_{red}]}{[A_{red}][B_{ox}]} \tag{11}$$

Dividing both sides by $n\mathcal{F}$ and changing signs (inverting logarithmic terms)

$$E = E^\circ + \frac{RT}{n\mathcal{F}} \ln \frac{[B_{ox}]}{[B_{red}]} - RT \ln \frac{[A_{ox}]}{[A_{red}]} \tag{12}$$

Since the reference electrode in equation (8) is the standard hydrogen half-cell, then both A_{ox} and A_{red} are at 1.0 M concentrations and

$$- RT \ln \frac{[A_{ox}]}{[A_{red}]} = 0 \tag{13}$$

so, equation (12) becomes $\quad E_h = E^\circ + \frac{RT}{n\mathcal{F}} \ln \frac{[B_{ox}]}{[B_{red}]} \tag{14}$

Equation (14), is the Nernst equation. It was developed at age 25 by the German physical chemist Walter Nernst (1864-1941). In the Nernst equation, E_h is the potential of the biological redox couple B_{ox}/B_{red} on the hydrogen scale of potentials (i.e. the reference half-cell is the hydrogen electrode) and $E°$ is the standard redox potential of B_{ox}/B_{red}. Since concentrations (not activities) are used in biochemical reactions, and since these reactions usually do not function at pH 0, $E°$ is replaced by E_m in biochemical versions of the Nernst equation. The subscript *m denotes midpoint potential*. The significance of the term *midpoint* is that if E_h is plotted versus the fraction of reduced component, i.e. Ox/red, on linear paper, then a curve of shape similar to that of the Henderson-Hasselbalch curve is obtained (Figure 9-1). At the midpoint of this curve, $E_h = E_m$ since the concentrations of oxidized and reduced forms are equal. That is, at the midpoint of Nernst curves, ln [oxidized]/[reduced] = ln 1=0 (Figure 9-2). In the biomedical sciences, the Nernst equation is written in two different, but related, forms. *The first form of the Nernst equation is*

$$E_h = E_m + 2.303 \frac{RT}{n\mathcal{F}} \log \frac{[\text{oxidized}]}{[\text{reduced}]}. \tag{15}$$

Regarding the second biomedical form of the Nernst equation, assume that instead of the standard hydrogen electrode and a biological redox couple, we had two vessels containing solutions of different concentrations $[c_1]$ and $[c_2]$ of the same salt. Then let us make the three additional assumptions shown graphically in Figure 9-3:

- that the two separated solutions are kept in electrical contact by a KCl bridge;
- that an identical metal electrode is placed in each vessel; and
- that changes in ionic concentration occur in the two solutions as electrons flow from c_1 to c_2 through the electrodes.

The free energy change of this reaction will be

$$\Delta G = RT \ln \frac{[c_1]}{[c_2]} \tag{16}$$

Therefore equation (15) may be transformed to (17), *the second form of the Nernst equation*:

$$E_h = 2.303 \frac{RT}{n\mathcal{F}} \log \frac{[c_1]}{[c_2]} \tag{17}$$

c_1 and c_2 = concentrations of a single ion between two compartments, 1 and 2.

Equation (17) is used to compute electrochemical potential given the concentrations of an ion across a cell membrane or, more generally, equation (17) may be used to calculate the concentrations of a single ion inside and outside of cells or cell compartments. Since equation (17) may be derived from equation (15) by assuming that E_h is to be calculated for two compartments which differ only in

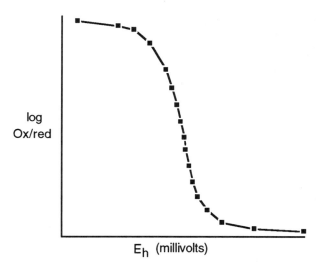

Figure 9-1.   A generalized potentiometric titration curve

ionic concentration of one component, then equation (17) may be regarded as a special case of equation (15) where the electrochemical reaction is the same on both sides. According to (15), a plot of E_h against log [oxidized]/[reduced] should yield a straight line, the slope of which depends only on n (at constant temperature) since 2.303 RT/$\mathcal{F}$ then is constant. When n = 1 (i.e., when one electron is transferred), the slope of the curve is twice as steep as when n = 2 (cf. Figure 9-2).

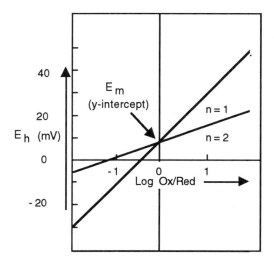

Figure 9-2 . Generalized Nernst curve for the determination of E_m

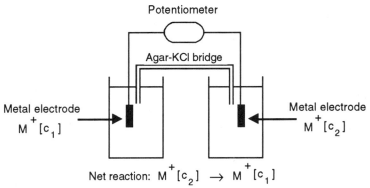

Figure 9-3. Diagram of apparatus for the measurement of redox potential

9-4. Midpoint potential is obtained from potentiometric titration

Equation (15), the version of the Nernst equation which is found in many textbooks of biochemistry, indicates that to obtain E_m graphically, E_h should be plotted against log [ox]/[red] as in Figure 9-2. Experimentally, therefore, our objective is to perform a titration to obtain several simultaneous measurements of E_h and of the ratios of concentrations of oxidant/reductant for the biological redox couple. E_h is measured with electrodes in a potentiometric apparatus, and the ratio of oxidized/reduced component is measured simultaneously by spectrophotometry or another physical technique. At the beginning of the titration, the biological redox couple is reduced to its maximum extent by stepwise addition of a reducing agent such as dithionite and the reaction (reduction) is monitored both potentiometrically and spectrophotometrically. After the biological couple has been maximally reduced, it is reoxidized by the stepwise addition (titration) of an oxidizing agent such as ferricyanide and again the reaction (this time oxidation) is monitored both potentiometrically and spectrophotometrically. The titration is continued until maximum oxidation is obtained (indicated by no additional change occurring in absorbance). Because the standard hydrogen half-cell (electrode) is inconvenient, secondary electrodes such as the calomel electrode are used instead. When a secondary electrode is used, its potential relative to the standard hydrogen electrode is added to each observed reading of E_h to obtain the true E_h value. Once E_h and the spectrophotometric measurement of proportions of oxidized to reduced forms of the biological redox couple are available, E_m can be determined from a Nernst semilog plot. The number of electrons, (n), transferred in the oxidation-reduction reaction also may be determined from Nernst semilog plots such as Figure 9-2.

9-5. E_m can clarify electron transport and drug detoxification

Biomedical scientists have used potentiometric titrimetric data, with other information, for several purposes, e.g., (1) to deduce the sequence of reactions in mitochondrial electron transport, (2) to study the light reaction of photosynthesis, (3) to decipher the central pathways of metabolism, and (4) to clarify processes by which drugs are detoxified. Such extensive use of oxidation-reduction data indicates that oxidation-reduction systems are ubiquitous in nature. This wide use of redox data suggests, moreover, that progress in understanding electron transport and the various processes coupled to electron transport, requires knowledge of the midpoint potentials of conjugate redox pairs. An interesting and much investigated case is that of *cytochrome P-450* in detoxification. This cytochrome, a hemoprotein of the cytochrome b class, is ubiquitous in nature; it is so common, in fact, that more than 150 different isoforms have been discovered among plants, animals, and microorganisms. Although P-450 is a cytochrome, it usually does not function as an electron carrier; instead, it often functions as an oxygenase, oxygenating naturally occurring substrates such as steroids, fatty acids, and retinoids. In addition however, cytochrome P-450 plays a role in the oxidation of *xenobiotic,* or man-made, foreign compounds including drugs, dyes, pesticides, carcinogens, petroleum products, anesthetics, and so on. The reviews of Guengerich, of Porter and Coon, and of Jakoby and Ziegler provide additional information for the interested reader.

We recall from section 6-7 that the oxidation of LDL may be involved in the development of atherosclerotic plaques. Biomedical scientists have obtained related data to support the hypothesis that those vitamins which can act as antioxidants (vitamins C, A, and E) may hinder, prevent, or protect against, heart disease, certain cancers, and some infectious diseases. The biochemical steps by which antioxidant vitamins confer such protective action are unknown; we may infer, however, that these vitamins work by protecting DNA, proteins, and the unsaturated fatty acids of cellular lipids from deleterious oxidative attack. Further support for a putative, protective role of the antioxidant vitamins comes from studying the human eye: the eye is exposed directly to light and oxygen and it accumulates antioxidant vitamins; when these vitamins are present at sufficiently high levels, they can prevent the growth of cataracts. These data therefore suggest that biomedical investigations of redox reactions and processes can have both fundamental and practical value.

9-6. Review questions

9-1. Define the term *standard oxidation-reduction potential.*
9-2. Define the terms *reducing* and *oxidizing biomolecule.*
9-3. Define the term *half-cell.*
9-4. Write an equation for the redox reaction of a donor and acceptor couple.
9-5. Outline the international standard for the electron affinities of redox couples.
9-6. Show how the free energy change of a reaction is related to the reaction quotient, the equilibrium constant and the oxidation-reduction potential.
9-7. What is the significance of the term *midpoint potential,* E_m?
9-8. Show how you would use potentiometric titration to determine E_m.
9-9. List some ways in which data from potentiometric titration have been used by biomedical scientists.
9-10. Show with the aid of a drawing how a generalized Nernst curve may be used to estimate midpoint potential.

9-11. Explain the statement that the Nernst equation is written in two different forms for biochemical purposes.

9-12. What do E_h, E_m and $E°$ represent?

9-13. What is the *standard hydrogen electrode*?

9-14. What is a *secondary electrode*?

9-15. Why is an understanding of oxidation-reduction important in studying energy metabolism?

9-16. Explain the principle of potentiometric titration to determine E_m.

9-17. If $T = 25°C$ and one electron is transferred, what is the value of $2.303RT/n\mathcal{F}$?

9-18. Summarize the major steps you would use to obtain the Nernst equation.

9-7. Additional problems

9-19. The potentiometric titration of a biological redox couple was carried out at pH 7.0 and 30°C. The totally oxidized redox couple (oxidized with ferricyanide) gave a reading of 100 arbitrary units. When μl quantities of buffered dithionite were added to reduce the totally oxidized solution, the following readings of oxidation-reduction and E_{obsd} were obtained with a calomel electrode and spectrophotometric apparatus:

Degree of reduction	14	29	41	56	82	95 (arbitrary units)
E_{obsd}	-122	-133	-140	-148	-168	-183 (mV)

If the standard potential of the calomel electrode at 30°C is 245 mV, calculate E_m of the redox couple and the number of electrons transferred per mole of reactant.
$R = 8.314$ J K^{-1} mol^{-1} $= 1.98717$ cal K^{-1} mol^{-1}
$\mathcal{F} = 96,487$ J Volt$^{-1} = 96,487$ coulombs mol$^{-1} = 23,062$ cal V^{-1} equiv^{-1}

9-20. The potentiometric titration of a biological redox couple was carried out at pH 7.0 and 25°C. The totally oxidized redox couple (oxidized with ferricyanide) gave a reading of 100 arbitrary units. When μl quantities of buffered dithionite were added to reduce the totally oxidized solution, the following readings of oxidation-reduction and E_{obsd} were obtained with a calomel electrode and spectrophotometric apparatus:

Degree of reduction	14	29	41	56	82	95 (arbitrary units)
E_{obsd}	348	337	330	322	305	287 (mV)

If the standard potential of the calomel electrode at 25°C is 245 mV, calculate E_m of the redox couple and the number of electrons transferred per mole of reactant.
$R = 8.314$ J K^{-1} mol^{-1} $= 1.98717$ cal K^{-1} mol^{-1}
$\mathcal{F} = 96,487$ J Volt$^{-1} = 96,487$ coulombs mol$^{-1} = 23,062$ cal V^{-1} equiv^{-1}

9-8. Spreadsheet solutions

	A	B	C	D	E	F	G	H	I	J
1	EXAMPLE 9-2									
2										
3										
4	INPUT		INPUT							OUTPUT
5	Redn	LogOx/Red	Eobsd	Eh	x^2	x*y	y^2			
6	14	0.7883704	348	593	0.6215	468	351649		SLOPE	29.7721
7	29	0.3888604	337	582	0.1512	226	338724		INTCP.	570.0068
8	41	0.1580682	330	575	0.025	90.9	330625		CORREL.	0.999867
9	56	-0.104735	322	567	0.011	-59.4	321489			
10	82	-0.658541	305	550	0.4337	-362	302500		SOLUTION:	
11	95	-1.278754	287	532	1.6352	-680	283024		Em	570
12										
13	Count	6								
14	Σ	-0.706731	1929	3399	2.8776	-317	1928011			
15	(Σx OR Σy)^2	0.4994692		1E+07						
16	Correl	499.15977	4.0946	121.92	499.23					
17										
18										
19										
20										

Plot of Question 9-2

Spreadsheet formulas

	A	B	C	D
1	EXAMPLE 9-2			
2				
3				
4	INPUT		INPUT	
5	Redn	LogOx/Red	Eobsd	Eh
6	14	=LOG10((100-A6)/A6)	348	=245+C6
7	29	=LOG10((100-A7)/A7)	337	=245+C7
8	41	=LOG10((100-A8)/A8)	330	=245+C8
9	56	=LOG10((100-A9)/A9)	322	=245+C9
10	82	=LOG10((100-A10)/A10)	305	=245+C10
11	95	=LOG10((100-A11)/A11)	287	=245+C11
12				
13	Count	=COUNT(B6:B11)		
14	Σ	=SUM(B6:B11)	=SUM(C6:C11)	=SUM(D6:D11)
15	(Σx OR Σy)^2	=B14*B14		=D14*D14
16	Correl	=B13*F14-B14*D14	=SQRT(B13*E14-B15)	=SQRT(B13*G14-D15)
17				

	E	F	G	H	I	J
1						
2						
3						
4						OUTPUT
5	x^2	x*y	y^2			
6	=B6*B6	=B6*D6	=D6*D6		SLOPE	=(B13*F14-B14*D14)/(B13*E14-B1
7	=B7*B7	=B7*D7	=D7*D7		INTCP.	=(D14*E14-B14*F14)/(B13*E14-B1
8	=B8*B8	=B8*D8	=D8*D8		CORREL.	=B16/E16
9	=B9*B9	=B9*D9	=D9*D9			
10	=B10*B10	=B10*D10	=D10*D10		SOLUTION	
11	=B11*B11	=B11*D11	=D11*D11		Em	=J7
12						
13						
14	=SUM(E6:E11)	=SUM(F6:F11)	=SUM(G6:G11)			
15						
16	=C16*D16					
17						

Suggested Reading

Brown, M.S. and J.L. Goldstein 1983. *Lipoprotein Metabolism in the Macrophage: Implications for Cholesterol Deposition in Atherosclerosis.* Annu. Rev. Biochem. 52: 223-261.

Chang, R. 1981. *Physical Chemistry With Applications to Biological Systems. 2d ed.,* Macmillan, New York.

Christian, G.D. 1980. *Analytical Chemistry, 3d ed.,* Wiley, New York.

Day, R. A. and A. L. Underwood 1980. *Quantitative Analysis, 4th ed.,* Prentice-Hall, New Jersey.

Dean, R.T., S. Gieseg and M.J. Davies 1993. Reactive Species and their Accumulation on Radical-damaged Proteins *Trends Biochem. Sci.,* 18: 437-441.

Dickerson, R. E., H. B. Gray, M. Y. Darensbourg and D. J. Darensbourg 1984. *Chemical Principles, 4th ed.,* Benjamin/Cummings, Reading, Massaachusetts.

Dutton, P. L. 1978. Redox Potentiometry: *Determination of Midpoint Potentials of Oxidation-Reduction Components of Biological Electron-Transfer Systems,* Methods in EnzymologyAcademic Press, New York, 54: 411-435.

Guengerich, F.P. 1991. Reactions and Significance of Cytochrome P-450 Enzymes. *J. Biol Chem.* 266: 10019-10022.

Jakoby, W.B. and D.M. Ziegler 1990. The Enzymes of Detoxication. *J Biol. Chem.* 265: 20715-20718.

Klotz, I. M. 1967. *Introduction to Biomolecular Energetics.* Academic Press, New York.

Krieger, M., S. Acton, J. Ashkenas, A. Pearson, M. Penman and D. Resnick 1993. Molecular Flypaper, Host Defense, and Atherosclerosis. *J. Biol. Chem.* 268: 4569-4572.

Lehninger, A. L. 1975. *Biochemistry, 2d ed.,* Worth, New York.

Lehninger, A. L., D. J. Nelson and M. M. Cox 1992. *Principles of Biochemistry,* 2d. ed., Worth, New York.

Marshall, A. G. 1978. *Biophysical Chemistry,* Wiley, New York.

Montgomery, R., and C. A. Swenson, 1976. *Quantitative Problems in the Biochemical Sciences, 2d ed.,* Freeman, San Francisco.

Morris, J. G. 1974. *A Biologist's Physical Chemistry, 2d ed.,* Arnold, Baltimore, Maryland.

Pecsok, R. L., L. D. Shields, T. Cairns and I. G. McWilliam 1976. *Modern Methods of Chemical Analysis, 2d ed.,* Wiley, New York.

Porter, T.D. and M.J. Coon 1991. Cytochrome P-450. Multiplicity of Isoforms, Substrates, and Catalytic and Regulatory Mechanisms. *J. Biol. Chem.* 266: 13469-13472.

Retsky, K.L., M.W. Freeman and B. Frei 1993. Ascorbic Acid Oxidation Product(s) Protect Human Low Density Liproprotein Against Atherogernic Modification. *J. Biol. Chem.* 268: 1304-1309.

Schottelius, B. A. and D. D. Schottelius 1978. *Textbook of Physiology, 18th ed.,* C. V. Mosby, St. Louis.

Wood, P. M. 1985. What is the Nernst Equation? *Trends Biochem. Sci.,* 10: 106-107.

PART IV.

DATA ANALYSIS

CHAPTERS 10 AND 11

Chapter 10

One- and Two-sample Analyses

10-1. Biostatistics provides an objective way to analyze biomedical data

This part of *Basic Biochemical Laboratory Procedures and Computing* deals predominantly with the statistical analysis of biomedical data. But, a few highly publicized scandals, and a report that researchers often encounter misconduct, are causing the general public to doubt the integrity of all science (e.g., see articles by Lawrence K. Altman, M.D. in *The New York Times* of November 22, 1993, and by Pamela Zurer listed in the references). Because of such reports of scientific misconduct, we suggest that *there is (or should be) both ethical, and technical dimensions, to the analysis and presentation of scientific data.* Zurer and Altman both report on the work of Dr. Judith Swazey and colleagues at the Acadia Institute in Bar Harbor, Maine. In a survey of 2 000 doctoral students and 2 000 faculty (cited by Zurer) Swazey found that 43% of faculty knew of "other faculty members making use of university resources for personal gain, 32% knew of authorship being assigned inappropriately, 22% reported their peers overlooking the use of flawed data, and 15% knew of instances where a fellow faculty member had not presented data that contradicted his or her previous research." Reports of the Swazey study also tell of scientists who used university resources to do commercial work, and of researchers who took paid vacations disguised as research trips. Not mentioned in these reports is other unethical behavior such as the inflation of achievements on applications for promotion; overstating the merit of friends' work while intentionally discrediting the work of rivals; or attacking colleagues viewed as competitors by giving poor evaluations to their graduate students. One issue which raises both statistical and ethical questions is when to discard experimental data. We present an objective criterion for data rejection in section 10-18. Earlier, in section 10-2, we list some guidelines for ethical scholarship; then we emphasize, in section 10-7, that deception is unacceptable in science; and in section 10-21 we provide some questions which examine various ethical issues. We shall not, however, attempt a broad review of ethical scientific conduct or of the best standards of personal and professional morality; neither shall we discuss the ethical principles which apply to the practice of biomedicine. For the latter, we suggest the books by Fletcher 1979, and Beauchamp and Childress, 1983.

The power of biostatistical analysis is that it provides objective criteria for interpreting experimental results. Chapters 10 and 11 review some principles, assumptions, and limitations of selected methods of data analysis, including the analyses outlined in Chapter 1, sections 12 to 14; from this review, we seek to cultivate a thoughtful approach to data analysis. We will present four sets of topics:

- Experimental design, some sampling distributions and hypothesis testing;
- Central tendency, dispersion and some one- and two-sample tests;
- Linear and nonlinear regression; and
- Analysis of variance.

Although Chapters 10 and 11 cover some basic statistical concepts they do not attempt to teach statistics. We recommend consulting suitable textbooks, or taking an introductory course, for a more complete grounding in biostatistics. Biostatistics is defined simply as the branch of statistics in which statistical methods are applied to data from biological and medical investigations. Three characteristics contribute to the need for a branch of statistics applied to the biomedical sciences:

- Biological values such as those for enzyme activity, are more variable than measurements of physical quantities, such as the absorption maxima of solutions of purines and pyrimidines. In biological material therefore, one finds ranges of "normal" values instead of single "true" values.
- Variability in biomedical research may arise from (a) limitations of methods used (b) inadequate skill, care or training of laboratory personnel, and (c) deficiencies in the maintenance, calibration, or precision of instruments.
- Experimental bias may, at times, be introduced inadvertently into a research plan because of the complexity of biological systems.

Some errors, notably the inaccuracies caused by instrumental, methodological or human factors can be eliminated, or controlled, by proper training and by preparing a good experimental plan; these *systematic errors* tend to yield inaccuracies by a constant amount. The detection and elimination of systematic error is, therefore, an experimental task. *Random errors*, however, usually cannot be controlled easily by an investigator. Biostatistics provides tools for minimizing the effects of uncontrollable errors and for helping to obtain unbiased answers to the two general types of hypotheses which are tested in biomedical research investigations: research hypotheses and statistical hypotheses.

Research hypotheses are the questions upon which one bases a clinical or laboratory investigation. *Statistical hypotheses* are the tentative, probability-based assumptions about the experimental results which one tests by appropriate statistical methods. In a properly designed experiment, the results should be consistent with *one research hypothesis* and should exclude alternative research hypotheses. Moreover, research hypotheses dictate how experimental results are collected and expressed. Therefore, the research hypotheses, laboratory methods, statistical hypotheses, and methods of statistical analysis, should all be defined during the conception or planning of an investigation. If statistical analysis is to provide answers with a high probability of being correct, the experiments first have to be designed properly. It is axiomatic that well-designed experiments save time, effort, and material. Furthermore, colleagues who devote time to repeating a statistically deficient experiment may "fail" (a) to reproduce the reported results and (b) to derive the reported conclusions. As a result, they may question whether the difficulty is attributable to intentional misconduct or to statistical naïveté.

10-2. Good experimental design depends on following simple rules

The principles of statistics provide guidelines for proper experimental design. An important principle is that biomedical investigators should plan controlled experiments. For example, an investigator may try to learn how a change in a single variable affects a sample or process. Or the goal may be to learn how a new procedure compares with a standard or well-tried procedure. In these, and in numerous other cases, *it is essential to design controlled experiments. A controlled experiment* is one in which samples which differ in only one way from a standard sample (or *control*) are run under the same conditions to evaluate the effects of the treatments on a measured quantity. Frequently, controlled experiments have one treated sample (a sample having a single change in experimental conditions) <u>and</u> one untreated (*control*) sample. But, it is possible also, to design controlled experiments in which more than one variable at a time, in more than one sample, is changed and compared with one or more controls. We assume, in the two- and multisample programs of *Basic Biochemical Laboratory Procedures and Computing*, that a controlled experiment produced the results. In addition to running controls, one may practice careful scholarship, and ensure statistical reliability, by adherence to the following:

To practice careful and ethical scholarship:

- Define the objectives of the work clearly.
- Select methods which have the appropriate sensitivity, accuracy, and precision. *Accuracy* refers to the closeness of a measured quantity to the true value. *Precision* is the degree of agreement between repeated measurements of the same item.
- Record all data in ink in a bound notebook; number all pages if your notebook does not have numbered pages; include date and time; and when necessary, have a witness countersign and date pages to affirm the authenticity of your work. Also, always date and sign computer or instrument generated data, graphs and calculations; then insert the signed printouts into a bound notebook. Follow all safety and environmental procedures explicitly. For example, do not foul the environment by flushing organic solvents or hazardous chemicals down the sink. Do not expose coworkers to toxic or infectious materials. Dispose of radioactive and microbial wastes (and broken glass) in approved containers or bags; and do not violate laws on the treatment of animals or human subjects. Always follow scholarly, legal, and ethical procedures for thorough and objective record-keeping; report data accurately; and keep your notebook in a secure place when it is not being used. In written and verbal presentations, acknowledge pertinent research and cite the work of others fairly; do not ignore, refuse to cite, spy upon, or disparage the work of rivals for personal gain. Do not present *curricula vitae,* or information required for job interviews, tenure or promotion, in a manner calculated to mislead or misinform. Respect all intellectual property rights. Do not abuse peer review to strangle competitors professionally. And do not fabricate, falsify, steal or plagiarize. Such behavior is inconsistent with the search for truth. Intentional misconduct does not occur in a vacuum. Senior personnel are responsible for sustaining a decorous, ethical climate. They should not condone, foster, engage in, wink at, or exploit improper behavior for ideology, for revenge, for political gain, or for prestige.

To ensure statistical reliability:

- Plan to use sufficient replication to obtain reliable data (section 10-13).
- Plan to use (and report) descriptive measures computed from the data, (e.g., measures of central tendency and dispersion, see section 10-4); such descriptive measures permit colleagues to evaluate the reliability of the results. Methods of organizing and summarizing data are a major segment of *descriptive statistics*.
- Plan experiments so that the expected results should be consistent with one research (working) hypothesis and should exclude alternative hypotheses.
- Plan to evaluate each treatment and interaction separately in experiments with more than one treatment. For example, in studying how increasing temperature and reaction time might affect an enzyme assay, evaluate (a) each factor individually and (b) the combined effect (the *interaction*) of higher temperature and increased time.
- Plan to take samples randomly from a population of interest to avoid bias, e.g., plan to use randomized, *double-blind trials* (where neither the subject nor the scientist knows who gets what drug or treatment) to evaluate drugs. A *population* is the entire set of objects in a clearly defined group. A *sample* is a small part of a population. The technique of making valid inferences about selected populations from small samples is a major segment of statistics known as *inferential statistics*.

10-3. Discrete and continuous data are evaluated differently

Discrete or discontinuous variables are those which contain intervals between any possible values they may assume. Examples are the number of defined products in the polymerase chain reaction; the number of erythrocytes in a sample of blood; the number of polyps on the colons of patients; or the number of nematodes infecting a plant. *Continuous variables*, conversely, are those which may have any conceivable value within a given range. Examples are the weights of laboratory animals (in kilograms) or the concentrations of protein in cellular homogenates (μg ml^{-1}).

For our purposes, the *probability* of an experimental result taking place is the relative frequency with which that result is estimated to occur in a defined population. Many biostatistical methods depend on estimating the probability of an experimentally obtained result in samples from a defined population (section 10-2). One makes inferences about the defined population from probability distributions; and, since there are different types of variables, there are also different types of probability distributions of these differing variables. Thus, the *probability distribution of a discrete variable* is a formula, table or curve which describes all possible values of the discrete variable and its corresponding probabilities. These probability values are all positive and less than 1; but their sum is equal to 1. The *probability distribution of a continuous variable* is a curve, or table, in which the total area under its curve and the x-axis = 1 square unit. In the probability distribution of a continuous variable, the area enclosed by the curve and perpendiculars to the x-axis provides the probability that an experimental, continuous variable X lies between the perpendiculars. It is generally agreed that, when a calculated statistic such as the *t*-statistic has an associated probability of 5% or less, it is unlikely that the experimental result is due to chance.

Not surprisingly, then, investigators use the probability distributions of discrete variables for data containing discrete variables and the probability distributions of continuous variables for data with continuous variables. We will discuss four continuous and two discrete probability distributions in this chapter.

The four continuous distributions are:
- the *normal*, the *t-* , the *F-* and the *chi-square* distributions;

The two discrete distributions are:
- the *Poisson* and the *binomial* distributions.

10-4. The normal curve shows the scatter of values around the mean

Scientists often group measurements into classes of varying size, e.g., the classes 1-5 or 6-10 as in histograms and tables. The distribution of the total number of measurements among the various classes is known as a *frequency distribution*. An important theorem, *The Central Limit Theorem,* states that the frequency distribution of means from any population, whether normal or nonnormal, will approximate the normal distribution. Such approximation improves as n, the sample size, increases. Therefore, *small samples, defined as those with n < 30,* frequently do not provide satisfactory approximations. Still, means from samples of n = 10, or even means from samples of n = 4 or 5, may approximate normality depending on how closely their original populations approximate the normal distribution. Since most quantitative biomedical data are expressed as means of replicate experiments, one may expect that when the size of n is adequate, the distribution of these sample means approximates normality. Unfortunately, there is no general procedure for estimating when n is sufficiently large for sample means to approximate the normal distribution. The graph of a normal distribution is a bell-shaped or Gaussian curve known as the *normal curve*:

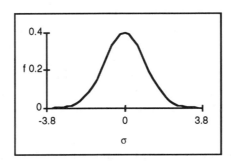

Figure 10-1.   The Normal distribution

Many curves from biomedical data are bell-shaped; but, not all bell-shaped curves are normal. Mathematically, the normal distribution is defined by the exponential equation:

$$f = \frac{1}{\sigma\sqrt{(2\pi)}} e^{-(x-\mu)^2/2\sigma^2}$$

where f is the frequency of the variable x; μ = the arithmetic mean (section 10-7); σ^2 = the variance = $\Sigma (x-\mu)^2/n$ (section 10-7) and σ = the standard deviation (section 10-7). The normal distribution, then, is determined completely by the two parameters μ, the mean, and σ, the standard deviation (section 10-7), both of which are measures of central tendency and dispersion. *Central tendency* refers to the clustering of values around the middle or average value; conversely, *dispersion* refers to the degree of variability or scatter of the data (i.e., to the lack of central tendency). Because both μ and σ can have different values, the normal curve is really a *family* of many curves, all of the members of the family being *symmetrical about* μ. Statisticians have established the wide applicability of the normal distribution and have used it to develop several statistical procedures which can be used throughout the physical, chemical, biomedical, and social sciences assuming that the data are drawn from a normal or approximately normal population.

Figure 10-2 shows that most of the values in the normal distribution lie within ± one standard deviation (± σ). More precisely, if we dropped perpendiculars from the normal curve to ±σ on the x-axis, approximately 68.26% of the total area will be included. The corresponding areas for ± 1.96 σ and ± 3.31 σ on the x-axis are approximately 95% and 99.95%, respectively. Each of the quantities 95% and 99.95% is known as a *confidence level or confidence coefficient.* The values

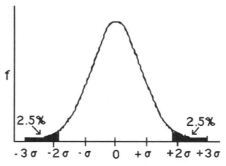

Fig.10-2. Normal curve with shaded areas (2.5% each) showing 5% of total area

-1.96 and +1.96 are known as *confidence limits* (lower, abbreviated L_1 and upper, L_2 respectively). By analogy, an estimate that a population mean or other parameter lies between two specified confidence limits is known as a *confidence interval* (abbreviated CI) or an *interval estimate*. Thus, the unshaded area of Figure 10-2 shows that the confidence interval corresponding to a probability of 0.95 is ± 1.96σ. This means that, 95% of the time, the means of populations are likely to lie within this interval (the unshaded area). The interval enclosed by the unshaded region is known as the *acceptance region*: this is the range within which the null hypothesis is true (section 10-5). Alternatively, Figure 10-2 indicates that only 5% of the time (or one measurement out of twenty) is it likely that the means will lie in the shaded area outside of the range ± 1.96σ (i.e., that the measurements will deviate more than approximately 2 standard deviations). The shaded area, in contrast to the unshaded area, is known as the *rejection region*. The null hypothesis is rejected in this region and test statistics which fall within the rejection area are said to be *significant*. A test statistic which separates the acceptance and rejection regions is known as the *critical value*. The critical value of a statistic provides an objective criterion for accepting or rejecting the null hypothesis: if the critical value is exceeded by a test statistic there is a high probability that there is a real difference between treated and control samples.

When $\mu = 0$ and $\sigma = 1$, the equation for the normal distribution reduces to

$$ f \; = \; \frac{1}{\sqrt{(2\pi)}} e^{-\frac{1}{2}(x)^2}, $$

the equation of the *standard normal distribution*. This gives the main curve in the family of normal curves. For any value of a variable X (the standard deviation) in this distribution we can define a quantity z known as the *standard normal deviate*:

$$ z \; = \; \frac{X-\mu}{\sigma}. $$

Thus z is the location of the variable X expressed as the number of standard deviations from the population mean. Some important values of z are those corresponding to probabilities of 50%, 5%, 1%, and 0.1%, where $z = 0.67$, 1.96, 2.58, and 3.31 respectively. Values of z may be used (a) to estimate the number of replica-tions needed to obtain results with desired levels of reliability (section 10-13) and (b) to evaluate the quality of data on the premise that points $\geq 2.58 \, \sigma$ from the mean probably are of doubtful validity (cf. section 10-18).

To summarize then, the normal curve (1) is determined by σ and μ, (2) is a family of curves, (3) is symmetrical about the mean, (4) has a total area of 1 square unit, (5) encloses 99.95 % of the total area at $\pm 3.31 \, \sigma$, and (6) suggests that data points $\geq 1.96 \, \sigma$ or $2.58 \, \sigma$ from the mean may be candidates for rejection because they fall within the 5% and 1% rejection regions of the standard normal curve.

10-5. Statistical testing requires the use of null and alternate hypotheses

Statistical analysis does not provide absolute proof that a research hypothesis is correct; it only shows whether experimental results support the research hypothesis. Statistical testing usually involves proposing two hypotheses: a *null hypothesis* (designated H_0) and an *alternate hypothesis* (designated H_A). Null hypotheses are stated in negative terms. Alternate hypotheses are inverse statements of null hypotheses; so, if the goal were to learn whether a new procedure is as good as a standard one then

the *null hypothesis* (H_0) for the experimental results may be stated as "there is no difference between the new and the standard procedures"; and

the *alternate hypothesis* (H_A) would state that "there is a difference between the new and the standard procedures."

Similarly, if one sought to learn whether high levels of lead in the blood of children had a negative effect on scholastic performance,

the *null hypothesis* might state that "there is no difference between children with an elevated level of lead and those within the normal range;" and

the *alternate hypothesis* would state that "there is a difference between children with an elevated level of lead and those within the normal range." It is good practice to specify explicitly, during the conception of a research project, the null hypothesis, the alternate hypothesis and the level of significance to be used.

Evaluation of the degree to which the experimental results support either the null hypothesis or the alternate hypothesis is accomplished as follows.

If the *null hypothesis is not true*, then the *research hypothesis is supported*. If the *alternate hypothesis is true* the *research hypothesis* also *is supported*.

If the null *hypothesis is true*, then the *research hypothesis* is *not supported*. Similarly if the *alternate hypothesis is not true*, then the *research hypothesis* also is *not supported*.

10-6. We may infer statistical significance from hypothesis testing

To make objective decisions on whether the null or alternate hypotheses are supported, we turn to *hypothesis testing*: we test whether a result is *significant* at selected levels of probability i.e. whether the result suggests that two or more populations are, in some way, statistically different or the same (cf. rejection or acceptance regions, section 10-4). Hypothesis testing to learn whether a test statistic is either significantly greater or smaller than a control value is known as a *two-tailed* or *two-sided* test since differences are tested in two directions (plus and minus). Similarly, a test which attempts only to learn whether a statistic is significantly greater than a control value is a *one-tailed* or *one- sided* test of significance since differences are tested in one direction only (plus). We show one-tailed and two-tailed tests by shading the appropriate side or sides on curves of probability distributions. In a one-tailed test of a result greater than a control (an *upper-tailed* test, A, Figure 10-3), we shade a segment indicating the appropriate probability, on the $+\sigma$ side of the curve. In a one-tailed *(lower-tailed)* test of a result less than a control, we shade the $-\sigma$ side. In a two-sided test, one shades segments of both the + and - sides of the curve. In shorthand notation we write:

Upper-tailed test.	H_0: $X \leq X_0$	versus	H_A:	$X > X_0$
Lower-tailed test.	H_0: $X \geq X_0$	versus	H_A:	$X < X_0$
Two-tailed test.	H_0: $X = X_0$	versus	H_A:	$X \neq X_0$

We show these possibilities in Figure 10-3 (A, B and C respectively).

Before microcomputers became widely available, biomedical investigators obtained levels of probability by interpolation from tables of probabilities because the computations of such probabilities can be both complex and tedious. Now, however,

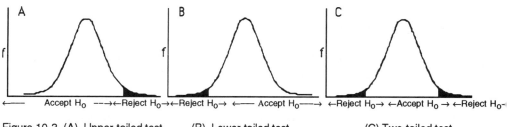

Figure 10-3. (A) Upper-tailed test.   (B) Lower-tailed test.   (C) Two-tailed test.

the computational power of the microcomputer has made the use of probability tables less necessary than before, and an investigator can *calculate* probabilities of experimental results (as in *Basic Biochemical Laboratory Procedures and Computing*) with such values as the *t*-statistic, the chi-square statistic, and *F*, the variance ratio.

We recall from sections 10-1 and 10-5 that it is wise to choose levels of significance for hypothesis testing during the conception of a research project. Unfortunately, there are no fixed rules for making such choices. The eventual choice of a level of significance should depend on such factors as whether it is necessary to make recommendations which have definitely the smallest margin of error (e.g., recommendations involving the safety of drugs used for treating patients) or whether the choices are less exacting. During the past sixty years, statisticians and biomedical scientists have used five percent (0.05) and one percent (0.01) levels of probability as appropriate levels for rejecting the null hypothesis. That is, if there is only a five percent or one percent chance that the null hypothesis is correct, then the investigator can feel safe in rejecting the possibility that the null hypothesis is true. The research hypothesis then is accepted. A concise notation has been developed to express such decisions. For example, if, as in section 10-4, the *t*-statistic were -7.78 for 24 degrees of freedom, then the probability (two-tailed) of *t* in this example is 5.9E-08 and this information is expressed concisely in the form $P(t = -7.78) = 5.98\text{E-}08$; the probability (P) of $t = -7.78$ is 5.98E-08.

10-7. Central tendency and dispersion are indices of variability

It is important, in the communication of research results, to provide information on the variability (or the opposite), the central tendency, of the results. A *measure of central tendency* (section 10-4) is a value around which a group of results tends to cluster. The *arithmetic mean* is a measure of central tendency and it describes the value around which the majority of individual measurements are found. The *arithmetic mean* of a sample, symbolized as $\overline{X}$ and read as "X bar," is the sum of all values in the sample divided by n, the number of values; that is, where "ΣX_i represents the sum of X's over all values of i" the mean is given by

$$\overline{X} = \frac{\Sigma X_i}{n}.$$

The sample *sum of squares, SS*, is an important measure of random variation. The sum of squares is defined as the sum of all deviations from the arithmetic mean squared, i.e., $\sum(X_i - \overline{X})^2$. The *geometric mean* and the *harmonic mean*, two other measures of central tendency, are not used extensively in the biomedical sciences; and, in *Basic Biochemical Laboratory Procedures and Computing*, when we speak of "the mean" we are referring to the arithmetic mean.

The *median* is another measure of central tendency. It is the value above, or below which, half of the measurements are found. For an odd number of measurements, the median is the middle value of the measurements, listed in ascending or descending order of magnitude. The median of an even number of measurements is the average of the two middle values. Generally, but not exclusively, the mean is a better measure of central tendency than the median.

In addition to reporting a measure of central tendency, it is desirable also to mention the extent to which data vary or are dispersed. Four measures of dispersion are the *range, standard deviation, coefficient of variation* and *standard error*.

The *range* of a group of measurements is the difference between the smallest and largest values in that group. The range is viewed as a poor estimate of dispersion since it is based only on two values and thus could be distorted if one of the two values is unreliable. It is therefore suggested that *the range* should be *accompanied by another index of dispersion such as the standard deviation* of the sample. The sample *standard deviation* (abbreviated SD) and symbolized s, is the square root of the sample variance. The *sample variance, s^2*, is the sum of squares divided by the number of independent observations. The number of independent observations is the number of observations, n, minus one. This value, n-1, is called the *degrees of freedom* and is abbreviated *DF*. The sample variance and sample standard deviation may be computed as

$$\text{Sample sum of squares, SS} = \sum(X_i - \overline{X})^2,$$

$$\text{Sample Variance, } s^2 = \frac{\sum X_i^2 - \dfrac{(X_i)^2}{n}}{n-1},$$

$$\text{Sample SS} = \text{Sample variance} * DF$$

$$\text{Sample standard deviation, } s = \sqrt{\frac{\sum X_i^2 - \dfrac{(X_i)^2}{n}}{n-1}}.$$

The *coefficient of variation*, designated by C (sometimes by CV or V), is the standard deviation of a group of values divided by the mean of the values x 100; that is

$$C = \frac{\text{the standard deviation}}{\text{the mean}} * 100$$

The coefficient of variation therefore expresses variability or dispersion relative to the mean as a percentage. Since the standard deviation and the mean have the same units, the coefficient of variation is a dimensionless value (see sections 1-1 and 1-2). Low values of s and C are desirable since they indicate high degrees of reliability.

The *standard error of the mean* is the variance divided by n, the number of values; that is, the standard error, abbreviated SE, is SD divided by $\sqrt{n}$ or,

$$SE = \frac{s^2}{n} = \frac{s}{\sqrt{n}}$$

Since the variance has square units, the standard error will have the same units as the original measurements, the standard deviation and the mean. We mentioned above that when the range is reported, it should be accompanied by another index of dispersion "such as the standard deviation." We may expand this quotation to state "such as the standard deviation or *the standard error.*" At this point, some brief comments about use of the standard deviation and the standard error are appropriate. Investigators who wish to indicate the *variability of the values or results which they obtained in an experiment* may report the standard deviation. When one reports the standard deviation, it is appropriate also to include the size of the sample (n), the mean, and the range. When, alternatively, one wishes to describe the *precision of calculated values such as means, coefficients, parameters, or estimates* one should report the standard error. As with SD, it is good practice to include the sample size, the mean and the range with SE. But it is not ethical practice to indulge in deceptive use of SE: for example, one should not, intentionally, use SE instead of SD to report the variability of research data simply because SE is a smaller number than SD. Illicit use of SE may make data "look better" than they are; but, just as society at large condemns deceptive advertising, deception is neither scientifically nor ethically acceptable. Deception is inexcusable in science!

The *confidence interval* for a *population mean* is obtained from

$$\text{Confidence interval} = \bar{x} \pm t \cdot \frac{s}{\sqrt{n}} ; t = \frac{(\bar{x} - \mu)}{s_{\bar{x}}}$$

The *confidence interval* for estimating the *difference between two means* is:

$$\text{Confidence interval} = (\bar{x}_1 - \bar{x}_2) \pm t \cdot \sqrt{\frac{s_p^2}{n_1} + \frac{s_p^2}{n_2}} ,$$

$$t = \frac{\bar{x}_1 - \bar{x}_2}{s_{\bar{x}_1 - \bar{x}_2}} ,$$

where
- $\bar{x}$ = the sample mean,
- t = Student's t (discussed in the next section),
- n = the size of a sample,
- s_p^2 = the pooled variance (SD squared), and
- s = the standard deviation.

10-8. Student's *t*-distribution is used for working with small samples

The *t*-distribution was developed by the head brewer at the Guinness brewing Co., W. S. Gosset (1908), who published his work under the pseudonym "Student." Gosset introduced the *t*-distribution to address the problem that, although the population parameters μ (the mean), and σ (the standard deviation), are needed for many statistical computations, both parameters are unavailable by direct calculation since the entire population is unknown. Moreover, in many biomedical experiments, n, the number of samples is *small, (i.e. less than 30)* and, the sample standard deviation, s, cannot, therefore, be considered an accurate estimate of σ, the standard deviation of the population. Gosset solved these problems by showing that, if the *t*-distribution is used, inferences about μ and σ may be made by employing the sample mean as an estimate of μ and the sample standard deviation as an estimate of σ. It should be emphasized, however, that, the *t*-distribution is valid only when the sample is taken at random from a normal distribution. Fortunately, the *t*-distribution is sufficiently "robust" to resist substantial deviations from normality. Like the normal distribution, the *t*-distribution is symmetrical about the mean and is, in effect, a family of curves. But, whereas the shape of the normal curve is always bell-shaped, the shape of the *t*-distribution varies with the number of degrees of freedom. The *t*-statistic is obtained from

$$t \quad = \quad \frac{(\bar{x} - \mu)}{s_{\bar{x}}}.$$

Thus, *t* is the difference between means divided by the standard error of this difference

$$t \quad = \quad \frac{\text{difference}}{\text{SE of difference}}.$$

The number of degrees of freedom varies with the type of statistical test. In the *t*-test the number of *degrees of freedom equals the number of measurements minus one*. For sample means, one would like to estimate whether a mean falls within a confidence interval. When n < 30, and s is not an accurate estimate of σ, confidence intervals may be calculated from

$$\text{confidence interval} \quad = \quad \text{estimator} \pm t * \text{standard error of estimator}.$$

In this equation, an *estimator* is defined as an estimate of a population parameter; for example, $\bar{X}$, the sample mean is an estimator of μ, the population mean.

Frequently a 95 percent probability (0.95 measured on a scale of 0 to 1) is considered to be sufficient to infer that a result is correct. But, when greater certainty is desired, 99% (correct 99 times out of 100), or higher, may be chosen.

When data from two samples are to be compared we could also use the *t*-distribution to test for the difference between the two means. The *t*-value for hypothesis testing is, therefore, the difference between the two means:

$$t = \frac{\overline{x}_1 - \overline{x}_2}{s_{\overline{x}_1 - \overline{x}_2}}$$

One assumes in a two-sample *t*-test that the variances of the control and treated populations, σ^2, are equal. The best estimate of σ^2 is the pooled variance of the two samples, s_p^2:

$$s_p^2 = \frac{SS_1 + SS_2}{DF_1 + DF_2}$$

$$s_{\overline{x}_1 - \overline{x}_2}^2 = \frac{s_p^2}{n_1} + \frac{s_p^2}{n_2}$$

$$t = \frac{\overline{x}_1 - \overline{x}_2}{\sqrt{\dfrac{s_p^2}{n_1} + \dfrac{s_p^2}{n_2}}}$$

We use a *goodness of fit* test to evaluate the significance of *t*: we compare the *t*-statistic with the theoretical distribution to learn whether *t* falls within the acceptance or rejection region (cf. sections 10-4 and 10-6). If *t* is greater than the critical value we reject the null hypothesis - and vice versa. Both one-tailed and two-tailed tests may be used.

10-9.   The *F*-test is used to compare the variances of two samples

The F-test compares the ratio of the variances of two samples. Named to honor its developer, the British statistician Sir Ronald Fisher, the *F*-statistic is defined as

$$F = \frac{s_1^2}{s_2^2}.$$

The *F*-distribution is calculated so that all values are greater than 1.0. Therefore, to be consistent with this definition, s_1^2 is always defined as the larger of the two variances. Since the *F*-test is performed on two groups of data, there are two variances and also two sets of degrees of freedom. The number of degrees of freedom corresponding to the larger variance is defined as DF_1 whereas the smaller is defined as DF_2. Biostatistical analysis with an *F*-test calls for four steps:

- setting up the null hypothesis,
- calculating *F*,
- evaluating the probability of *F*, and
- testing for significance.

In *Basic Biochemical Laboratory Procedures and Computing* the user sets up and tests the null hypothesis, the programs carry out steps (2) and (3). The null hypothesis for testing with *F* assumes that the two samples consist of randomly drawn variates taken from normally distributed populations having the same variance σ^2. Evaluation is based on goodness of fit procedures. As indicated in section 10-8, a *goodness of fit* test is one in which an observed frequency distribution is compared with a theoretical distribution. Here, if the calculated *F* is greater than the critical value, then one accepts the alternate hypothesis and rejects the null hypothesis; if the calculated *F* is less than the critical value, one rejects the alternate hypothesis and accepts the null hypothesis. Critical values of the *F*-distribution require consideration of the degrees of freedom for the numerator, the degrees of freedom for the denominator and the corresponding confidence level.

10-10.   The chi-square test is a nonparametric procedure

Since the *t*-test and many other biostatistical procedures seek to test hypotheses about parameters of a population, such tests are referred to as *parametric tests*. The use of tests which do not make inferences about population parameters (*nonparametric tests*) is called for under other circumstances such as

- when results are expressed as counts or frequencies,
- when the experimental data consist of ranked measurements which do not permit the computations necessary for parametric testing,
- when the form of the sampled population is not known, or
- when tests which do not refer to the parameters of populations are used.

At this point, we should make a distinction between methods classified as nonparametric and those classified as distribution-free. *Nonparametric methods* are those which test hypotheses about populations without making inferences about the parameters of the population. *Distribution-free methods* are those which do not make assumptions either about sampled populations or their parameters.

One widely used nonparametric method, *chi-square analysis*, is appropriate when results are expressed as counts. The use of chi-square analysis is particularly appropriate when data are expressed as frequencies and when an investigator wants to test for differences between observed and expected frequencies. Such data usually consist of counted items which have certain attributes or which fall within certain categories. Examples of data expressed as counts include: the numbers of children born to women taking fertility drugs; the numbers of patients who fall into each of several blood groups; or the numbers of saturated, monoenoic, dienoic or polyunsaturated fatty acids esterified to the *sn*-1, 2, or 3 positions of glycerol in triacylglycerols. The chi-square statistic is calculated from the relation:

$$\chi = \Sigma \frac{(O - E)^2}{E},$$

where χ = the chi-square statistic,
         O = the observed frequency,
and E = the expected frequency.

We use a *goodness of fit* test to learn whether the null hypothesis (that the observed and expected frequencies are the same) is true: we compare the calculated chi-square statistic with the theoretical chi-square distribution. If χ derived from the experimental data is greater than the critical value, then we reject the null hypothesis - and vice versa.

The theoretical chi-square distribution is a continuous distribution. But, chi-square values from experimental data are discrete. In spite of this difference, the approximations of experimentally derived chi-square values to the theoretical distribution are generally good except for 1 degree of freedom. To overcome this discrepancy, it is usually recommended that the *Yates correction for continuity* be used. This correction decreases the absolute value of each difference between observed and expected values by 0.5 unit; that is, the Yates correction modifies the chi-square relation to:

$$\chi = \frac{\Sigma(|O - E| - 0.5)^2}{E}.$$

Thus, in this book, we use parametric procedures in:
- the *t-test* (for linear regression, for the correlation coefficient, for the mean, for the difference between means, and for the difference between two slopes); and
- the *F-test* (for comparing slopes and variances; and for evaluating Single Factor, Randomized Block, and Two-Factor Analyses of Variance).

The program on Chi-square as a test of goodness-of-fit is based on a nonparametric procedure whereas the Direct Plot of Eisenthal and Cornish-Bowden is classified as a distribution-free method.

10-11. The Poisson distribution is used with discrete data

The Poisson distribution is used to evaluate discrete data which occur at random in space or time. The Poisson distribution is a special case of the binomial distribution; thus, when, the distribution of an experimental variable follows the Poisson distribution, only two results are possible for a given observation: success or failure. Success may, for example, be taken as an α-particle or a γ-ray making a fatal *hit* on a sensitive structure in an organism, failure as the absence of a lethal hit.

In experiments which obey the Poisson process, it is assumed (a) that successive results are independent, that one result does not influence the chance of success of another, (b) that the probability of more than one occurrence in a small segment of a space is small, (c) that the process is random, and (d) that an infinite number of occurrences is possible in the specified space or interval.

The supposition that an experimental variable belongs to the Poisson distribution should be based on the experimental plan rather than on the experimental results. Examples of experimental plans which would seem to suggest

Poisson distribution of populations are plans which involve counting the number of microbial cells in batch culture, which call for counting the number of phage particles attached to bacteria, or which call for counting the number of nematodes infecting parts of a plant. In the microbial culture, we can think of the liquid medium as being divided into a large number of small droplets and we can then define success as the occurrence of an organism in a droplet. Similarly, in the nematode-infected plant, we can view the plant as being constructed of several small segments and then define success as the presence of a nematode in a segment and failure as the absence of a nematode.

The Poisson distribution is expressed by the equation:

$$P(x) = \frac{e^{-\gamma}\gamma^x}{x!},$$

where $\quad$ P(x) $\quad$ = $\quad$ the probability of an occurrence x in a sample,

and $\quad\quad$ γ $\quad$ = $\quad$ the mean number of occurrences per unit of space.

Usually, a *goodness of fit* test (section 10-10) is used to determine whether an observed distribution is consistent with that expected from the Poisson distribution. This fit is tested by chi-square. The variance of a Poisson process is equal to the mean i.e. $\sigma^2 = \mu$; thus, in measuring radioactivity (a Poisson process), when the mean number of counts is N, the standard deviation is $\sqrt{N}$ (cf. section 8-3).

10-12. The binomial distribution is also used with discrete data

The binomial distribution, like the Poisson distribution, is used to evaluate discrete data. According to the binomial theorem, when a single trial is conducted, a successful event may occur with a probability p and a failure may occur with a probability q = 1-p.

If we have a population of two groups (for example two phases in countercurrent distribution), and these two phases contain solute weighing a total of n grams of which r grams of the solute is in the mobile phase, and n - r grams is in the stationary phase, then the probability of r is

$$P(r) = \frac{n!}{r!(n-r)!}\, p^r q^{n-r}.$$

The binomial coefficient is $\frac{n!}{r!(n-r)!}$. This expression for the coefficient is equivalent to the number of combinations of n items taken r at a time, that is, $_nC_r$.

In *Basic Biochemical Laboratory Procedures and Computing* we use the binomial theorem to calculate the fraction of solute found in each tube of a countercurrent apparatus (section 3-2). The relevant equation for computing the fraction of solute with the binomial coefficient is

$$T_{n,r} \quad = \quad \frac{n!}{r!(n-r)!} \left[\frac{1}{K+1} \right]^n K,$$

where $T_{n,r}$ =   the total number of transfers,

               r =   tube number,

and k =   the partition coefficient.

As in the Poisson distribution, once binomial probabilities have been calculated, a goodness of fit test may be used to determine whether an observed distribution is consistent with that expected from the binomial distribution. This fit also may be tested by chi-square.

10-13. Estimating the number of replications needed for reliable results

In planning their investigations, biomedical scientists frequently want to know how many times they should repeat an experiment to obtain a reliable mean value for such measurements as the amount of a biomolecule, or the rate of a process. If too many replications are carried out, the sample mean, although acceptable, is evaluated at too high a cost of time, effort and material. If too few replications are carried out, the precision of the mean will be too low and the investigation might yield inconclusive results of little, if any value. An approximate method for calculating the number of replications for the satisfactory estimation of a population mean is given by

$$n \quad = \quad (z^2 * s^2) \div d^2, \qquad\qquad (1)$$

where n =   the number of replications,

               z =   the standard normal deviate (section 10-4),

               s =   the standard error (section 10-7),

and d =   confidence interval, mean $\pm$ d units (section 10-4).

When equations such as (1) are used, calculated values of n are rounded to the next higher whole number if the calculated values are not integers. Use of equation (1) is demonstrated in Example 1.

Example 1. You estimate that the standard error for an enzyme assay is 7 μg substrate reacting/min/mg protein. How many replications should you carry out for (a) a 95% and (b) a 0.99 confidence interval corresponding to 15% of the average specific activity reported in the literature (40 μg substrate/min/mg protein)?

Solution.

(a) n = $(z^2 * s^2) \div d^2,$

               = $(1.96^2 * 7^2) \div (0.15*40)^2,$

               = $188.24 \div 36,$

   n = $5.23 \sim 6$ replications with 95% confidence.

(b) n = $2.58^2 * 7^2 \div (0.15*40)^2,$

               = $326.16 \div 36,$

   n = $9.06 \sim 10$ replications with 99% confidence.

In addition to estimating a population mean, investigators often want to know how many times they should repeat experiments to estimate differences between two means. If narrow confidence intervals with high reliability are sought, many replications will be required. If the expected differences between sample means are small, a large number of replications also will be required. We now provide answers to two relevant questions:

1. For a specified number of replications and an expected coefficient of variation, *what percentage difference between means would be significant at specified probabilities?*

2. For differences between sample means of an expected size, *how many replications are required for statistical significance at a given level of probability?*

We assume that estimates of sample means, coefficients of variation or standard deviations are available. Approximate solutions to questions 1 and 2 above then may be obtained from equations (2) and (3) respectively:

$$D\% = (1.96*C*\sqrt{2}) \div \sqrt{n}, \tag{2}$$

where
D = difference between means as a percent,
C = coefficient of variation (section 10-7),
n = number of replications,

and
$$n = [(1.96*C*\sqrt{2}) \div D\%]^2, \tag{3}$$

where n, D, and C are defined as above.

We demonstrate the use of these two equations in Examples 2, 3 and 4.

Example 2. Four replicates gave a 20% coefficient of variation for the phospholipid content of a cellular fraction. Calculate the percent difference between the mean content of control and treated fractions which would be significant at (a) the five and (b) the one percent levels.

Solution.

(a) D% = $(1.96*C*\sqrt{2}) \div \sqrt{n}$,
    = $(1.96*20*\sqrt{2}) \div \sqrt{4}$,
    = $55.4372 \div 2$,
    D% = 27.7% ~ 30% difference between the means.

(b) D% = $2.58*C*\sqrt{2} \div \sqrt{n}$,
    = $(2.58*20*\sqrt{2}) \div \sqrt{4}$,
    = $72.9734 \div 2$,
    D% = 36.49 ~ 40% difference between the means.

Example 3. You expect the mean enzymatic activity of a cytosolic enzyme to differ approximately by 15% at two stages of growth. If the coefficient of variation for the enzyme assay is 13%, how many replications do you need to learn whether the 15% difference is significant at the 5 percent level? If the difference between mean activities were 10%, how much would the number of replications have to be increased to show, with 95 percent probability, that the 10% change did not occur by chance?

Solution.

(a)   n $= [(1.96*C*\sqrt{2}) + D\%]^2,$
           $= [(1.96*13*\sqrt{2}) + 15]^2,$
           $= [36.0342 + 15]^2,$
           $= 5.77 \sim 6$ replications.

(b)   n $= [(1.96*13*\sqrt{2}) + 10]^2,$
           $= [36.0342 + 10]^2,$
           $= 12.99 \sim 13$ replications,
Therefore, an addiditonal 7 replications would be needed.

Example 4. You expect the amounts of a biomolecule in treated cells to increase 10% over the control value of 32.00 ± 2.25 ug/mg protein (mean ± standard deviation). Without converting the standard deviation to the coefficient of variation, calculate the number of replications required for detecting whether the 10% change is significant at the five percent level.

Solution.

n $= [(1.96*SD*\sqrt{2}) + (0.10*mean)]^2,$
      $= [(1.96*2.25*\sqrt{2}) + (0.10*32.00)]^2,$
      $= [6.2367 + 3.2]^2,$
n $= 3.798 \sim 4$ replications are needed.

10-14. The method of least-squares is used to fit linear regression

The term *simple regression* refers to the relation between two variables in which the magnitude of one variable (known as the *dependent* variable) is a function of the second variable (known as the *independent* variable). A relationship wherein one variable changes as the second changes but neither of the variables depends on the other is referred to as *correlation*. Thus, the dependence of one variable on another is a consequence of regression, but not of correlation. In simple regression, the independent variable usually is denoted by x, the dependent variable by y. In *linear regression,* the relation between the two variables is a straight line expressed by the *slope-intercept form of the equations of a straight line.* In the slope-intercept equation, the parameters b and a express the relation between the variables y and x as shown in equation (1) of section 10-15:

	y	=	a + bx,
Here	y	=	the dependent variable,
	x	=	the independent variable,
	b	=	the slope, the x-coefficient,
and	a	=	y-intercept, a constant.

If we plot the values of the two variables y and x, we find, usually, that we can draw a straight line through the points but all of the points do not fall exactly on the line. For the most reliable results, we would like to draw the *best-fit* line through the points. The best fit is achieved by the *method of least-squares* in which the sum of squares (section 10-7) of the deviations of each data point from the best-fit line is a minimum. This sum of squares is referred to as the *residual or error sum of squares*. Mathematically, the principle involved in finding the point at which the sum of squares is a minimum is explained in the following way. First, express the sum of squares as a function of b then as a function of a and set each function equal to 0. For each of the two functions, if the first derivative = 0 and the second derivative > 0 then the minimum is reached. The two functions, each set equal to 0, lead to two equations with two unknowns, b and a. Solving for the two unknowns simultaneously yields two equations with unique values for b the slope, and a, the y-intercept. These equations for the slope and the y- intercept, (equations 2 and 3 in section 10-15) give the desired minimum values for the squares of the deviations of each data point from the best-fit line.

Four computational steps are needed to apply the method of least-squares to obtain the best-fit line in linear regression. These steps are (a) calculate the quantities listed at the beginning of section 10-15; (b) calculate the slope and y-intercept from equations (2) and (3); (c) substitute the slope b and intercept a into equation (1) to obtain the best-fit line and (d) calculate the least-squares parameters of equations (4) to (11) to evaluate the quality of the fit. Many investigators use the correlation coefficient as the sole criterion for evaluating the quality of a least-squares fit. However, Davis, Thompson and Pardue (1978) suggest that complete reliance on the correlation coefficient may be unwise. Davis et al. suggest that, instead of relying solely on the correlation coefficient, investigators should use additional parameters and tests. Davis et al., and standard statistics textbooks, suggest that, in addition to the correlation coefficient, the following be used to evaluate least-squares fits:

- the standard error of estimate;
- the standard error of the slope;
- the standard error of the intercept; and
- *t*-tests and analysis of variance testing (cf. Chapter 11).

We define these parameters and various tests of significance in equations (4) to (11) of section 10-15, and we outline their properties in section 10-16.

In statistics, and in other branches of mathematics, it is customary to distinguish related values from one another by attaching diacritical marks such as bars (recall the use of $\bar{\upsilon}$ to represent the average number of molecules of L bound in section 5-10, and the use of $\bar{X}$ to distinguish the mean from the variable X in section 10-7). In regression analyses, estimated values for the best-fit line are often distinguished from experimental values by attaching *hats* e.g. $\hat{y}$ to the estimated variables and parameters. When diacritical marks are used in regression, the best-fit line is usually represented as:

$$\hat{y} = \hat{a} + \hat{b}x$$

Here, $\hat{y}$, $\hat{a}$ and $\hat{b}$ are referred to as *"y hat"*, *"a hat"* and *"b hat"* respectively.

Once we have obtained the best-fit line, we can predict values of y for given values of x. *An important rule* in making such predictions is that *one should not attempt to predict beyond the values of x which were used in the data*; that is, *one may interpolate but one may not extrapolate.* Some underlying assumptions of the linear regression model are that

- the values of x are chosen or fixed by the investigator;
- the error in measuring x is small; and
- the y values are random and do not depend on each other.

Linear regression is widely used in the biomedical sciences. But, the popularity of this method is not matched by recognition of its limitations and underlying assumptions. We encourage readers who need more information to consult suitable textbooks or articles on the interpretation and weaknesses of least-squares analyses such as those by Duggleby and Leatherbarrow, *Trends Biochem. Sci.*, listed in the references at the end of this chapter.

10-15. Quantities and equations for least-squares computations

Quantities (sums) related to parameters and variables defined in section 10-14:

Sums: Σx, Σy, $\bar{x}$, $\bar{y}$, Σx^2, Σxy, Σy^2, $(\Sigma x)^2$, $(\Sigma y)^2$, $\Sigma x \Sigma y$, n.

Equations:

Straight line:
$$\hat{y} = \hat{a} + \hat{b}x; \tag{1}$$

Slope:
$$\hat{b} = \frac{n\Sigma xy - \Sigma x \Sigma y}{n\Sigma x^2 - (\Sigma x)^2}; \tag{2}$$

Intercept:
$$\hat{a} = \bar{y} - \hat{b}\bar{x}; \tag{3}$$

SE of estimate, S_e
$$= \left[\frac{\Sigma y^2 - \dfrac{(\Sigma y)^2}{n} - \dfrac{(\Sigma xy - \bar{x}\,\Sigma y)^2}{\Sigma x^2 - \bar{x}\,\Sigma x}}{n - 2} \right]^{\frac{1}{2}}; \tag{4}$$

SD of slope, s_b
$$= S_e \left[\frac{1}{\Sigma x^2 - \bar{x}\Sigma x} \right]^{\frac{1}{2}}; \tag{5}$$

SD of intcp., s_a
$$= S_e \left[\frac{\Sigma x^2}{n\Sigma x^2 - (\Sigma x)^2} \right]^{\frac{1}{2}}; \tag{6}$$

Correlation coeff, r $\quad = \quad \dfrac{n\sum xy - \sum x \sum y}{[n\sum x^2 - (\sum x)^2]^{\frac{1}{2}} \, [n\sum y^2 - (\sum y)^2]^{\frac{1}{2}}}$; $\qquad$ (7)

SD of r, $\qquad s_r \quad = \quad \left[\dfrac{1 - r^2}{n - 2}\right]^{\frac{1}{2}}$; $\qquad$ (8)

t (slope) $\qquad = \quad \hat{b}/s_b$; $\qquad$ (9)

t (intcp) $\qquad = \quad \hat{a}/s_a$; $\qquad$ (10)

t (r) $\qquad = \quad \left[\dfrac{r^2(n-2)}{(1-r^2)}\right]^{\frac{1}{2}} \quad = \quad r/s_r.$ $\qquad$ (11)

10-16.   Properties of least-squares coefficients and parameters

Slope b.   The x-coefficient. It expresses quantitatively the dependence of y on x. Alternatively, the slope is defined as the change in y for a given change in x, or $\Delta y/\Delta x$.

y-Intercept a. The constant in the slope-intercept equation. It expresses the distance of y above or below the x-axis at x = 0. Stated another way, the y-intercept is the ordinate of the point at which a curve intersects the y-axis.

$s_{y.x}$.   Standard error of estimate or standard error of y-estimate, is an estimate of the error in fitting the regression line. It measures the scatter of the experimental values of y about the best-fit line; i.e., it shows the accuracy with which the computed regression equation predicts the dependence of y on x. The sums of squared differences between experimental and best-fit values of y, that is $(y - \hat{y})^2$ are known as *residuals.*

s_b.   Standard error of the slope. Indicates the accuracy with which the regression equation predicts the slope.

s_a.   Standard error of the intercept (i.e., of the constant). Indicates the accuracy with which the regression equation predicts the intercept.

r^2.   The coefficient of determination. This indicates the proportion of the total variation due to the fitted regression; i.e., $r^2 = SS_{regression}/SS_{\,total}$.

SS$_{total}$.   The variance (section 10-7) of the experimental y values/(n-1).

SS$_{regression}$. $\Sigma\Sigma_{total}$ * r^2.

r. The correlation coefficient, also known as *Pearson's correlation coefficient,* ranges from -1 to +1; it measures the degree of linear relationship between y and x. Karl Pearson introduced r in an acclaimed study during 1902-1903. Spreadsheets usually give built-in formulae for the slope and SD, thus permitting r to be computed from $r = slope*(SD_x/SD_y)$.

t_b *and* t_r. These are identical values for testing whether there is a linear relation between y and x. The table for *r* can be calculated from the *t*-table. In a good fit, t_b and t_r are significant; that is, the relation between y and x is not due to chance alone.

t_a. This is equal to (â - maximum rate)/ s_a. It is used for testing whether the value of the intercept obtained from a regression equation is the same or different from the intercept obtained experimentally. In a good fit, t_a is not significantly different from the maximum rate.

10-17. There are many procedures for fitting nonlinear equations

As we saw in section 10-14, the term regression refers to the relation between two variables in which the magnitude of the dependent variable is a function of an independent variable. *Linear regression* is so called because it describes *equations or mathematical models in which the parameters are linear.* Thus, linear regression is described by equations of the type $y = a + bx$ in which the parameters are a and b. Linear regression also may be described by *bilinear equations* of the type

$$y = a + b_1x_1 + b_2x_2,$$

in which the parameters are a, b_1 and b_2.

Nonlinear regression, conversely, is described by equations in which the parameters are nonlinear. For example, nonlinear equations may have exponential parameters such as k, the rate constant, in *first order decay curves*:

$$[A] = [A_0] e^{-kt}.$$

Exponential growth curves also have nonlinear parameters, but unlike decay curves, the parameter k (or its equivalent) is positive; for example:

$$y = A e^{bt}.$$

Nonlinear equations, additionally, may have hyperbolic parameters as, for example, in the *Michaelis-Menten equation,* where the parameters are K_m and V_{max}:

$$v = \frac{V_{max}.[S]}{K_m + [S]}.$$

The following points are usually adduced to support use of nonlinear regression:

- It provides more accurate estimates of parameters than are possible from linear regression of transformed data or from graphical procedures;
- It tests whether sets of experimental data fit selected mathematical models or equations;
- It provides good estimates of a dependent variable for chosen values of the independent variable; and
- It provides estimates of the goodness-of-fit and reliability of the data.

The general procedure for performing a nonlinear regression may be divided, for convenience, into three steps:
- Obtain preliminary estimates of the parameters of the nonlinear equation.
- Compute the "best" estimate of the parameters by *iteration* (a repetitive process of adjusting successive estimates to obtain progressively better estimates).
- Compute the standard errors of the best estimate of the parameters.

In step 1, good preliminary estimates of parameters which are of interest to biomedical scientists often may be obtained from least-squares analyses of linearized forms of the original nonlinear equation.

In step 2, progressively better estimates of the parameters are calculated by adjusting or refining the preliminary estimates obtained in step one. Several optimization or adjustment procedures have been developed for conducting step two. Among the best known are the *Gauss-Newton procedure*, the *method of steepest descent,* and the *Marquardt method.* The latter is a combination of the first two optimization methods. In *Basic Biochemical Laboratory Procedures and Computing*, we use the Gauss-Newton optimization procedure. Wallace Cleland, who has published several FORTRAN programs for investigations of enzyme kinetics, also has used the Gauss-Newton procedure for his programs. The reason for this choice is that, when good preliminary estimates are available, (as from linear transformations of the Michaelis-Menten equation), the Gauss-Newton procedure arrives at the best estimates of the desired parameters rapidly. Another advantage of the Gauss-Newton method (an advantage which is common to other optimization procedures) is that it permits investigators to calculate measures of variability for obtaining the best estimates. As a result, an investigator can estimate not only the parameters of the equations but can also determine how well the parameters have been estimated. An appropriate equation is needed for conducting step 2. If such an equation is not available from the literature, it may (for the Gauss-Newton method) be derived by expanding the original nonlinear equation as a Taylor series. Then, by discarding the high order terms of the Taylor series, one obtains an approximately linear equation from which the parameters may be estimated by least-squares (cf. the method of Wilkinson, section 5-8). Estimation of parameters using the linear equation derived from the Taylor series expansion is conducted with as many iterations (successive approximations) as are needed to obtain progressively better estimates and, eventually, arrive at the best estimate of the parameters. In step 2 then, the preliminary estimate is first adjusted to obtain a better or refined estimate. If necessary, the refined estimate is adjusted again to obtain an even better estimate and, as many cycles of iteration as necessary are conducted to reduce the differences between succesive estimates to negligible levels.

After the iterative procedure has been stopped, an investigator needs objective criteria for evaluating the results. Three criteria are useful in this regard:

- The calculations should *converge*.
- The best fit curve should agree closely with the mathematical model.
- The standard errors for estimating the parameters should be small.

Convergence is evaluated by calculating and comparing the *relative change in the sum of squares* for the parameters. For example if, in successive iterations, the relative change in the sum of squares of the parameters is less than a predetermined small number (such as less than one thousandth of a percent change, i.e., 10^{-5}) convergence is considered to have occurred, and the iterations are stopped. The relative change in the sum of squares is defined as the sum of squares at a given iteration / the observed value of the parameter for that iteration. That is, where θ is the value of a parameter, the relative change is $(\theta_{obsd} - \theta_{calcd})^2 / \theta_{obsd}$.

Since, in the Gauss-Newton procedure, the equations derived from Taylor series expansion are approximately linear, they may be fitted by the method of least-squares. The least-squares fit will yield linear parameters for estimating the best fit in addition to "correction factors" for computing standard errors. Some examples of equations of biomedical interest which may be treated by the Gauss-Newton procedure are those for: (a) exponential and radioactive decay; (b) exponential growth; (c) hyperbolic functions and (d) logistic functions such as those used to describe bacterial growth and the growth of human populations.

Step 3 of the nonlinear regression procedure is the computation of standard errors. Such computation is possible because the best fit is achieved by least-squares methods in which the sum of squares of the vertical deviations from the best-fit line is minimized with respect to the parameters. Once the sum of squares is available, standard errors can be calculated (section 10-7). A small value of the standard errors is equivalent to a good fit; conversely, a large value of the standard errors is equivalent to a poor fit. These calculations of the best fit parameters and standard errors are based on the following assumptions which may not be exactly correct for each set of data:

- the x values are chosen (fixed) by the investigator;
- the errors in the x values are negligible;
- the y values are independent (one value of y does not depend on the value of another value of y);
- the y values are normally distributed and are sampled randomly;
- the variances of y populations corresponding to any value of x are equal.

It is worth reemphasizing here that no procedure for fitting nonlinear equations, (Gauss-Newton, the method of steepest descent or the Marquardt method) can improve bad data. Consequently, it is wise to subject data from nonlinear regression analyses to graphical evaluation to detect gross deviations from the kinetic equation or mathematical model. In *Basic Biochemical Laboratory Procedures and Computing*, we have made provisions for graphical evaluation by providing for users of the nonlinear regression program the opportunity to plot the independent variable against the best-fit values of the dependent variable.

It also may be instructive to review a short paper by Duggleby, *Trends Biochem. Sci.* 16: 51-52 (1991), an article by Beechem, *Methods in Enzymology* 2 37-54 (1992) and an earlier article by Mannervik, *Methods in Enzymology* 87: 370-390 (1982).

Duggleby, a pioneer in the use of nonlinear regression, wrote his *TIBS* paper in response to Leatherbarrow's earlier *TIBS* article on pitfalls in the biochemical use of linear and nonlinear regression. Duggleby explored the view that "most of the supposed advantages of nonlinear regression are illusory and the results obtained from a hand-drawn line after linear transformation may be almost as useful." After he explored his dissenting view on the advantages of nonlinear regression, Duggleby said that although he has argued that the usual justifications for nonlinear regression have little merit, he does not wish to abandon such analyses. Duggleby offered to send readers "copies of a more detailed exposition of his views" than was possible in *TIBS*.

Beechem, on the other hand, stated in Methods in Enzymology that:
"for greater than 90% of all of the data collected in the fields of biochemistry and biophysics, the proper technique to analyze the data is nonlinear least-squares. No attempt will be made to justify the preceding statement; it has simply been arrived at from personal experience with a variety of data analysis projects and by communications with other biophysical data analysis groups."

But Mannervik, also in Methods in Enzymology, stated earlier that:
"By using statistical methods, a certain degree of objectivity is ascertained insofar as all investigators should get the same analytical results once they have agreed on the techniques to use. However, it should be borne in mind that the choice of a statistical method (such as the least-squares algorithm) is not necessarily unbiased and will often affect the results."

This type of scholarly debate indicates, as we suggested in section 10-1, that a prudent investigator will approach data analysis thoughtfully in the context of the requirements of each project. Unenlightened use of any method or software package can lead to inconclusive or anomalous results.

10-18. Questionable values may be rejected with objective criteria

When an investigator makes an error in measuring a value, that value should be rejected; this is an easy and clear-cut choice. But, most persons who conduct laboratory investigations are faced, at some time, with the problem that the correctness of one value in a set of data is suspect because that value differs markedly from others in the set even though all of the measurements were, apparently, made correctly. The difficulty, in this case, is to decide whether it is valid (and ethical) to retain or reject that suspect measurement. The statistical difficulty arises because if the standard chosen for rejection is too low then data which should be rejected are retained; alternatively, if the standard is too high, then data which should be retained are rejected.

Statisticians have proposed many criteria for rejecting suspect data. We saw (section 10-4) that if data are approximately normal, then points $\geqslant 1.96\sigma$ or 2.58σ from the mean (where σ is the standard deviation) may be candidates for rejection. In this section, we present another criterion for data rejection, the *Q test* of Dean and Dixon.

The Q test (where Q refers to *rejection quotient*) takes into account the fact that most laboratory investigations are conducted with small numbers of measurements. The Q test is carried out according to the following steps:

- Arrange the data in order of increasing magnitude.

- Calculate the range of values, that is, the difference between the largest and the smallest values.

- Calculate the difference between the suspect value, x_2 and its nearest neighbor, x_1.

- Calculate $\quad Q = \dfrac{x_2 - x_1}{range}$

- Compare the calculated Q with the tabulated values of Q given in the table below. If the calculated Q is greater than the tabulated value for the appropriate number of measurements, then reject the questionable value with 90% confidence. Conversely, if the calculated Q is less than the tabulated value for the given number of values then retain the suspect value.

Tabulated values of Q, as computed by Dean and Dixon, are:

Number of observations:	2	3	4	5	6	7	8	9	10	∞
Rejection quotient, Q:	0.94	0.76	0.64	0.56	0.51	0.51	0.47	0.44	0.41	0.00

10-19. Review questions

10-1. What problems do you face in deciding whether to discard experimental data?

10-2. Define the terms *dispersion* and *central tendency*.

10-3. What is the *central limit theorem*?

10-4. Define the term *acceptance region*?

10-5. Explain the terms *confidence interval, confidence limit* and *confidence level*.

10-6. Define the terms *accuracy* and *precision*.

10-7. What are the main assumptions of hypothesis testing?

10-8. Why is there a need for a branch of statistics applied to the biomedical sciences?

10-9. Define *standard deviation* and *standard error*.

10-10. Under what conditions should the *standard deviation* and *standard error* be used?

10-11. On what assumptions is nonlinear regression based?

10-12. Define the terms *dependent* and *independent* variable.

10-13. What are *research* and *statistical hypotheses*?

10-14. List the major statistical assumptions behind the method of least-squares.

10-15. Distinguish between *parametric, nonparametric* and *distribution-free methods*.

10-16. What is a *goodness of fit* test?

10-17. Define the terms *controlled experiment* and *double-blind experiment*.

10-18. Explain, in as many ways as you can, the term *statistically significant*.

10-19. What are some guidelines for ethical scientific scholarship?

10-20. How would you ensure statistical reliability in your investigations?

10-20. Additional problems

Measures of Central Tendency and Dispersion

10-21. The protein content of four bacterial cultures (mg/ml of cell suspension) was determined in six replicates by the method of Lowry et. al with the following results:

> (a)   111, 110, 115, 115, 148, 147
> (b)   330, 322, 345, 331, 323, 318
> (c)   200, 183, 183, 185, 210, 220
> (d)   195, 185, 220, 213, 190, 180

Calculate the mean, standard deviation, standard error, coefficient of variation and 95% and 99% confidence limits for each set of data.

Percent Calculations

10-22. (a)   The current yearly salaries of your three laboratory associates are $40,000, $44,700 and $55,500 respectively. Members of the administrative staff attached to the laboratory make $21,000 and $23,000 per year, respectively. If the new contract under which your institution operates calls for increases, based on current salaries, of 6 % in the first year and 4.5 % in the second year, what dollar increases would your five colleagues be receiving in each of the two years of the new contract?

(b)   You have been told that, if you and your colleagues combine to purchase twenty or more computers at a time, you can purchase the computers which normally cost $5,500 at $3,500. What percent decrease are you being offered?

(c)   Given a = 15 and b = 75, what are the results of adding a% of b to b, subtracting a% of b from b, multiplying b by a% of b, and dividing b by a% of b.

(d)   The following are indices of unsaturation for four samples of a fungal phospholipid at three different ages. Calculate the percent change between times 1 and 2 and between times 2 and 3 for each sample.

Age	1	2	3
Sample 1	5.90	11.90	7.80
Sample 2	16.90	36.00	5.70
Sample 3	1.26	1.08	1.02
Sample 4	1.69	1.23	0.94

Chi-Square Goodness of fit

10–23.   The following are calculated and expected values (mol %) for eight fatty acids:

Observed	1.9	15.8	7.1	2.7	62.0	9.2	1.1	0.2
Expected	3.0	14.8	7.6	4.3	61.8	8.1	0.3	0.1

Compare the two sets of data by chi-square to learn whether or not the observed and expected values are statistically different.

t-Test for Two Independent Samples

10-24.   Determinations of levels of blood urea (mg ml^{-1}) were conducted on two groups of laboratory animals with the following results:

Group 1:   111, 110, 115, 115, 148, 147, 124, 195, 185, 220, 213, 190, 180, 183
Group 2:   320, 213, 227, 293, 211, 330, 322, 345, 331, 323, 318, 343.

The working hypothesis of the experiment was that animals from group 2 generally have higher levels of urea than those of group 1. Test this hypothesis by computing and evaluating *t*.

Paired-Sample *t*-Test

10-25. Twenty flasks containing equal amounts of the same growth medium were inoculated in an identical fashion to grow cultures of mushrooms for making "Fabulous" brand mushroom soup. After the cultures had grown for an appropriate time, ten of the cultures were treated with "supermiracle," a compound which stimulates growth of such cultures; the other cultures were treated with the buffered medium used to dissolve "supermiracle." Test the hypothesis that growth of the cultures (in g dry wt/ flask) is the same with or without the miracle compound.

With   miracle: 46.3, 43.1, 44.8, 42.9, 46.1, 45.8, 41.6, 44.3, 41.6, 44.7
Without miracle: 48.4, 46.0, 41.6, 44.4, 45.2, 41.8, 43.3, 42.9, 42.2, 46.4

The *F*-Test

10-26. Measurements of the diameters (mm) of noncancerous, intestinal polyps were carried out on two groups of patients with the following results:

Group 1:   7, 18, 5, 8, 3, 10, 3, 5, 8, 21, 11, 6, 8, 5, 6, 4, 10, 7, 15, 3
Group 2:   3, 8, 5, 7, 5, 10, 3, 15, 8, 5, 7, 3, 5, 3, 5, 7, 8

Do these data provide adequate reason to conclude that the sizes of polyps differ between the two groups?

Poisson Goodness of Fit

10-27. The DNA which codes for a mouse antibody was amplified by the polymerase chain reaction and diluted with the DNA of a mutant cell line. Then, ten 1-μg samples of the diluted DNA, were transferred to a nylon membrane and subjected to Southern hybridization with a ^{32}P-labeled oligonucleotide probe, specific for the amplified DNA. The data are expressed in terms of the numbers of successful hybridizations in the equivalent of 10^7 cells.

Number of Occurrences	Number of Hybridizations
0	8
1	9
2	19
3	17
4	15
5	13
6	10
7	9
8	7
> 9	0

Test whether the amplified and diluted DNA was distributed randomly.

Poisson Goodness of Fit continued

10-28. Infection by nematodes causes a red-colored ring to be formed in the trunks of certain tropical trees. Felling the trees and cutting the trunks into sections gave the following data on the numbers of nematodes infecting each cubic meter of the trees, including the ring:

Number of Occurrences	Number of nematodes $* 10^2$
0	4
1	12
2	16
3	23
4	3
5	0

Test whether the red-colored ring is caused by localized infection or whether the nematodes are distributed randomly.

Linear Regression and Correlation

10-29. The oxygen content of distilled water was measured in a procedure involving the reaction of N-methyl-phenazonium methosulfate with increasing levels of NADH in the presence of catalase and oxygen dissolved in 3.4 ml of distilled water. Oxygen uptake was measured polarographically with a Clark electrode with the following results:

Amount of NADH, μmol	:	0.128	0.256	0.384	0.512	0.768
Recorder deflection, %	:	8.0	16.0	23.75	30.75	46.0

How good is the correlation between the two variables?

10-30. The following results were obtained from an experiment in which the abilty of varying amounts of an enzyme extract to catalyze the incorporation of a radioactive precursor into trichloracetic acid/deoxycholate precipitable material was measured:

Volume of enzyme extract, μl	:	5	10	15	20	25
cpm of precipitate	:	350	450	600	1030	4400

Do you conclude that the incorporation occurred in a linear fashion?

10-21. Some questions which raise ethical issues

Questions 10-31 to 10-42 may be used in four ways:
- For discussion in class;
- For written responses;
- For written responses followed by class discussion; and
- For individual reflection on the issues involved.

10-31. You have been given such a high final grade in one of your courses that you think there must be an error in your favor. Perhaps, you conjecture, an assistant made an incorrect entry into the program which your professor uses for grading. What would you do about the suspected error, and why?

10-32. You are about to graduate with an M.D. degree and you are being interviewed for the residency which is your first choice. You have a good record; but it seems from your interviewer's comments that in one important board score, someone in the medical school erred and reported a higher score than you received. Would you tell the interviewer that your score was lower? Explain your answer.

10-33.   Members of your research group are asked to review a grant proposal. The proposal is excellent. But the head of your group tells you, without reading the proposal, to give it as low a ranking as possible because he thinks that the principal investigator has enough money. Would you comply? Explain your answer.

10-34.   A senior colleague tells you that you "should not say anything bad" about any grant proposal or research paper you are asked to review because if you make negative comments, the authors may find out that you did and then will retaliate if subsequently they review your work. What will be your response? Explain.

10-35.   You have referred to papers from a competitor's laboratory in your research. But, after you have prepared the first draft of the completed research for publication, your supervisor tells you not to cite his competitor's references because he does not want to increase the recognition or citation count of the competitor. How would you respond, and why? Justify your response.

10-36.   You are asked to review a research paper in your specialty. It turns out that you are, independently, working on a closely similar experiment which is not yet completed. Would you follow the advice of colleagues who tell you to delay returning your review (a) to prevent the authors from receiving publication priority and (b) to increase your chance of being awarded a research grant?

10-37.   A colleague is reviewing a grant application, sees an idea which is applicable to your research, shows it to you , and suggests that you "use" (i.e., steal!) the idea for your research. Would you? How do you explain your answer?

10-38.   A professor with impeccable academic credentials brags that he will not pay money for any software. This professor also gloats that he has all of the major software in several categories. Moreover, colleagues have seen this individual copying software manuals openly in a busy departmental office in the presence of students for whom he should be setting a good example. If this professor offered you copies of any two software packages from his laboratory for a copy of one of your legally obtained software packages, what would you do? Does it surprise you that a highly educated individual does not seem to have a sense of propriety? Is it pertinent that this professor mocks and ridicules colleagues who refuse to conspire with him to gain political advantage? Is it germane that he will walk past colleagues and refuse to acknowledge their presence by looking at them or saying "hello" or "good morning?" Explain your views, and say whether the university administration should be told that the professor shows flagrant disregard for its software policy. Is the professor's violation of copyright a form of professional misconduct or is it a form of "campus crime?" Would you be less likely to report him if his piracy were less flagrant and if he observed ordinary rules of personal and civil conduct? Would you fear reprisal in your exams for reporting such a person?

10-39.   A secretary left the original of the biochemistry final exam for your class of 105 on the glass plate of a copier to which students have access. One of your classmates found the exam, made a copy, and left the original on the copier so that the secretary could retrieve it. Before long, this"entrepreneurial" classmate started selling copies of the final exam to members of your class for $25.00 each, insisting that he is justified in charging that amount since (a) he is assuming a big risk and (b) he is providing "a useful service." Assume that you had to obtain a high grade in biochemistry to improve your chance of admission to medical school; would you purchase a copy of the stolen exam? Explain your choice. Should someone report the activities of the entrepreneur? What percent of the class do you think would purchase the stolen exam? Give reasons.

10-40.   The duties of many academic scientists are such that they no longer have the time to work at the lab bench. The reason is that, in addition to teaching and supervising undergraduate, graduate and postdoctoral research, science professors devote considerable time to writing papers, writing grant proposals, writing books, attending meetings and preparing and managing budgets. Most science professors also keep abreast of new developments by reading and studying the scientific literature. Professor Stealth, on the contrary, does not subscribe to, or read specialized journals. As a result, he usually is unaware of how to implement new physicochemical methods. Stealth however, takes pride in the length of his publication list and tries to publish an average of one paper a week i.e. more than fifty papers a year. This professor avoids technical obsolescence in two ways: (a) by instructing his graduate students to spy on the laboratories of other faculty members who work out newer techniques and (b) by arranging for his graduates to work in well-equipped industrial labs. Another point: no one in the department gets any help from Professor Stealth's laboratory. How should students and professors in other laboratories respond to Professor Stealth's behavior? What ethical issues are involved in sharing or refusing to share information with Professor Stealth? Be prepared to justify your response during a class discussion.

10-41.   Columns for use with high-performance liquid chromatographs can be expensive. You and two other undergraduate students were left in the laboratory to complete an experiment while your professor and her more experienced research students were attending a major scientific meeting. Your professor was a little uneasy to leave inexperienced students to complete their senior research projects on the new HPLC column. But, it is the spring semester before you graduate, and your professor did not want to hold up your graduation. Perhaps, through nervousness, you damage the new HPLC column. How would you handle this? Would you admit that you damaged the expensive column? Or, would you keep quiet and say nothing to your partners or the professor so that when she discovers the damage the other two students could be suspect as well as you and perhaps even get you "off the hook"? Does the fact that you will graduate in a few weeks and perhaps never see the professor again make a difference in what you would do?

10-42.   There are five graduate students in your laboratory. All five of you work in the same research area, so you use many of the same supplies, some of the same glassware, growth media for microorganisms, radioisotopes etc. The general rule in the laboratory is that when someone empties or uses nearly all of any material, that person should get an order for replacement, if necessary. This rule works well, in general, except for one of the five students: he will never stop his work to replace dwindling supplies. How does one deal with this type of person?

Suggested Reading

Abelson, P.H. 1992. Integrity of the Research Process. *Science* 256: 1257.

Beechem, J.W. 1992. *Global Analysis of Biochemical and Biophysical Data* Methods in Enzymology 210: 37-54.

Blaedel, W.J., V.W. Meloche and J.A. Ramsay 1951. A Comparison of Criteria for the Rejection of Measurements. *J. Chem. Educ.*, 28: 643 - 647.

Beaauchamp, T.L. and J.F. Childress 1983. *Principles of Biomedical Ethics*. 2nd ed. Oxford University Press, New York.

Campbell, R.C. 1974. *Statistics for Biologists*. 2nd ed. Cambridge University Press, London.

Causton, D. R. 1987. *A Biologist's Advanced Mathematics*, Allen & Unwin, Boston.

Cleland, W.W. 1979. *Statistical Analysis of Enzyme Kinetic Data*. Methods in Enzymology 63A: 103 - 138.

Daniel, W.W. 1983. *Biostatistics: A Foundation for Analysis in the Health Sciences*. 3rd. ed. John Wiley & Sons, N.Y.

Davis, R.B., J.E. Thompson and H.L. Pardue 1978. Characteristics of Statistical Parameters Used to Interpret Least-Squares Results. *Clinical Chemistry* 24: 611 - 620.

Dean, R,B. and W.J. Dixon 1951. Simplified Statistics for Small Numbers of Observations. *Anal. Chem.* 23: 636-638.

Draper, N. and H. Smith 1981. *Applied Regression Analysis*. 2nd. ed. John Wiley & Sons, NY.

Duggleby, R.G. 1991. Analysis of Nonlinear Data by Nonlinear Regression: Is it A Waste of Time? *Trends Biochem. Sci.* 16: 51-52.

Duggleby, R.G. 1981. A Nonlinear Regression Program for Small Computers. *Anal. Biochem.* 110: 9-18.

Dunnett, C.W. 1964. A Multiple Comparison Procedure for Comparing Several Treatments With a Control. *J. Amer. Statistical Assn Biometrics* 20: 482 - 491.

Fletcher, J. 1979. *HUMANHOOD: Essays in Biomedical Ethics*. Prometheus Books, Buffalo, NY.

Johnson, M.L. and L.M. Faunt 1992. *Parameter Estimation by Least-Squares Methods*. Methods in Enzymology 210: 1-37.

Johnson, M.L. and S.G. Frasier 1985. *Nonlinear Least-Squares Analysis*. Methods in Enzymology 117: 301 - 342.

Katz, M. A. 1976. *Calculus for the Life Sciences*. Marcel Dekker, NY.

Leatherbarrow, R.J. 1990. Using Linear and Nonlinear Regression to Fit Biochemical Data. *Trends. Biochem. Sci.* 15: 455-458.

Lee, J.D. and T.D. Lee 1982. *Statistics and Numerical Methods in BASIC for Biologists*. Van Nostrand Reinhold, NY.

Mannervik, B. 1982. *Regression Analysis, Experimental Error, and Statistical Criteria in the Design and Analysis of Experiments for Discrimination Between Rival Kinetic Models*. Methods in Enzymology 87C: 370 - 390.

Marquardt, D.W. 1963. An Algorithm for Least-Squares Estimation of Nonlinear Parameters. *J. Soc. Indust. Appl. Math.* 11: 431-441.

Nimmo, I.A. and G.L. Atkins 1979. The Statistical Analysis of non-normal (real?) data. *Trends Biochem. Sci.* 4: 236 - 239.

Motulsky, H.J. & L.A. Rasnas 1987. Fitting Curves to Data Using Nonlinear Regression. *FASEB. J.* 1: 365 - 374.

Ridgman, W.J. 1975. *Experimentation in Biology*. Halsted Press, NY.

Sagnella, G.A. 1985. Model Fitting, Parameter Estimation, Linear and Nonlinear Regression. *Trends Biochem. Sci.* 10: 100 - 103.

Snedecor, G.W. & W.G. Cochran 1967. *Statistical Methods*. 6th. ed., Iowa State Univ. Press.

Straume, M. and M.L. Johnson 1992. *Analysis of Residuals: Criteria for Determining Goodness-of-fit*. Methods in Enzymology 210: 106-117.

Taubes, G. 1993. Misconduct: Views From the Trenches. *Science* 261: 1108-1111.

Youden, W.J. 1951. *Statistical Methods for Chemists*. John Wiley & Sons.

Zar, J.H. 1974. *Biostatistical Analysis*. Prentice-Hall, Inc., Englewood, NJ.

Zierler, K. 1989. Misuse of Nonlinear Scatchard Plots. *Trends Biochem. Sci.* 14: 314–317.

Zurer, P.S. 1993. Survey Finds Researchers Often Encounter Scientific Misconduct. *Chem & Engineering News* Pages 24-25, November 22.

Chapter 11

Analysis of Variance

11-1. Analysis of variance is used to test three or more samples

Analysis of variance, abbreviated ANOVA, is used for comparing the means and variances of three or more experimental groups or samples. We cover analysis of variance in *Basic Biochemical Laboratory Procedures and Computing* because biomedical investigations frequently include more than two samples. For those experiments designed to investigate just one or two samples, the appropriate procedures from Chapters 1 and 10 should be used.

Analysis of variance may be defined as a procedure in which the *total variance is divided into individual variances so that the contributions of each variance to the total variance may be estimated and compared.*

The comparisons obtained from an analysis of variance reveal how much of the observed variation in a multisample experiment is due to differences between the samples and how much is due to random errors. From these comparisons, an investigator can estimate the relative importance of the differences between the samples.

An example of a simple multisample experiment is one designed to evaluate the effects of four different buffers on the activity of a given enzyme; here, in essence, the investigator studies the effects of multiple versions or levels of one factor (type of buffer) on one variable (enzyme activity). Stated another way, an individual buffer in this experiment is a level of the factor "type of buffer."

There are several types of ANOVA, but the basic procedure was developed in 1925 by Sir Ronald Fisher in England. At first, the analysis of variance was applied to agricultural experiments; during the past sixty years, however, tests of the procedure in many disciplines have demonstrated its validity and broad applicability. When the use of ANOVA is considered, it is good practice to define the sources of variation during the planning of the research. The variations often arise from two sources: from the experimental treatments and from uncontrollable biological variation.

In this chapter, we will discuss briefly, three analyses of variance:

- single-factor,
- randomized block, and
- two-factor analysis of variance.

Performing ANOVA by hand calls for tedious arithmetic computations. To decrease this tedium we have written computer programs to perform the analyses. In the following sections, we will simplify our ANOVA presentation by taking three steps: we will describe principles and assumptions in words; we will outline algorithms in qualitative statements; and we will omit the mathematical symbols and steps required for describing manual calculation.

11-2. The basic assumptions and calculations are similar for all ANOVA

There are three important assumptions behind every analysis of variance:

- that all of the samples are taken from normal populations;
- that the populations are randomly distributed; and
- that (for hypothesis testing) the variances of the populations are equal.

Not only are there similar underlying assumptions behind every analysis of variance, but, the sequence of computational steps used for all of the analyses is fundamentally similar, varying mainly in detail. These computational steps, based on the principles developed by Fisher for the simplest ANOVA, are

- Calculate the *total sum of squares*; this is the sum of squares of all of the values.
- Calculate the *treatments sum of squares* and treatments variance.
- Calculate the *residual or error sum of squares and variance*; the error sum of squares equals the total sum of squares minus the treatments sum of squares.
- Use the *F*-test to determine whether the treatment and error variances are significantly different from each other.

One completes an analysis of variance by preparing a standardized table which lists the total variation, the treatments variation, the residual variation and related data. Since a major reason for undertaking a multisample investigation is to compare the means of individual samples, it is often of interest to extend the results reported in the standardized ANOVA table to ask which means differ from the control. One may ask (1) which individual means differ significantly from the control or (2) which are the largest differences among the various means and the control. To answer the first question, Student's *t* may be used; to answer the second, one should use Dunnett's *t* tables. We compute Student's *t* but not Dunnett's *t* in *Basic Biochemical Laboratory Procedures and Computing*. Usually, an ANOVA table has five headings:

- the source of variation;
- the sums of squares;
- the degrees of freedom;
- the mean square, and
- the variance ratio, F.

The heading *source of variation*, as the name implies, identifies sources of variability; the heading *sums of squares* identifies the individual contributors to the total sums of squares; *degrees of freedom* refers to the DF attributable to the sums of squares; the *mean square* is the sum of squares divided by the corresponding degrees of freedom; and the *variance ratio, F* is an estimate of the significance of selected sources of variation. We now describe the three types of ANOVA covered in *Basic Biochemical Laboratory Procedures and Computing*.

11-3. Single-factor ANOVA tests three or more levels of one variable

Single-factor analysis of variance is so called because it tests more than two levels of *one factor* (e.g., type of buffer). Thus, data to be analyzed by a single-factor analysis of variance are usually arranged in columns and are classified according to one criterion. Example 11-21 contains data from a single-factor experiment.

To perform analyses of variance, we need usually, to compute the following values:

1. the number of groups and the number of values, n, in each group
2. the sum of the values in each group
3. the mean of each group of values
4. the sum of each sum in item 2
5. the square of item 2 divided by n
6. the sum of the square of each value
7. item 4 squared, ÷ by the total number of values, N (the sum of each n)
8. the sum of squares of the deviations of each value from the sample mean
9. the mean square (or variance) as in Tables 11-1 to 11-3, and
10. the variance ratio, F, as in Tables 11-1 to 11-3.

After performing these calculations, we proceed as follows for single-factor ANOVA. Let SS = the sum of squares, DF = degrees of freedom, MS = the mean square, n = the number of observations; and let t, c and e refer to total, columns and error respectively. Then, to obtain the ANOVA table, we perform the following steps:

- Calculate the *total sum of squares*, SS_t.
- Calculate the *treatments (columns) sum of squares*, SS_c, and mean square, MS_c.
- Subtract SS_c from SS_t to calculate *residual or error sum of squares*, SS_e and MS_e.
- Calculate and test F for the treatments. These calculations yield:

Table 11-1 Summary of Calculations for a Single Factor ANOVA

Source of variation	SS	DF	MS	F
Total	SS_t	n-1	$\dfrac{SS_t}{}$	
Treatments (columns)	SS_c	c-1	$\dfrac{SS_c}{c-1}$	$\dfrac{MS_c}{MS_e}$
Residual (Error)	$SS_t - SS_c$	n-c	$\dfrac{SS_t - SS_c}{n-c}$	

In many analyses of variance, the investigator chooses specific treatments e.g. phosphate, tris and HEPES buffers. The treatments in this case are called a *fixed effects* or Model I ANOVA. If the buffers were chosen at random, e.g. by pointing at random in a catalog they would have been called a *random effects* or Model II ANOVA.

11-4. Randomized block tests two or more factors without replication

Randomized block analysis of variance is also known as *two-factor analysis of variance without replication*, the term *randomized* signifying that treatments are assigned at random to experimental groups. In this design, data are usually arranged in columns and rows and are classified according to two criteria. Data from the criterion or factor under investigation are usually arranged in columns, whereas data for the second factor are placed in rows and are used for removing the effects of components which contribute to biological variability. These components may include time, age, species, size, location, light or other environmental factors. The description "two-factor without replication" in this design refers to the fact there is only one reading for each cell or combination of columns and rows. Examples 11-22 to 11-24 show tables of data arranged for a randomized block analysis of variance. To obtain the randomized block ANOVA table, we proceed as follows.

Let SS = sum of squares, DF = degrees of freedom, MS = mean square, n = the number of observations; and let t, c, r and e refer to total, columns, rows and error, respectively. Then, to obtain the ANOVA table, we perform the following steps:

- Calculate the *total sum of squares*, SS_t.
- Calculate the *treatments (columns) sum of squares*, SS_c, and mean square, MS_c.
- Calculate the *blocks (rows) sum of squares*, SS_r, and mean square, MS_r.
- Subtract $SS_c + SS_r$ from SS_t to calculate SS_e and MS_e.
- Calculate and test F for the treatments. These calculations yield:

Table 11-2. Summary of Calculations for Randomized Block ANOVA

Source of variation	SS	DF	MS	F
Total	SS_t	$cr-1$		
Treatments (columns)	SS_c	$c-1$	$\dfrac{SS_c}{c-1}$	$\dfrac{MS_c}{MS_e}$
Blocks (rows)	SS_r	$r-1$	$\dfrac{SS_r}{r-1}$	$\dfrac{MS_r}{MS_e}$
Residual (Error)	$SS_t - (SS_c + SS_r)$	$(c-1)(r-1)$	$\dfrac{SS_t - (SS_c + SS_r)}{(c-1)(r-1)}$	

The term block comes from agricultural experiments with which the two-factor ANOVA without replication was developed. In those early experiments, a *block* consisted of plots of land which received treatments such as the addition of different types of fertilizer. Blocks permitted the agricultural scientist to separate, from the total variation, that source of variation caused by differences among the blocks. Today, the term block is used much more broadly than in its original agricultural context. Thus, blocks may include, for example, groups of microbial cultures in an incubator, groups of microtiter plates used in enzyme-linked immunosorbent assay (ELISA), or groups of laboratory animals kept in animal cages.

11-5. Two-factor ANOVA with replication is a factorial design

Two-factor analysis of variance with replication is the name given to experiments in which an investigator studies, simultaneously, both the individual effects, and the interactions, of two factors. The factors are usually organized into categories or *levels* and, because two factors are tested simultaneously, this type of design is known as a *factorial experiment*. Data for this type of analysis of variance are organized according to two criteria and are arranged in columns and rows; but, unlike the randomized block design, there are replicate observations (levels) for each combination of factors in a two-factor analysis of variance. Examples 11-25 and 11-26 show data for a two-factor analysis of variance with replication. We can extend a standardized ANOVA table from data such as in examples 11-25 and 11-26 to ask whether means differ significantly from a control or from each other. We show such a comparison of means for a single-factor ANOVA in Table 11-4. Generally, to obtain a standardized two-factor ANOVA table with replication, we proceed as follows.

Let SS = sum of squares, DF = degrees of freedom, MS = mean square, n = the number of observations; and let t, c, r , cr and e refer to total, columns, rows, interaction and error respectively. We then obtain entries for the ANOVA table from performing the following steps:
- Calculate the *total sum of squares*, SS_t.
- Calculate the *treatments (columns) SS, (SS_c)* and mean square, MS_c.
- Calculate the *blocks (rows) sum of squares*, SS_r, and mean square, MS_r.
- Calculate the *interaction sum of squares*, SS_{cr}, and mean square, MS_{cr}.
- Subtract $SS_c + SS_r + SS_{cr}$ from SS_t to calculate SS_e and MS_e.
- Calculate and test F for the treatments. These calculations yield:

Table 11-3. Calculations for Two Factor ANOVA With Replication

Source of variation	SS	DF	MS	F
Total	SS_t	ncr -1		
Treatments (columns)	SS_c	c-1	$\dfrac{SS_c}{c-1}$	$\dfrac{MS_c}{MS_e}$
Blocks (rows)	SS_r	r-1	$\dfrac{SS_r}{r-1}$	$\dfrac{MS_r}{MS_e}$
Interactions (cr)	SS_{cr}	(c-1) (r-1)	$\dfrac{SS_{cr}}{(c-1)\,(r-1)}$	$\dfrac{MS_{cr}}{MS_e}$
Residual (Error)	SS_t - (SS_c + SS_r + SS_{cr})	cr(n -1)	$\dfrac{SS_e}{cr(n-1)}$	

The two-factor ANOVA confers many advantages, among the most important being the simultaneous evaluation of more than one factor so that one experiment answers more than one question. This saves time and money.

11-6.   An ANOVA table and a comparison of means with a control

Table 11-3   Solution of Question 11-21

Source of variation	SS	DF	MS	F
Total	462999.6	23		
Treatments	69868.25	3	23289.42	1.184816
Residual (Error)	393131.4	20	19656.57	

P for F from Treatments   = 0.3406533

Comparison of Treatment Means With the Mean of a Control, #4

#	Mean	Difference	t Value	P (two-tailed)	P (one-tailed)
1	3.04E+02	1.424E+02	1.759	9.38E-02	4.69E-02
2	1.86E+02	2.435E+01	0.301	7.67E-01	3.83E-01
3	2.10E+02	4.805E+01	0.594	5.59E-01	2.80E-01
4	1.62E+02	0.000E+00	0.000	0.00E+00	0.00E+00

11-7.   Review questions

11-1.   Why is analysis of variance important in biomedical investigations?
11-2.   Define the terms *fixed effects model* and *random effects model*.
11-3.   What is a factorial experiment?
11-4.   Define the term *analysis of variance*?
11-5.   Explain the difference between *single-factor* and *randomized block* ANOVA.
11-6.   Explain the term *two-factor ANOVA with replication*.
11-7.   What are the main assumptions of analyses of variance?
11-8.   Outline the basic computational steps in analyses of variance.
11-9.   Explain the term *block* in randomized block analysis of variance.
11-10. What are the headings in a standard ANOVA table?
11-11. Explain how the error SS is calculated in single-factor ANOVA.
11-12. Explain how the error SS is calculated in randomized block ANOVA.
11-13. Explain how the error SS is calculated in two-factor ANOVA with replication.
11-14. List three assumptions behind analyses of variance.
11-15. Define the term *factor*.
11-16. Define the term *level*.
11-17. Define the term *mean square*.
11-18. What is the meaning of the term *randomized* in randomized block ANOVA?
11-19. What is an *interaction* in analysis of variance?
11-20. What source of variation is used for hypothesis testing in every ANOVA of Chapter 11?

11-8.   Additional problems

Single factor ANOVA

11-21.   In a study of the effects of biochemical buffers on the activity of mitochondrial fractions, mitochondrial activity was measured polarographically in four assay media. One medium ("sucrose") did not contain any buffer. The remainder contained Hepes, Tris, or phosphate. Subject the resulting data to single-factor analysis of variance to learn whether there was a difference in activity (expressed as n atoms oxygen/min/mg protein) among the four assay media.

ACTIVITY OF MITOCHONDRIAL FRACTIONS IN FOUR ASSAY MEDIA

Hepes	Tris	Phosphate	Sucrose
478.7	172.1	369.2	390.4
116.2	85.3	85.3	76.3
560.8	389.8	334.3	186.9
209.5	87.4	153.4	61.2
311.9	276.6	216.7	190.0
148.2	105.8	100.3	66.1

Randomized block ANOVA

11-22. The malonaldehyde content of control (sample 1) and three treated samples was measured with the following results (n mol/g dry wt):

Block	36	48	72	96	120
			Time (hr)		
1	823	309	81	93	487
2	293	242	109	113	221
3	300	164	55	90	446
4	355	85	109	101	334

Perform a randomized block analysis of variance.

11-23. The effective hydrophobic lengths of fatty acids from control (sample 1) and three treated samples was measured with the following results:

Block	36	48	72	96	120
			Time (hr)		
1	13.2	13.4	13.3	13.2	12.8
2	13.1	12.8	12.7	12.8	13.1
3	12.5	12.4	13.0	12.9	13.3
4	12.4	12.8	13.1	13.0	13.4

Perform a randomized block analysis of variance.

11-24. The ratio of unsaturated to saturated fatty acids of control (sample 1) and three treated samples was measured with the following results:

Block	36	48	72	96	120
			Time (hr)		
1	3.92	3.17	3.79	3.59	2,59
2	1.38	1.49	3.06	3.17	1.38
3	2.45	2.82	3.06	3.30	1.65
4	3.05	2.93	3.53	3.72	2.79

Perform a randomized block analysis of variance on these data.

Two factor ANOVA

11-25. You wanted to learn whether a supplement, αT, can modulate changes in lipid content induced by growth of an organism at different temperatures. The lipids were separated by HPLC and lipid content was expressed as n mol/mg of protein. Values are means of four independent experiments.

		Factor B	
		(Presence [-] or absence of [+] αT)	
Factor A			
(Temperature)	Block	$-\alpha T$	$+\alpha T$
22°C	S1	44.7	75.9
	S2	23.8	16.1
	S3	31.5	8.0
28°C	S1	70.1	45.3
	S2	13.7	49.5
	S3	16.2	5.3

Perform a two factor analysis of variance on these data.

11-26. It was desired to learn whether a supplement, αT, can modulate changes in lipid content induced by growth of an organism at different temperatures. The lipids were separated by HPLC and lipid content was expressed as n mol/mg protein. Values are means of four independent experiments.

		Factor B	
		(Presence [-] or absence of [+] αT)	
Factor A			
(Temperature)	Block	$-\alpha T$	$+\alpha T$
20°C	S1	44.7	65.9
	S2	43.8	66.1
	S3	45.5	68.0
30°C	S1	70.1	85.3
	S2	73.7	89.5
	S3	76.2	83.3

Perform a two-factor analysis of variance on these data.

11-9. Questions which call for planning statistical analyses

11-27.   The mayor of your home town has asked your research team for help because the rate of breast and prostate cancer in the town is 25% higher than the state average. Your spouse (who is a patient), and many residents, fear that an environmental factor, e.g. pesticides, or electromagnetic fields, may be the cause. You may make your own assumptions about the population of the town, ages of the patients, their nutritional habits, the types of pollution (air, water, toxic dumps, etc.), the cancer you want to investigate, the way the disease develops, the geology of the town, and type of industry, etc. Design a study to explain the high cancer rates. Then describe the clinical procedures, laboratory (especially instrumental) methods, statistical methods, and methods of computation where appropriate.

11-28.   Describe how you would go about preparing a suitable three-buffer system for studying the pH dependence and Michaelis-Menten constants of an enzyme that may have commercial value. Assume that you will test ten different microorganisms for the enzyme and that your preliminary work suggests that enzyme activity should be tested between pH 3.5 and 8.0. Since your preliminary work has provided data which will allow you to estimate how many replicates you should carry out to obtain reliable results, describe the statistical procedures you would use in the study.

11-29.   The link between the oxidation of LDL and atherosclerosis is suggestive but not proven. Review the literature on this topic, and then design a study to answer one segment of the problem that currently is unanswered. Describe clinical procedures, laboratory methods, statistical procedures and methods of computation.

11-30.   Suppose that dogs are beginning to develop a mysterious, rabies-like disease which the supermarket tabloids say is transmitted by bites from strange, Martian animals quietly brought to earth in flying saucers and then returned to Mars at speeds several times the speed of light. You do not have any way of proving that Martians are involved; but, it is possible to obtain sick dogs for study. The disease is following the path being taken by the so-called *killer bees,* i.e, from South to North in the USA. Public health authorities have already developed adequate precautions to protect the citizenry, so only dogs are at risk. Assume that you are the head of a six-person team consisting of a virologist, a pathologist, a veterinarian, a cell biologist/electron microscopist, a biochemist and one other specialist of your choosing. Work with five other students on an interdisciplinary project to show how you would isolate and characterize the suspected virus so that you may eventually develop a vaccine. You may change the type and number of specialties (students) involved as you see fit; but, be sure to describe clinical procedures, laboratory methods, statistical procedures and methods of computation.

11-31.   Choose two spreadsheet examples from *Basic Biochemical Laboratory Procedures and Computing* (developed with Microsoft Excel®); one example should be solved by linear regression the other without linear regression. Then translate the two problems for use with Lotus® 1-2-3®, and show the results on printed copies.

11-32.   Describe what issues you would consider in choosing whether to use nonlinear regression to analyze data from any three procedures of your choice. Regardless of whether you choose nonlinear regression or not, justify your choice.

11-33.   The news media have reported that electromagnetic fields generated by electric power lines, microwave ovens, police radar guns, electric razors, electric blankets, hair dryers, computer screens, cellular telephones, and more, can lead to increased rates of leukemia and other cancers. There is, however, little scientific evidence to support such reports. Try to learn as much as you can about one or more of the reports, then design an investigation to attempt clarification. Describe all relevant instrumental, clinical, statistical, biochemical or other methods. Could the absence of reliable information have several causes such as: the great complexity of the problem; that some supposedly "positive results" are caused by artifacts such as "shocks, vibration, noise or ozone in the experimental system;" that some physical scientists with little biomedical training do not know enough about the biomedical sciences; that some biomedical scientists with inadequate knowledge of physics do not know enough about physics; that judges, juries and the general public often are poorly informed about the facts, principles and conduct of science; that some reporters are too anxious to be first with a big story; and that some trial lawyers "smell" big money if they could link electromagnetic radiation to cancer? What training should scientists have to work effectively on such problems? More generally, is there perhaps too much fragmentation and specialization in the training of scientists who may have to work on such practical problems? The article *Health Effects of Electromagnetic Fields Remain Unresolved* by Bette Hileman on Pages 15 to 29 in the November 8, 1993 issue of Chemical & Engineering News is a good place to start.

11-10.   Questions related to contemporary research problems

11-34.   Write a spreadsheet template to calculate the volumes (ml) required to dilute stock solutions 100,000 times in steps of 100.

11-35.   You have isolated a soluble protein of molecular weight 20, 000 by chromatographic procedures and now you would like to characterize it by spectroscopic methods. You may choose any single spectroscopic technique or combination of techniques. Please state which technique(s) you would choose, why you made that choice, and what are the shortcomings which made you exclude other available methods. Include in your answer principles of the method(s) you choose, description of instrumentation, and methods of computation or analysis where appropriate.

11-36.   Biomedical scientists have made good progress in understanding the functions of macromolecules through the tools of biochemistry and molecular genetics. But, as others have noted, progress in deducing the structures of such macromolecules has been much slower than understanding their functions. Make a survey of modern methods of determining the structures of macromolecules and explain why it is so difficult to obtain detailed information on macromolecular structure.

11-37. You are being considered for a faculty position. Among the factors which will determine whether you are chosen is that you have to submit a proposal to the NSF for the instruments you will need for a new laboratory for studying the structures of proteins and nucleic acids. List and justify your choice of instruments.

11-38. Explain your selection of chemicals, supplies, and instruments for the separation, purification, detection, and assay of antibodies to a mitochondrial protein.

11-39. Computers are now essential components of analytical instruments, i.e., they are used not just for data analysis but also to control the instruments. Why?

11-40. Protein kinases and phosphatases regulate various aspects of growth, differentiation and gene expression. Usually, no single research group is able to study all aspects of protein phosphorylation/dephosphorylation. Thus, some research groups isolate and clone genes, some determine three-dimensional structure, others use recombinant DNA procedures to obtain large quantities of nonabundant protein kinases and phosphatases, and so on. Review one area of this field, describe the cellular processes which are regulated, the chemical techniques used, difficulties involved in the area of study, and any specialized techniques used such as the generation and use of antibodies to phosphoamino acids.

11-41.   Describe how you would conduct a Fourier transform infrared spectroscopy experiment to determine protein structure. Explain the factors which determine the positions of the various absorption bands.

11-42.   Review, then write a paper on, the roles of cytokines, phospholipases, and various lipids in the regulation of cellular processes.

11-43.   Enormous quantities of information are now generated monthly in virtually all active fields of biomedical investigation. Among the results of this information explosion are that it is extremely difficult for a single individual to keep up with all the new developments in his or her field; for libraries to afford all of the new journals that keep appearing; or for individuals to afford to purchase the number of journals which they would like to have at their fingertips. Online computer databases (cf. section 4-23) can help the busy scientist to find information in several areas: DNA, RNA, and protein sequencing, immunoglobulins, kinetics, metabolic pathways, research papers, product literature, patents, review articles, regulatory laws etc. But, both for individuals and institutions, these databases sometimes have certain drawbacks including cost, incomplete or limited coverage, and difficulty of use. In spite of the drawbacks, however, computerized bibliographic retrieval from online databases is an essential tool for today's biomedical scientists. With this in mind, write a set of instructions to show a novice how to use one of the following: STN International®, DIALOG Information Services®, MEDLINE® or PaperChase® on MS-DOS or Macintosh® computers. Remember that one may gain access to these scientific databases through membership in CompuServe® or other services.

11-44. Write a review of the use of nuclear magnetic resonance spectroscopy to determine the structure of proteins of molecular weight up to approximately 12 000 in solution. Include in your survey an evaluation of how the resolution of such nuclear magnetic resonance spectroscopy compares with that of high resolution x-ray crystal structure and comment on where you think current efforts to improve and extend NMR spectroscopy may lead.

11-45. In this question you are being requested to write a computer program (or a group of programs) in a language of your choice to print the following questions and answers for a class of 30 students:

Assume that you plan to ask the 30 students to carry out calculations for preparing buffer solutions of 5 different pH values for each of

(a) *phosphate buffer* and
(b) one buffer chosen from *acetic acid, Mes, Pipes, Mops, Tes, Hepes, Tricine, Tris, Bicine and Caps*. These buffers provide both conjugate acid and conjugate base.

For phosphate buffer your program should give each student the pK_a', the desired molarity, the volume of buffer required, the molecular weights of the conjugate acid and conjugate base and a pH increment of 0.1 unit within the range pH 5.6 to 8.0. The range pH 5.6 to 8.0 in increments of 0.1 pH times 6 different molarities yields 30 sets of 5 buffer solutions.

For the second buffer the program should give each student the pK_a', the desired molarity, the molecular weight of the buffer, the normality of acid or base required, the volume of buffer required, and a pH increment of 0.5 unit for a range of 2 pH units around the pK_a' of the selected buffer. The range of 2 pH units in increments of 0.5 pH around the pK_a' for each of 10 buffers at 3 different molarities yields 30 sets of 5 buffer solutions. Thus the only data input which your program needs to change are the pH values for each student, the 6 molarities for phosphate and the 3 molarities for the second buffer. You may keep the volume of buffer required constant by asking every student to prepare, say, 0.1 or 0.5 l.

The program, therefore, should allow you to give each student two numbered sets of data (questions) for preparing two sets of buffers and provide you with the answers to each set of two questions. Each student will calculate the amount of each component needed to obtain buffer solutions of the specified volumes, pH and molarity, and each will present answers in tabular form, making sure to include the number of moles of conjugate acid and conjugate base, the weights of the two conjugate partners where appropriate, and the volume of acid or base of specified normality when required. The students also should should be asked to include intermediate values such as the fraction of conjugate acid or base dissociated, to consult one or more handbooks of biochemistry to see how such information is presented, and to prepare reports with a word processing program.

11-46. One of your colleagues is planning to modify a published method to conduct a study of a newly discovered enzyme. The published method, which called for the use of a three-buffer mixture (pK_a' values of 3.4, 6.15 and 8.35 respectively), was used for studying pH-activity profiles of a set of five microbial enzymes. Your friend, however, is planning to conduct his experiments at the single pH value of 6.2 but he still plans to use the original three-buffer system. Would you question your friend's choice of three buffers and advise a change? Explain your answer.

Suggested Reading

Campbell, R.C. 1974. *Statistics for Biologists.* 2nd ed. Cambridge University Press, London.

Daniel, W.W. 1983. *Biostatistics: A Foundation for Analysis in the Health Sciences.* 3rd. ed. John Wiley & Sons, NY.

Dunnett, C.W. 1955. A Multiple Comparison Procedure for Comparing Several Treatments With a Control. *J. Amer. Statistical Assn.* 50: 1096 - 1121.

Dunnett, C.W. 1964. New Tables for Multiple Comparisons With a Control. *Biometrics* 20: 482 - 491.

Hileman, B. 1993. Health Effects of Electromagnetic Fields Remain Unresolved. Pages 15-29, *Chem. & Engineering News,* Nov. 8.

Lee, J.D. and T.D. Lee 1982. *Statistics and Numerical Methods in BASIC for Biologists.* Van Nostrand Reinhold, NY.

Nimmo, I.A. and G.L. Atkins 1979. The Statistical Analysis of non-normal (real?) data. *Trends Biochem. Sci.* 4: 236 - 239.

Ridgman, W.J. 1975. *Experimentation in Biology.* Halsted Press, NY.

Snedecor, G.W. and W.G. Cochran 1967. *Statistical Methods.* 6th. ed. Iowa State University Press.

Youden, W.J. 1951. *Statistical Methods for Chemists.* John Wiley & Sons.

Zar, J.H. 1974. *Biostatistical Analysis.* Prentice-Hall, Inc., Englewood, NJ.

Appendix 1. Properties of some concentrated acids and ammonia

	Molecular weight	Specific Gravity	Percent (w/w)	Approximate Molarity	Approximate Normality
Acetic acid	60.0	1.05	100	17.4	17
Ammonia	17.0	0.91	25	14.3	14
Formic acid	46.0	1.20	89	23.6	24
Hydrochloric acid	36.5	1.18	37	12.0	12
Nitric acid	63.0	1.42	70	15.7	16
Perchloric acid	100.5	1.54	70	9.2	12
Phosphoric acid	98.0	1.75	85	14.7	15
Sulfuric acid	98.1	1.83	98	18.3	37

Appendix 2. Physical properties of some buffer acids and bases

Buffer Acid or Base	Molecular Weight	pK_a at 25°C	ΔpK_a/°C
Phosphoric acid	177.98	2.12 (pK_{a1})	0.0044
Glycine	75.07	2.34 (pK_{a1})	
Citric acid	192.12	3.06 (pK_{a1})	
Citric acid	192.12	4.74 (pK_{a2})	-0.0016
Acetic acid	60.05	4.76	0.0002
Citric acid	192.12	5.40 (pK_{a3})	0.0
Succinic acid	118.09	5.57 (pK_{a2})	0.0
MES	195.23	6.15	-0.011
ADA	212.15	6.60	-0.011
PIPES	342.26	6.80	-0.0085
ACES	182.20	6.90	-0.020
BES	213.20	7.15	-0.016
Phosphoric acid	177.98	7.21 (pK_{a2})	-0.0028
TES	229.25	7.50	-0.020
HEPES	238.11	7.55	-0.014
Tricine	180.18	8.15	-0.021
Tris	121.14	8.30	-0.031
Bicine	163.17	8.35	-0.018
Glycylglycine	132.12	8.40	-0.028
Boric acid	43.82	9.24	-0.008
Glycine	75.07	9.60 (pK_{a2})	
CAPS	221.32	10.40	0.032
Phosphoric acid	177.98	12.32 (pK_{a3})	-0.026

ACES	=	N-(Acetamide)-2-aminoethanesulfonic acid
ADA	=	N-(2-Acetamide) iminodiacetic acid
BES	=	N,N-Bis(2-hydroxyethyl)2-aminoethanesulfonic acid
CAPS	=	3-(Cyclohexylamino)propanesulfonic acid
HEPES	=	N-2-Hydroxyethylpiperazine-N'-2-ethanesulfonic acid
MES	=	2-(N-Morphilino)ethanesulfonic acid
PIPES	=	Piperazine-N,N'-bis(2-ethanesulfonic acid)
TES	=	N-Tris(hydroxymethyl)methyl-2-aminoethanesulfonic acid

Appendix 3. Some radioisotopes used in biomedical work

Element (mass number)	Type of Radiation	Half-life	Energy of β-particle or γ-ray (MeV)
^{3}H	β⁻	12.1 yr	0.0185
^{11}C	β⁺	20.5 min	0.95
^{14}C	β⁻	5100 yr	0.156
^{22}Na	β⁺, γ	2.6 yr	0.58
^{24}Na	β⁻, γ	14.8 hr	1.39
^{32}P	β⁻	14.3 days	1.71
^{35}S	β⁻	87.1 days	0.169
^{36}Cl	β⁺, κ, β⁻	2×10^6 yr	0.64 (β⁻)
^{45}Ca	β⁻	152 days	0.260
^{59}Fe	β⁻, γ	46.3 days	0.46
^{60}Co	β⁻, γ	5.3 yr	0.31 (β); 1.16(γ)*
^{125}I	γ	60 days	0.035 (γ)
^{131}I	β⁻, γ	8.1 days	0.605 (β); 0.637 (γ)*

* Not all energies are shown

Appendix 4. Values of some constants used in bioresearch

Absolute temperature	273.15 K
Acceleration of gravity	980.7 cm s^{-2}
Avogadro constant (N_o)	6.0225×10^{23}
Boltzmann's constant (k)	1.3805×10^{-23} J K^{-1}
Faraday constant ($\mathcal{F}$)	96 486.7 C mol^{-1}
Gas constant (R)	8.314 J K^{-1} mol^{-1}
Planck's constant (h)	6.256×10^{-34} J s
π	3.14159
e	2.30258

Appendix 5. Some useful conversion factors

1 Ångstrom	=	10^{-8} cm = 0.1 nm 10^{-10} m
1 atm	=	760 mm Hg = 760 torr = 101 325 Pa = 101 325 N m^{-2}
1 cal	=	4.184 J = 4.184×10^7 ergs
1 coulomb	=	2.9979×10^9 esu
1 eV	=	23.06 k cal mol-1 = 1.602×10^{-12} erg = 1.602×10^{-19} J
1 $\mathcal{F}$	=	96 487 C mol^{-1} = 23 062 cal V^{-1} equiv^{-1}
1 R	=	8.314 J K^{-1} mol^{-1} = 1.987 cal K^{-1} mol^{-1}

Basic Biochemical Laboratory Procedures and Computing reviews the theoretical basis of many biophysical chemical techniques used in the biochemistry laboratory, and emphasizes the usefulness of computer spreadsheets in solving quantitative problems related to these methods. The approach taken is multidisciplinary. Topics discussed are found in textbooks, biomedical research journals, annual review publications, and monographs. By pointing out the applications of the topics discussed to current research, *Basic Biochemical Laboratory Procedures and Computing* reflects the thinking, methods, instruments, and experimental strategies of today's practicing biomedical scientists. The text is ideal for upper level undergraduates and graduate students taking courses on laboratory methods and techniques in departments of biological sciences, biochemistry, biophysics, and molecular biology.

About the Author

R. Cecil Jack is retired, former Professor and Chairman, Department of Biological Sciences, St. John's University and former Associate Biochemist, Boyce Thompson Institute for Plant Research. Professor Jack has extensive experience teaching biochemistry to undergraduate and graduate students.

Cover design by Ed Atkeson/Berg Design

90000

9 780195 078978

ISBN 0-19-507897-7